High Voltage Engineering:
Theory and Practice

High Voltage Engineering: Theory and Practice

Edited by
Drake Shaw

WILLFORD PRESS

www.willfordpress.com

Published by Willford Press,
118-35 Queens Blvd., Suite 400,
Forest Hills, NY 11375, USA

ISBN: 978-1-64728-527-2

Cataloging-in-Publication Data

High voltage engineering : theory and practice / edited by Drake Shaw.
    p. cm.
Includes bibliographical references and index.
ISBN 978-1-64728-527-2
1. Electric power distribution--High tension. 2. High voltages. 3. Electrical engineering.
I. Shaw, Drake.
TK3144 .H54 2023
621.319 13--dc23

For information on all Willford Press publications
visit our website at www.willfordpress.com

Contents

Preface

High voltage engineering is the science of planning, testing and operating high voltage electrical devices, as well as designing insulation coordination to ensure the steady operation of the power network. It ensures that the consumers situated far from the power generation units have access to electrical energy. A higher voltage results in improved efficiency and reduces loss of electricity during transmission over long distances. High voltage engineering optimizes and develops the operating properties of internal and exterior insulators. It is concerned with the proper operation, application and the practical use of high voltages and high fields. Furthermore, it studies the properties of materials used for electrical insulation for estimating their performance under high stress during their life span under various physical conditions. The stress on the insulation caused by the electric field serves as the foundation for the design of high voltage equipment, where magnitude of the stress depends on the type of voltage. This book contains some path-breaking studies in the field of high voltage engineering. Researchers and students in this field will be assisted by it.

Significant researches are present in this book. Intensive efforts have been employed by authors to make this book an outstanding discourse. This book contains the enlightening chapters which have been written on the basis of significant researches done by the experts.

Finally, I would also like to thank all the members involved in this book for being a team and meeting all the deadlines for the submission of their respective works. I would also like to thank my friends and family for being supportive in my efforts.

Editor

Calculation of Ion Flow Field of Monopolar Transmission Line in Corona Cage Including the Effect of Wind

Zhenyu Li and Xuezeng Zhao *

Department of Mechatronics Control and Automation, School of Mechatronics Engineering, Harbin Institute of Technology, Harbin 150001, China; constantlzy@163.com
* Correspondence: zhaoxz@hit.edu.cn.

Abstract: In this work, the ion flow field of a monopolar transmission line inside the corona cage of a square cross-section is iteratively calculated concerning the effects of wind. The electric field distribution is solved analytically using the charge simulation method (CSM). Meanwhile, the upwind finite volume method (UFVM) with 2nd order accuracy is presented for the distribution of space charge density. Additionally, a dual mesh grid is established in the calculation domain, the interlaced geometric construction of the mesh assures a quick and effective convergence rate. In the final part, a reduced-scaled experiment is designed to examine the feasibility and accuracy of this approach, electric field and ion current density on the bottom side are measured by field mills and Wilson plates. The data numerically computed fits well with that acquired by measurement.

Keywords: corona discharge; electric field analysis; ion flow field; space charge density; UFVM

1. Introduction

In operating HVDC transmission lines of a power system, the phenomenon of corona discharge is a leading cause of radiation interference (RI), noise interference (NI), and corona loss (CL) [1]. Thus, investigation on the ion flow field distributed around the conductors receives considerable attention in the design of HVDC transmission lines.

Commonly, one of the main obstacles in solving the ion flow field is the nonlinearity between the electric field and space charge density. The vast majority of solutions calculate the electric field and the space charge density iteratively and the iteration process ends once the criteria are met. In the meantime, wind flow affects the distribution of the electric field around transmission lines to certain degree as well. All the above-mentioned issues increase the difficulty of calculating the ion flow field.

In the past few decades, research on the ion flow field calculation has varied in terms of the methods utilized to calculate the electric field and the space charge density in the domain of interest.

With regard to the electric field, Janischewskyj and Gela [2] introduced finite element method (FEM) to solve the electric field numerically; afterwards, this method was frequently adopted and well-developed. In 1983, Takuma et al. [3] applied CSM to calculate the nominal electric field without space charge, while FEM was used to solve the electric field induced by space charges. Since then, this approach has been broadly applied in electric field calculation [4,5]. CSM offers satisfactory accuracy, whereas the calculation domain is restricted to an infinite field above the ground or axisymmetric structure. Simultaneously, a drawback of FEM is that the accuracy of the electric field close to the conductor surface is not as expected because of the steep gradient.

In regard to the calculation of space charge density, method of characteristics (MOC) is diffusely utilized [6–10], space charge density is calculated along electric field lines with given initial charge density on the conductor surface. This approach relies on the Deustch's assumption, which assumes the space charge affects the amplitude of an electric field rather than its direction. Xiao calculated the ion flow field around a cross-over transmission line in the 3D domain with the MOC method. However, the effectiveness of this method is unsatisfactory if the influence of wind flow is under consideration. Lu et al. [11] proposed an upwind FEM, which avoids non-physical instability of the numerical calculation. Zhou et al. [12] induced upwind weighting function to FEM for the purpose of eliminating the oscillations in simulation of charge conservation. Levin [13] established dual mesh based on the triangulation grid in calculation domain; the new mesh is called donor cell and the space charge density is hereby solved in accordance with Gauss' Law. Then, upwind the FVM method were used in several research projects [5,14–17] in which the numerical stability, effectiveness, and accuracy of the solution process was improved substantially. Yang et al. [18] proposed an upstream meshless method to solve the current continuity equation.

For the purpose of implementing an indoor experiment and control of environmental parameters, a corona cage is designed in where the phenomenon of corona discharge initializes on a relatively lower voltage level. Bian et al. [19] and Lekganyane et al. [20] investigated the ion flow field in a square cross-section cage and compared the result with that of an indoor test line; Zhou et al. [8] presented a comprehensive study on the ion flow field distribution in a cylindrical cage employing a mesh-based method and MOC. However, there is paucity of published research concerning the effect of wind flow on the corona discharge of corona cage.

However, solution of the ion flow field generally concerns the numerical stability, calculation accuracy and the impact of wind flow. The referenced articles barely meet these requirements at the same time. Therefore, it is necessary to develop a method that offers a quick, stable, and accurate solution of ion flow field.

In this paper, the calculation domain is tessellated in the form of a dual mesh. Next, CSM is utilized for a nominal electric field in the absence of space charges. Simultaneously, electric fields generated by space charges is available if the space charges density is known, by this means, the accuracy of the calculated field is guaranteed even on the conductor surface. The 2nd order upwind FVM is employed to calculate the space charge density distribution. Eventually, the calculated result is validated with that derived by experiments.

The importance and originality of this study consists of the nominal electric field in a square cross-section being solved by means of proper placement of the simulation charges, the more accurate solution of space charge density distribution involving the impact from wind flow, as well as the applicability of this approach in presence of wind flow. The calculation process provides rapid convergence rate as the analytically calculated electric field is less time-consuming compared to the traditional method using FEM.

2. Mathematical Description

2.1. Governing Equations and Simplifying Assumption

Generally, the ion flow field in the ambient of conductor is governed by Poisson's equation and the current density conservation equation [3]:

$$\begin{cases} \nabla E = -\rho^- / \varepsilon_0 \\ \nabla \cdot J^- = 0 \\ J^- = \rho^- (bE + W) \end{cases} \tag{1}$$

where,

b is the ion mobility, 1.5×10^{-4}, m^2/V/s;

E is the electric field, V/m;

ρ is the negative space charge density, C/m^3;

J^- is the negative ion current density vector, A/m^2;

W is the wind velocity vector, m/s; and
ε_0 is the permittivity of air equals 8.854×10^{-12}, F/m.

Further, certain assumptions are proposed in advance in order to reduce the complexity of calculation and acquire satisfactory precision:

(a)   The thin ionization layer close to the conductor surface is neglected;
(b)   The ion mobility remains unchanged throughout the solution process;
(c)   Influence exerted by ion diffusion is ignored;
(d)   Kaptzov's assumption [21] which presumes the electric field on conductor surface remains constant after the applied voltage reaches the onset value is adopted.

2.2. Boundary Conditions

Before proceeding to the solution process. Boundary conditions and initial conditions of the calculation domain are listed in Table 1.

Table 1. Boundary conditions and initial conditions.

Distribution Variables	Conductor Surface	Cage Wall
Electric potential	V_{app}	0
Space-charge density	ρ_s	$\frac{\partial \rho}{\partial n}$
Electric field	E_{on}	$\frac{\partial E}{\partial n}$

where,

V_{app} is the voltage supplied on the conductor, V;
ρ_s refers to the space charge density on the conductor surface, C/m^3;
E_{on} is the onset electric field, V/m; and
E_{on} is assumed to be constant on the conductor surface according to Kaptzov's assumption, the explicit value is attained using Peek's empirical formula [22]:

$$E_{on} = 30m(1 + \sqrt{\frac{0.0906}{r}}) \tag{2}$$

where,

m is the roughness factor set to 0.65; and
r is the radius of the conductor, m.

An appropriate guess of the initial value of the charge density on conductor surface determines the accuracy of the result and diminishes the iteration process. The empirical formula introduced in [23] is referred to in this paper:

$$\rho_s = \frac{E_g}{E_c} \frac{8\varepsilon_0 V_c(V_{app} - V_c)}{rH_{con}(5 - 4V_c/V_{app})} \tag{3}$$

where,

E_g is the ground level electric field under the conductor, V/m;
E_c is the nominal electric field on the conductor, V/m;
V_c is the onset voltage of the conductor, V;
V_{con} is the conductor voltage, V; and

H_{con} is the height of the conductor, m.

Takuma [3] assumes that the initial charge density is evenly distributed on the conductor surface, yet it is not suitable for the situation where the transmission line is placed inside corona with a square cross-section cage. Because distances between the conductor surface to the cage wall are diverse, which differs from the situation above the ground or in the coaxial cage. Hence, in this work, charge density is set to be linearly dependent on E_c. To be specific, for each node on the conductor surface, E_c is calculated by CSM, neglecting space charges. Thus, the initial charge density of this node is achieved by substitute corresponding E_c into Equation (3).

3. Solution Process

The distributions of electric field and space charge density are solved iteratively, in this process, the initial charge density on the conductor surface is updated in each iteration in case that the condition of convergence is not satisfied. The detailed procedure is organized and illustrated in Figure 1.

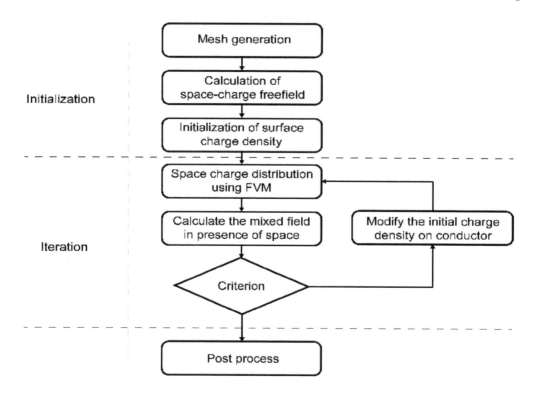

Figure 1. Flowchart of the method.

3.1. Discretization of Calculation Domain

The calculation is conducted in the 2D cross-section of the cage, and the calculation domain is divided in form of dual mesh. Specifically, a Delaunay Triangular mesh is generated in the first place; after that, polygon cells are constructed via connecting the barycenter and the midpoints of triangular cell edges, which share common vertexes [24].

The meshing of the calculation domain is demonstrated in Figure 2. Meshing of the area in the vicinity of the conductor surface and the cage wall are finer in the cause of the need of more accurate calculation results. As a result of the grid independency test, the difference of the calculated result is less than 1%, while the domain is tessellated into 1879 cells.

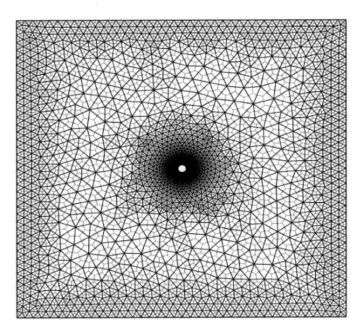

Figure 2. Tessellation of calculation domain.

Since the electric field is calculated analytically, the electric field in each single point can be calculated defectively. Regarding to the space charge density, they are stored in the nodes of the triangulation. The effectiveness of the solution process is ensured resulting from the interlaced meshing scheme.

3.2. Calculation of Electric Field

The calculation of electric field composes of two parts: the nominal field and the field induced by space charges.

Regarding the nominal field to be calculated by CSM, there are 40 simulation charges equally distributed inside the conductor.

In contrast to the circumstance of outdoor lines above the ground, extra simulation charges are placed outside the cage wall to maintain the zero potential of the grounded cage wall; as a result, as demonstrated in Figure 3, there are 160 simulation charges equidistantly arranged with interval d_c, the perpendicular distance between the cage wall and simulation charges are two times of d_c. Image charges are placed symmetrically in the opposite sides of the ground. The match points are placed right on the conductors and corona walls, respectively.

According to the principle of CSM, values of the simulation charges must ensure the potential on the conductor surface and cage wall to be V_{app} and zero. Thus, coefficient equations are listed below:

$$\begin{cases} P_{cond} \cdot Q_{cond} + P_{cage} \cdot Q_{cage} = V_{app} \\ P'_{cond} \cdot Q_{cond} + P'_{cage} \cdot Q_{cage} = 0 \end{cases} \tag{4}$$

where Q_{cond} and Q_{cage} are simulation charges of the conductor and cage, P_{cond}, P_{cage} are the potential coefficients regarding the conductor surface, and P'_{cond}, P'_{cage} are the potential coefficients for the cage wall.

The nominal electric field is therefore obtained by superposing the field caused by simulation charges. The space charges-induced field can be calculated if the space charges are known. Accordingly, the ion flow field is available as follows:

$$E_m = \sum_i \frac{Q_i}{4\pi\varepsilon_0}\left(\frac{\vec{r_i}}{r_i^2} - \frac{\vec{r_i'}}{r_i'^2}\right) + \sum_j \int\limits_{s_j} \frac{\rho_j}{4\pi\varepsilon_0}\left(\frac{\vec{r_j}}{r_j^2} - \frac{\vec{r_j'}}{r_j'^2}\right) ds_j \tag{5}$$

where,

s_j is the area of the polygon cell, m³;

r_i and r_j are the distances between the observation point to the source, m;

r'_i and r'_j are the distances between the observation point to the image points, m;

Q_i are the simulation charges consist of Q_{cond} and Q_{cage}, C; and

ρ_j is the charge density on triangulation nodes, C/m³.

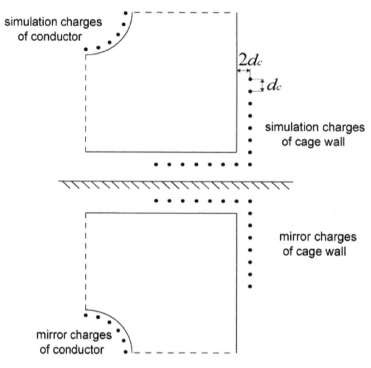

Figure 3. CSM method.

3.3. Calculation of Ion Current Density

FVM ensures the maximum principle of charge density on cell boundaries resulting in numerical stability in the calculation process. Additionally, the upwind scheme satisfies the physical fact that the migration of space charges is affected by upwind stream only [25]. Particularly, this method is feasible for calculation in the presence of wind flow.

By substituting Equation (3) into Equation (2), following equation is obtained:

$$\nabla \cdot [\rho^- (b\mathbf{E} + \mathbf{W})] = 0 \tag{6}$$

Afterwards, Equation (6) is converted to integral form:

$$\iint_s \rho(b\mathbf{E} + \mathbf{W})dl = 0 \tag{7}$$

where,

s and l are the area and boundary of the cell.

Next, Equation (7) is rewritten in the form of linear equations in the light of Figure 4:

$$\sum_{n=1}^{n} \rho V_n L_n = 0 \tag{8}$$

where,

$V_n = bE_n + W_n$, E_n and W_n are the outward normal component of electric field and wind speed on cell edges, and

L_n is the length of the ith cell; and

n presents the serial number of the edges.

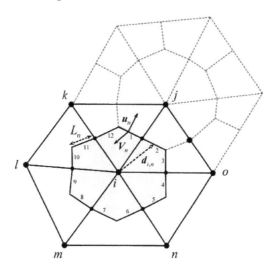

Figure 4. Control volume of UFVM.

Finally, a system of linear equations is established and the distribution of charge density is therefore achieved.

Normally, charge density is defined as the average value of adjacent nodes [15], or determined by the upwind scheme. Nevertheless, these methods are only of 1st order accuracy which is insufficient for engineering requirements.

With an aim to promote the accuracy of space charge density, a 2nd order upwind scheme is utilized. Charge densities on the edges are solved directly. Apparently, the charge densities on cell edges can be expressed according to the Taylor series expansion as Figure 4:

$$\begin{cases} \rho_{i,n} = \rho_i + \nabla\rho_{i,n} \cdot d_{i,n} & if \quad V_n \cdot u_n > 0 \\ \rho_{i,n} = \rho_n + \nabla\rho_{n,i} \cdot d_{n,i} & if \quad V_n \cdot u_n < 0 \end{cases} \tag{9}$$

where,

$\rho_{i,n}$ is the charge density on the edge, C/m^3;

$d_{i,n}$ is the vector direct from the ith node to the corresponding neighboring nodes; and

$\nabla\rho_{i,n}$ and $\nabla\rho_{n,i}$ are the gradient of corresponding upwind node.

For the known charge densities on the ith node and its adjacent nodes, the following over determined matrix equation can be established; thus, the gradient of charge density on the ith node is obtained:

$$\begin{vmatrix} \Delta x_1 & \Delta y_1 \\ \Delta x_2 & \Delta y_2 \\ \dots & \dots \\ \Delta x_n & \Delta y_n \end{vmatrix}_{A_{n\times2}} \times \begin{vmatrix} \frac{\partial\rho_i}{\partial x} \\ \frac{\partial\rho_i}{\partial y} \end{vmatrix}_{X_{2\times1}} = \begin{vmatrix} \rho_i - \rho_1 \\ \rho_i - \rho_2 \\ \dots \\ \rho_i - \rho_n \end{vmatrix}_{B_{n\times1}} \tag{10}$$

3.4. Terminal Criteria and Initial Charge Density

The iteration procedure will terminate while the following conditions are met.

$$\frac{E_c - E_0}{E_0} < \delta_E \tag{11}$$

$$\frac{1}{N}\sum_{i=1}^{N}\frac{\left|\rho_{m,i}-\rho_{m-1,i}\right|}{\left|\rho_{m-1,i}\right|}<\delta_{\rho} \tag{12}$$

where,

δ_E and δ_ρ are relative terminal criteria;

E_c is the electric field on conductor surface, V/m; and

$\rho_{m,i}$ and $\rho_{m-1,i}$ are consecutive space charge densities of the iteration process in ith cell, C/m^3.

However, the initial charge densities on the conductor surface need to be modified in each iteration so as to maintain the electric field on conductor surface as E_0, the principle abides by the following equation:

$$\rho_{m-1}=\rho_m\left(1+\mu\cdot\frac{E_c-E_0}{E_c+E_0}\right) \tag{13}$$

where,

ρ_{m-1}, ρ_m are charge densities of two consecutive iteration on conductor surface, C/m^3; and

μ is the acceleration factor equals to two.

Distributions of the space charge density of different wind speed are indicated in Figure 5. The convergence rate of the calculation under $-120\,kV$ voltage supply and 10 m/s wind speed is shown in Figure 6; the iteration process has a good convergence and ends after about 45 times of iteration. It takes fewer than 10 min since the coefficient matrix of the electric field is prepared once and for all.

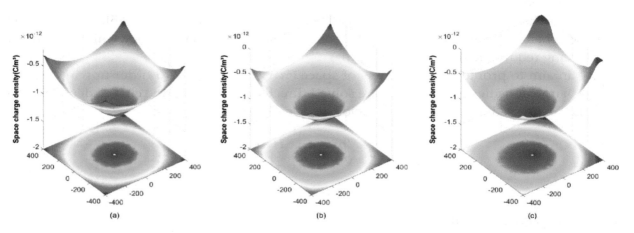

Figure 5. Space charge density under 100 kV supply voltage. (**a**) space charge density under 0 m/s wind speed; (**b**) space charge density under 5 m/s wind speed; and (**c**) space charge density under 10 m/s wind speed.

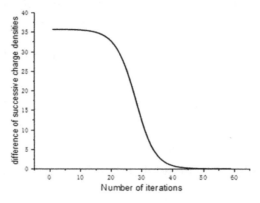

Figure 6. Convergence to terminal criteria.

4. Validation

4.1. Design of the Experiment

The corona cage is frequently used, which offers much more intensive field strength on relatively low voltage level. The reduced size of experimental facility enables indoor experiments.

The dimension of the main section of the corona cage is $1500 \times 800 \times 800$ (mm), there are two guard sections size at $300 \times 800 \times 800$ (mm) in order to eliminate the impact results from end effect [26]. The conductor of 10 mm radius is hung along the center of the cross-section, and the two ends of the conductor are fixed on a steel frame with two composite insulators as a connection. The schematic diagram and experimental equipment are indicated in Figures 7 and 8.

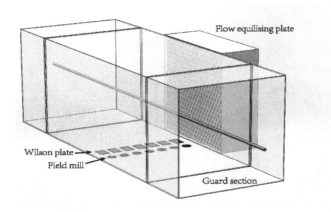

Figure 7. Schematic diagram of the experimental setting.

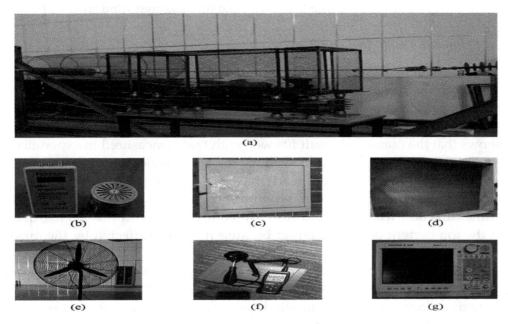

Figure 8. (**a**) The general view of the corona cage; (**b**) The electric field mill; (**c**) The Wilson plate; (**d**) Flow equalizing plate; (**e**) adjustable speed fan; (**f**) digital high precision anemometer; and (**g**) scope coder.

Seven Wilson plates and field mills are parallelly laid on the bottom side of the cage equidistantly for the measurement of ion current density and electric field.

Yokogawa scope coder DL 850 is used to record sampling signals of the ion flow current collected by Wilson plates, this equipment provides high sampling rate which is up to 100 Ms/s and enables continuous synchronous measurement of multi-channel.

The voltage was supported by a DC source with fluctuation less than 5%. The experiment was conducted at standard air pressure in a high voltage laboratory. The relative humidity was 20% to 30% and the temperature ranged from 10 °C to 15 °C.

An adjustable speed fan with four gears is placed next to the cage, facing one side of the cage, and it provides lateral wind speed of 0–10 m/s. The impact of flow divergence is moderated by a flow equalizing plate, the wind speed is measured by a digital high precision anemometer (AS-8336). The desired wind speed is achieved by adjusting the gear and distance between the fan and the equalizing plate. Since even distribution of the wind speed is the prerequisite of the experiment, wind speed values of 16 points are measured in the cross-section of the cage. The measurement shows that the standard deviation of wind speeds is less than 5%. The average measured result is demonstrated in Figure 9.

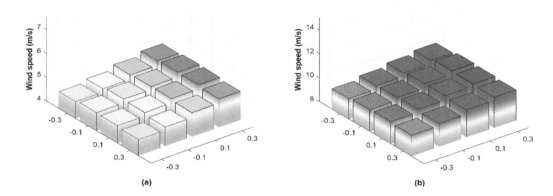

Figure 9. (**a**) measured wind speeds of 5 m/s; and (**b**) measured wind speeds of 5 m/s.

4.2. Discussion of Numerical and Measured Results

For the purpose of investigating the impact caused by varied supply voltage and wind speed, the voltage applied on the conductor is adjusted to −80 kV, −100 kV, and −120 kV, the wind speed is specified at 0 m/s, 5 m/s and 10 m/s. Both the numerically resolved and practically measured results of the electric field and the ion current density on the bottom side of the cage are demonstrated in Figure 10. It shows that the numerical result fits well with that is measured in experiment. As a result, the method used in this work is qualified to evaluate the ion flow field in a corona cage.

The electric field and ion flow current reach their peaks right under the conductor and decrease with the distance from the center point when the wind speed is zero. As the wind speed rises, the curve shifts in the same direction with the wind flow. In comparison with the electric field, the degree of the shift on the ion current density is larger because it is more affected by the movement of the space charges.

In order to illustrate in a more intuitive manner, the cross-section of interest is bisected by vertical center line which is illustrated in Figure 11, the section in where the wind flow and the electric field move in opposite directions is defined as upwind; on the contrary, the other section is downwind.

In the upwind section, electric field and ion current density weakens because the wind flow reduces the density of space charges. This effect decreases as approaching the conductor due to the increasing electric field force.

For both electric field and ion current density, absolute values in the upwind section is slightly lower than that in the downwind section. The absolute value declines along with the increasement of wind speed. In addition, the degree of the impact results from wind speed mounts as the electric field strength decreases. Additionally, the ion current density on lower voltage level is more sensitive to the wind speed owing to weaker field strength.

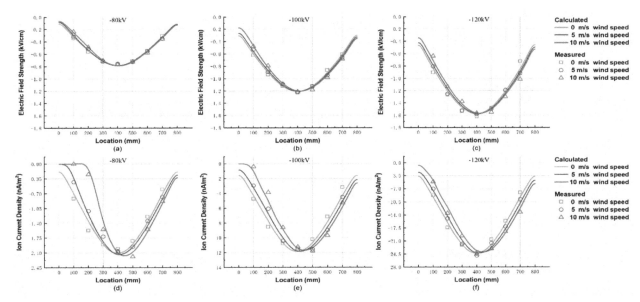

Figure 10. (**a**) Electric field on the bottom side under −80 kV supply voltage; (**b**) electric field on the bottom side under −100 kV supply voltage; (**c**) electric field on the bottom side under −120 kV supply voltage; (**d**) ion current density on the bottom side under 80 kV supply voltage; (**e**) ion current density on the bottom side under 100 kV supply voltage; (**f**) ion current density on the bottom side under 120 kV supply voltage.

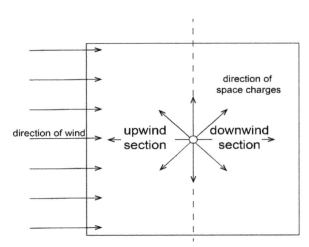

Figure 11. Schematic diagram of the cross-section.

In upwind and downwind section, the wind flow exerts contrarily influence on the electric field and the ion flow current. To be explicit, increased wind speed weakens the electric field intensity and ion current density in the upwind section. For the downwind section, the result is opposite. The explanation is that the charge density in the upwind section is larger than that of the downwind section which is rather obvious, as shown in Figure 5.

5. Conclusions

The method combining CSM and upwind FVM, which concerns the impact of wind flow, is proposed in this paper, experiment is designed to examine the numeric results. The results indicate that such a method provides a relatively accurate evaluation of the ion flow field under a corona discharge conductor. For the merits of the presented method, the analytically calculated electric field offers a more precise electric field in the vicinity of the conductor than that is solved in traditional FEM; simultaneously, it reduces calculation time and enhances numerical stability to a large extent by means of avoiding iterative calculation of two numeric methods. Moreover, 2nd-order UFVM improves the accuracy of space charge density and is competent for the conditions, including wind impact. The solution process converges effectively and stably. The dependence of the ion flow field on the wind is studied, which approves the influence of wind flow on the ion flow field.

Author Contributions: Conceptualization, Z.L. and X.Z.; Methodology, Z.L.; Software, Z.L.; Validation, Z.L.; Formal Analysis, Z.L.; Investigation, Z.L.; Resources, Z.L. and X.Z.; Data Curation, Z.L.; Writing-Original Draft Preparation, Z.L.; Writing-Review & Editing, Z.L.; Visualization, Z.L.; Supervision, X.Z.; Project Administration, X.Z.; Funding Acquisition, X.Z.

References

1.   Maruvada, P.S. *Corona Performance of High-Voltage Transmission Lines*; Research Studies Press Baldock: Herfordshire, UK, 2000.

2.   Janischewskyj, W.; Cela, G. Finite Element Solution for Electric Fields of Coronating DC Transmission Lines. *IEEE Trans. Power Appar. Syst.* **1979**, *PAS-98*, 1000–1012. [CrossRef]

3.   Takuma, T.; Ikeda, T.; Kawamoto, T. Calculation of ION Flow Fields of HVDC Transmission Lines By the Finite Element Method. *IEEE Trans. Power Appar. Syst.* **1981**, *PAS-100*, 4802–4810. [CrossRef]

4.   Abdel-Salam, M.; Farghally, M.; Abdel-Sattar, S. Finite element solution of monopolar corona equation. *IEEE Trans. Electr. Insul.* **1983**, *EI-18*, 110–119. [CrossRef]

5.   Li, X.; Ciric, I.R.; Raghuveer, M.R. Highly stable finite volume based relaxation iterative algorithm for solution of DC line ionized fields in the presence of wind. *Int. J. Numer. Model. Electron. Netw. Devices Fields* **1997**, *10*, 355–370. [CrossRef]

6.   Davis, J.L.; Hoburg, J.F. HVDC transmission line computations using finite element and characteristics method. *J. Electrost.* **1986**, *18*, 1–22. [CrossRef]

7.   Fortin, S.; Zhao, H.; Ma, J.; Member, S.; Dawalibi, F.P. A New Approach to Calculate the Ionized Field Transmission Lines in the Space and on the Earth Surface. In Proceedings of the 2006 International Conference on Power System Technology, Chongqing, China, 22–26 October 2006.

8.   Zhou, X.; Cui, X.; Lu, T.; Fang, C.; Zhen, Y. Spatial distribution of ion current around HVDC bundle conductors. *IEEE Trans. Power Deliv.* **2012**, *27*, 380–390. [CrossRef]

9.   Guillod, T.; Pfeiffer, M.; Franck, C.M. Improved coupled ion-flow field calculation method for AC/DC hybrid overhead power lines. *IEEE Trans. Power Deliv.* **2014**, *29*, 2493–2501. [CrossRef]

10. Zhang, B.; Mo, J.; He, J.; Zhuang, C. A Time-domain Approach of Ion Flow Field around AC-DC hybrid Transmission Lines Based on Method of Characteristics. *IEEE Trans. Magn.* **2015**, *52*, 7205004. [CrossRef]

11. Lu, T.; Feng, H.; Cui, X.; Zhao, Z.; Li, L. Analysis of the ionized field under HVDC transmission lines in the presence of wind based on upstream finite element method. *IEEE Trans. Magn.* **2010**, *46*, 2939–2942. [CrossRef]

12. Zhou, X.; Lu, T.; Cui, X.; Zhen, Y.; Liu, G. Simulation of ion-flow field using fully coupled upwind finite-element method. *IEEE Trans. Power Deliv.* **2012**, *27*, 1574–1582. [CrossRef]

13. Levin, P.L.; Hoburg, J.F. Donor Cell-Finite Element Descriptions of Wire-Duct Precipitator Fields, Charges, and Efficiencies. *IEEE Trans. Ind. Appl.* **1990**, *26*, 662–670. [CrossRef]

14. Zhang, B.; He, J.; Zeng, R.; Gu, S.; Cao, L. Calculation of Ion Flow Field Under HVdc Bipolar Transmission Lines by Integral Equation Method. *IEEE Trans. Magn.* **2007**, *43*, 1237–1240. [CrossRef]

15. Long, Z.; Yao, Q.; Song, Q.; Li, S. A second-order accurate finite volume method for the computation of electrical conditions inside a wire-plate electrostatic precipitator on unstructured meshes. *J. Electrost.* **2009**, *67*, 597–604. [CrossRef]

16. Yin, H.; He, J.; Zhang, B.; Zeng, R. Finite volume-based approach for the hybrid ion-flow field of UHVAC and UHVDC transmission lines in parallel. *IEEE Trans. Power Deliv.* **2011**, *26*, 2809–2820. [CrossRef]

17. Yin, H.; Zhang, B.; He, J.; Zeng, R.; Li, R. Time-domain finite volume method for ion-flow field analysis of bipolar high-voltage direct current transmission lines. *IET Gener. Transm. Distrib.* **2012**, *6*, 785–791. [CrossRef]

18. Yang, F.; Liu, Z.; Luo, H.; Liu, X.; He, W. Calculation of ionized field of HVDC transmission lines by the meshless method. *IEEE Trans. Magn.* **2014**, *50*, 7200406. [CrossRef]

19. Bian, X.; Yu, D.; Meng, X.; Macalpine, M.; Wang, L.; Guan, Z.; Yao, W.; Zhao, S. Corona-generated space

charge effects on electric field distribution for an indoor corona cage and a monopolar test line. *IEEE Trans. Dielectr. Electr. Insul.* **2011**, *18*, 1767–1778. [CrossRef]

20. Lekganyane, M.J.; Ijumba, N.M.; Britten, A.C. A comparative study of space charge effects on corona current using an indoor corona cage and a monopolar test line. In Proceedings of the 2007 IEEE Power Engineering Society Conference and Exposition in Africa, Johannesburg, South Africa, 16–20 July 2007.

21. Kaptzov, N.A. *Elektrische Vorgänge in Gasen und im Vakuum*; VEB Deutscher Verlag der Wissenschaften: Berlin, Germany, 1955; pp. 488–491. ISBN 978-3-446-42771-6.

22. Peek, F.W. *Dielectric Phenomena in High Voltage Engineering*; McGraw-Hill Book Company, Inc: New York, NY, USA, 1920.

23. Abdel-salam, M.; Al-hamouz, Z. A finite-element analysis of bipolar ionized field. *IEEE Tans. Ind. Appl.* **1995**, *31*, 477–483. [CrossRef]

24. Zhou, X.; Cui, X.; Lu, T.; Zhen, Y.; Luo, Z. A time-efficient method for the simulation of ion flow field of the AC-DC hybrid transmission lines. *IEEE Trans. Magn.* **2012**, *48*, 731–734. [CrossRef]

25. Li, X. Numerical Analysis of Ionized Fields Associated with HVDC Transmission Lines Including Effect of Wind. Ph.D. Thesis, The University of Manitoba, Winnipeg, MB, USA, 1997.

26. Urban, R.G.; Reader, H.C.; Holtzhausen, J.P. Small corona cage for wideband HVac radio noise studies: Rationale and critical design. *IEEE Trans. Power Deliv.* **2008**, *23*, 1150–1157. [CrossRef]

2

Is the Dry-Band Characteristic a Function of Pollution and Insulator Design?

Maurizio Albano *, A. Manu Haddad and Nathan Bungay

School of Engineering, Cardiff University, The Parade, Cardiff CF24 3AA, UK; Haddad@cardiff.ac.uk (A.M.H.); BungayN@cardiff.ac.uk (N.B.)

* Correspondence: AlbanoM@cardiff.ac.uk.

† This paper is an extended version of our paper published in 2018 IEEE International Conference on High Voltage Engineering (ICHVE 2018), Athens, Greece, 10–13 September 2018; 0-TM5-5.

Abstract: This paper assesses the dry-band formation and location during artificial pollution tests performed on a 4-shed 11kV insulator with conventional and textured surface designs in a clean-fog chamber and with the application of a voltage ramp-shape source. The different designs present the same overall geometrical dimensions, but the textured ones are characterized by the application of a patented insulator surface design. Three pollution levels, extremely high, high and moderate, were considered. A newly developed MATLAB procedure is able to automatically recognize the perimeter of the insulator, the trunk and shed areas on infra-red recordings. In addition, using the vertical axis identification, all trunks are subdivided into zones and into left and right areas, significantly increasing the capability of abnormalities detection. Any temperature increase within these areas enables to detect the appearance and the extension of dry bands. The results of the analysis of the statistical location and extension development over time of the dry bands during these set of comparative tests show a clear distinction between designs and pollution levels. These results may offer interesting design guidelines for dry-band control.

Keywords: insulator design; dry band; pollution

1. Introduction

Pollution is a key aspect for high voltage insulator design and selection. Pollution deposition and wet conditions can initiate discharge activity and can lead to flashover and, consequently, the availability of a high-voltage transmission line is compromised. The adoption of silicone rubber insulators is justified by their improved performance under polluted environments in comparison with traditional ceramic and glass insulators, and particularly when the pollution level is severe. However, the superior performance of the composite material is compromised when discharge activity is established on the insulator area and, consequently, dry bands appear. This activity determines a hydrophobic reduction of the silicone rubber, and if it persists, no hydrophobicity recovery can be initiated. Moreover, further partial arcs can initiate localized erosion, permanently weakening the silicone rubber insulator surface, as presented in several research works [1–5].

Numerous works [6–11] have been published recently on dry-band formation and pollution estimation, showing the importance of this topic. The papers [6,7] propose an adaptive technique to predict the contamination level increase in order to warn the utility before reaching high levels, conditions that are favourable to extensive dry-band activities. In [8,9], dry-band formation is studied applying controlled non uniform areas of pollution on flat samples. In [10,11], the influence of multiple dry bands on flashover characteristics under various environmental and pollution conditions was investigated on rectangular silicone rubber test samples.

Previous research works [12,13] performed by the authors showed the valuable information obtained by the analysis of dry-band and infrared data. However, the dry-band location was not fully automatically detected along the full insulator, but always with user intervention requiring considerable amount of time.

In this research work, high voltage tests have been performed on 4-shed 11kV insulators with various surface designs in a clean-fog chamber test facility applying a wide range of artificial pollution levels [14]. The more common locations of the dry bands have been identified by analysing the infrared recordings. Further characteristics, such as extension and temperature profile, have been investigated during the time frame of the tests, applying a newly developed algorithm. The MATLAB procedure is able to identify the insulator physical boundaries, the vertical axis and trunk sub-areas, in order to evaluate, with good accuracy, the extension of each dry band and to monitor the variation of these hot zones continuously during the entire test. The series of clean-fog tests have been performed adopting conventional and three textured designs. In addition, after the application of the artificial pollution layer and a resting period of 24 hours, the equivalent salt deposit density (ESDD) and the Non-Soluble Deposit Density (NSDD) have been determined on all selected insulator designs. The ESDD and NSDD levels can contribute to the dry-band characteristics. These results confirm the interesting design advantage of texturing for dry-band control.

2. Experimental Set-Up

The experimental set-up adopted for this investigation was based on a clean-fog test chamber and a programmable voltage ramp-shape source to energise the insulators, previously artificially polluted. The high voltage source was provided by a Hipotronics 150 kVA transformer, and the primary circuit was fed via a Paschen transformer (0–960 V) providing a programmable voltage variation up to 75 kV. In this series of tests, the voltage control unit is set up to provide a constant rate of increase of 4 kV per minute to achieve the ramp shape until the flashover event; this permits to assess the insulator performance under increasing stress in a limited period of time. This methodology has been proposed by the Cardiff University research team, and it is described in detail in [15]. A schematic diagram of the measurement set-up used in the experiment is shown in Figure 1. Previous investigations focused on evaluating the impact of fog rate and pollution level applied as ESDD and NSDD on the insulator surface, and each test insulator was subjected to a ramp series (between 4 and 12 successive flashovers, with a 5 min interval). These comparative laboratory tests, based on voltage ramp shape and an artificially polluted insulator in a clean fog chamber, facilitated the evaluation of the impact on withstand levels of textured design compared with plain surface insulators. A clear indication of improved flashover withstand performance of textured insulators in comparison with conventional surfaces was demonstrated in the research investigation [16].

In this work, only the first ramp was selected in order to have a uniform temperature distribution over the insulator surface at the start of each test. Further tests have been performed using selected constant voltage levels to observe the stability of the dry-band extension over time and the most appropriate methodology to identify the dry-band formation.

The insulator designs selected for this investigation share an identical axial length (175 mm), trunk diameter (28 mm) and shed diameter (90 mm), as illustrated in Table 1. The increased creepage length offered by texturing the surface of the insulator (TT) is shown in the first column of Table 1. Additional increments of the creepage length is offered in the TTS design, where a logarithmic spiral double-ridge pattern is applied under the sheds. The insulator with an all plain surface is referred to in this work as the conventional design (CONV).

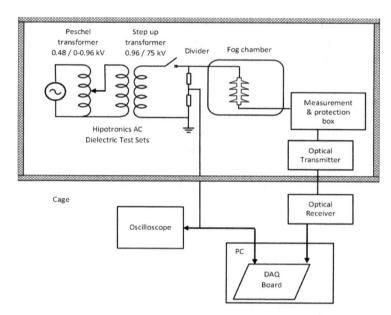

Figure 1. Schematic diagram of the measurement set-up used in the experiment [15].

Table 1. Insulator designs.

Design	Creepage Length (mm)	Dimple Size Radius (mm)	Under Shed Profile (mm)	Axial Length (mm)	Trunk Diameter (mm)
CONV	375	0	0	175	28
TT4	471	4	0	175	28
TTS4	503	4	4	175	28
TT6	471	6	0	175	28
TTS6	503	6	6	175	28

During the laboratory tests, video and images were recorded with a high-definition camcorder and a still camera to obtain signs of the discharge phenomena in the trunk areas. Unfortunately, the faint discharges were not clearly visible, and they could be observed only using long-time exposure techniques. Simultaneously, a FLIR A325 camera was adopted and fixed on the same tripod of the visual camera in order to achieve a close focal point between the two recordings. IR pictures were recorded using a 1 Hz frame rate to minimize memory storage requirement. The IR camera offers the capability to monitor the temperature at a frequency rate up to 60 Hz and with a precision of 0.5 °C. These capabilities are clearly not sufficient to record the instantaneous maximum temperature of the discharge. However, the IR records can show with good precision the maximum temperature of any area of the insulator. The camera does not offer any triggering signal, and only by using the time stamp associated with each IR frame recorded was it possible to synchronize the temperature with all the other data recorded during the test, such as electrical parameters and visual records, with a precision of 1 ms [14].

3. Experimental Results and Dry-Band Detection Procedure

A long series of laboratory tests were performed on the five insulator designs after the application of extremely high, high and moderate pollution levels, 1.15, 0.64 and 0.42 mg/cm^2, respectively, and a constant NSDD of 0.1 mg/cm^2.

In the initial tests, selected constant voltage levels were applied and some discharge activity was identified in visual recordings only using a long-exposure digital stills camera. These small discharges were not visible in the high-definition (HD) digital video camera recordings but only in the long-exposure photo. These discharges, of the corona/streamer type, are characterized by a purple-blue colour as shown in Figure 2b. The associated dry bands were localized only with the help of the IR camera, as shown in Figure 2c. The visual photos were post-processed in order to enhance the spark visibility. A second type of discharge, the streamer discharges, are shown by a red/orange colour crossing an extended area than that of the dry band.

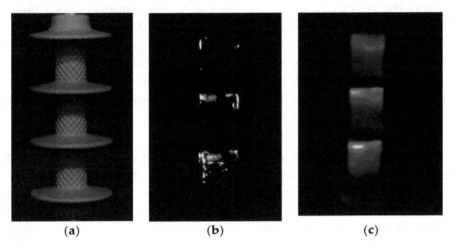

(a)   (b)   (c)

Figure 2. Visual frame before fog application (**a**), visual frame using long-exposure acquisition (**b**) and the IR frame (**c**) during the constant voltage test on sample TT4; equivalent salt deposit density (ESDD) was 0.64 mg/cm^2 with a fog spray rate of 3 L/h.

The synchronized IR images present areas of increased temperature (Figure 2c) where the visual images showed the presence of fully formed dry bands (Figure 2b). It is important to observe that no significant streamer activity is present in the upper trunk. However, the presence of a dry band in each trunk is confirmed by the IR photos, with a lower temperature magnitude on the upper one.

A differential temperature between the insulator surface and the surrounding fog along a vertical segment has been calculated for each test. After preliminary tests, using constant voltage, a fixed threshold of 5.5 °C appeared to be a quick identification method. However, if this methodology would be simply applied to the all extended ramp voltage tests the results would be not correct. In fact, a lower extension of dry bands would have been detected, with the temperature profile presented in Figure 3 as example. Moreover, the identification of the dry band is very time consuming if performed by the user by manually examining a significant number of frames and series of tests and this suggested the need for developing an automatic procedure.

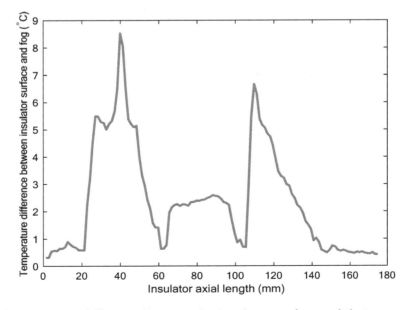

Figure 3. Temperature difference between the insulator surface and the surrounding fog.

The maximum average temperature on the whole insulator surface during the tests and the average values on identified areas can be evaluated from the thermal recordings. The extraction of the maximum temperature recorded during the test and the observation of its temporal variation can help to highlight only the timestamp of the spark/dry bands events but no information about their location, size or number. These limitations suggested the authors to develop a MATLAB computer procedure to acquire and process the temperature variation at different areas of the insulator. Preliminary investigations showed higher frequency of occurrence of dry-band and corona discharge events on the trunks in comparison with the shed areas. Consequently, increased surface temperature or spark events appeared mainly on these areas before flashover, suggesting limiting the analysis mainly on these areas.

3.1. Computer Technique for Identification of Dry-Band Boundaries

The precise identification of the insulator location and its perimeter on the thermal frame is a fundamental feature of the proposed analysis. The authors developed a MATLAB procedure to automatically detect the profile boundary of the insulator. This procedure is based on the analysis of the small variations between the ambient air surrounding the insulator and the insulator surface. Since the temperature difference between the insulator and the fog is quite small at the start of each test, the automatic identification is not a trivial task. In addition, possible reflections on the object can cause significant error in the outline identifications. A schematic diagram of the procedure adopted is shown in Figure 4.

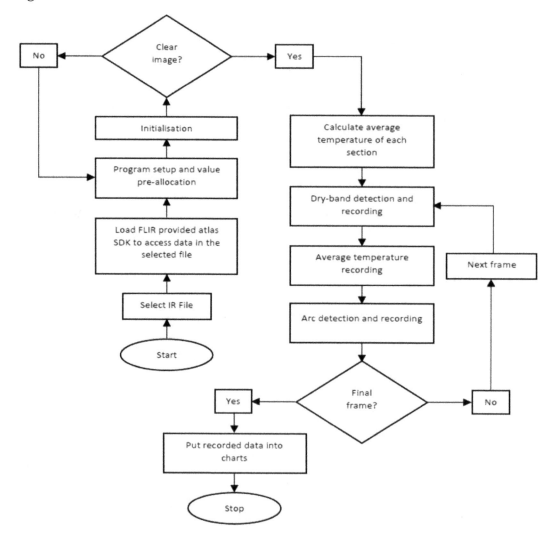

Figure 4. Schematic diagram of the procedure adopted.

The boundary detection procedure is based on Canny Edge detection, but other algorithms, such as Sobel, Prewitt and Roberts, were also tested. The Canny algorithm showed the best accuracy when applied to this environmental condition. The two threshold levels required by the Canny algorithm are automatically selected using two key areas of the frame, assuming the left corner as representation of the fog temperature and the hotter area indicating the insulator position within the frame. Once the perimeter is extracted, it is subdivided in two sections, the left- and right-hand boundaries, as can be seen in Figure 4 with the red and blue colours, respectively. Then, they are plotted over the frame for final acceptance by the user. As shown in Figure 5b, the boundaries are not always identified completely, and the user can decide to modify the thresholds or introduce some corrections. However, despite some uncertainty on sheds areas, the areas of interest, the trunks are correctly identified in most cases.

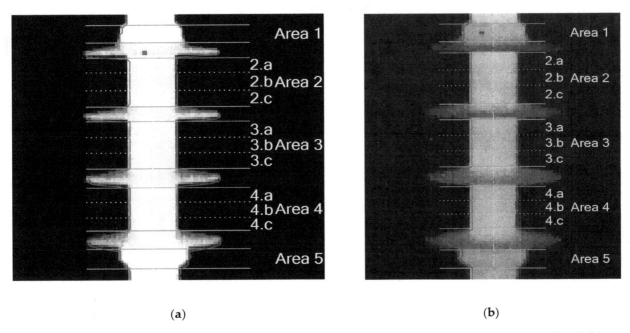

(a)   (b)

Figure 5. Automatic boundaries recognition and area selection applied to (**a**) a conventional and (**b**) a textured design.

A second step of the procedure aims to identify the shed and trunk areas, since only the trunk areas are processed on the temperature analysis. The trunk surfaces have been subdivided into three zones (a, b, c), identified by segmented white lines in Figure 5. The presence of the shed characterizes the type area a, meanwhile the type area b is located on the central area, more exposed to the fog and prone to a stronger wetting action. The remaining one, area type c, is located on the trunk near the top shed zone.

Assuming the insulator is always on an almost vertical position, the average of all the horizontal coordinates of the trunks permits to identify the position of the vertical axis of symmetry, drawn on Figure 3 as a dot–dash white line.

3.2. Average Temperature Calculation

The precise identification of the boundaries of the insulator enables the extraction only of the values related to each area from the IR temperature data, neglecting any measurement related to the fog. The procedure calculates the average temperature of each area for each frame. This permits to identify different trends related to the design. Figure 6 shows the two end-fitting areas during the whole test for the conventional and textured TT4 insulators.

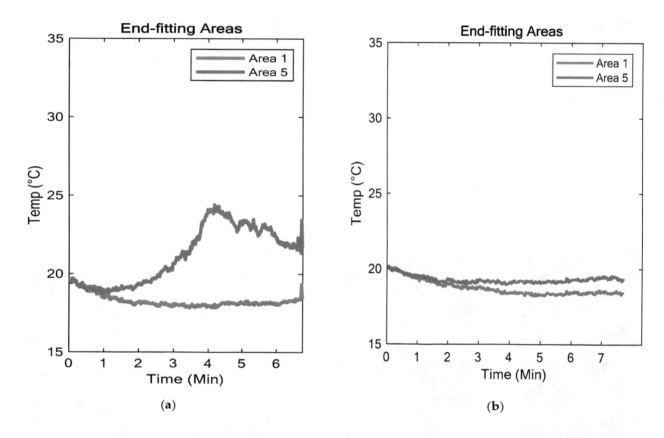

Figure 6. Average temperature of end-fitting areas during the whole test, on (**a**) conventional and (**b**) textured TT4 insulators.

In the first two minutes, the fog formation determines a cooling effect on the initial period. Afterwards, a steady increase in dry-band extension is observed as the voltage and current increase. The polluted insulator previously stored at laboratory temperature is placed in the fog chamber where, as the test begins, the fog cools the insulator surface until the discharge activity heats up the surface.

In particular, the temperature of area 5, which is the trunk close to the ground termination, is observed to increase in many tests of the conventional design. The increment of temperature in the graph confirms the presence of a dry band and discharge activity. This trend is not followed for the textured design, indicating no significant activity in this area. The area 1, which is the trunk area close to the high voltage termination, shows no discharge activity for both designs.

Figure 7 shows the average temperature of each zone of the three trunk areas for a TT4 design, with a pollution level of 0.64 mg/cm^2. The temperature trends are aggregated in two subplots to facilitate the comparison between the left- and right-hand areas. The comparison between left and right areas confirms that the artificial pollution has been applied correctly to achieve uniform ESDD on the surface and no significant variation is observed.

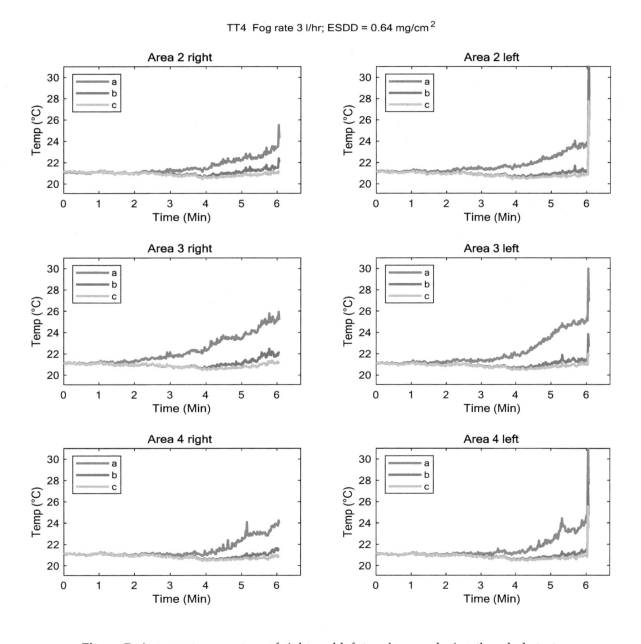

Figure 7. Average temperature of right and left trunk areas during the whole test.

From the plots, the different temperature trends are clearly visible for zone a, suggesting that the formation of dry bands is facilitated by the presence of the shed, shielding the wetting on this zone. Another clear indication from the average temperature of subareas is that subarea c always presents the lowest average temperature in each trunk. This suggests that the wetting is enhanced in the other two subareas, meanwhile in the area closer to the shed, the wetting is slightly reduced, creating better conditions for dry-band formation.

It is noted that the textured design tends to initiate dry bands on all the three areas, creating a more uniform distribution. This is not observed on the conventional design, exhibiting a more localized formation, as shown in the Appendix A in Figure A1. In fact, the middle trunk, area 3, does not show any significant increment, meanwhile the other two trunks show temperature increments in zone a. A rapid increase in temperature at around 5 to 6 min before the flashover event is observed.

3.3. Dry-Band Analysis

Another useful indication is the localization and progression of any dry band during the test. The identification of dry bands using a fixed selected threshold cannot be detected correctly, as described in the temperature profile reported in Figure 3. In addition, using a variable level function only on the first IR frame is still not always applicable because of the cooling trend by the fog, as described previously.

The proposed algorithm of dry-band detection is based on identifying the points/area where the minimum average value is exceeded within the trunk zone. This permits to take into account any cooling effect in the initial period or overall variation not caused by a localized event. Each dry-band width is then evaluated calculating the average temperature of each left- and right-hand horizontal row of data given by the individual IR pixels. If these values exceed the threshold, the row is flagged and counted. The total dry band extension along the vertical axis for each frame is calculated converting the number of pixels in mm, given the distance between the terminals of the insulator; and in this case, it is equal to 175 mm.

The sum of all dry-band extensions along the insulator axis for all the duration of the test (Design CONV, ESDD 0.64 mg/cm^2, fog rate 3 L/hr) is presented in Figure 8. An analogue test using a lower pollution level (ESDD 0.42 mg/cm^2) is shown in the Appendix A as Figure A3.

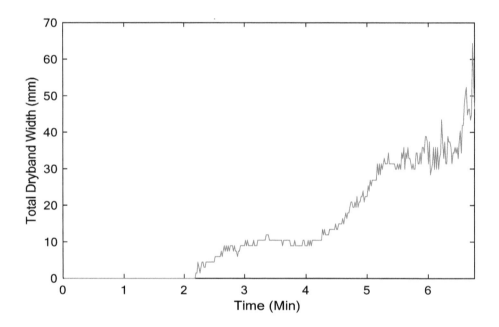

Figure 8. Cumulative dry-band extension on all the trunks. Ramp-test: Design CONV, ESDD = 0.64 mg/cm^2, and a fog rate of 3 L/hr.

In order to gain a deeper understanding of the growth of each dry band, the calculated dry-band location and duration were plotted on a single graph taking advantage of the contour facility introducing time as the x-axis value, the vertical position as the y-axis value and the temperature of the dry band as the colour level. In addition, the last frame of the IR image is automatically cropped according to the identified boundaries and rescaled along the vertical axis of symmetry, facilitating the user to localize the dry band on the insulator. The resulting plot is shown in Figure 9. This graph offers a valuable overview of the full test since each dry-band width is presented as function of time and as its dry-band temperature range. The graph allows the comparison of the dry-band temperature distributions for a specific design very readily. The individual dry-band extension function of time and the dry-band temperature range for the conventional design (Figure 9a) and textured TT4 insulator (Figure 9b) are also presented.

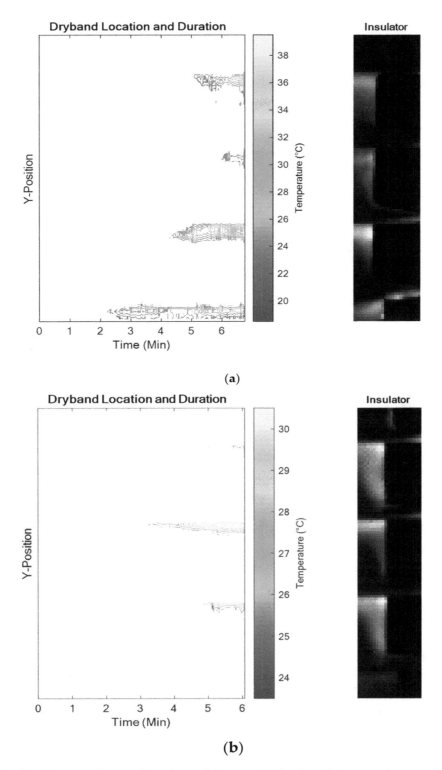

Figure 9. Dry-band extension as function of time and dry-band temperature range for the (**a**) conventional design and (**b**) textured TT4 insulator examples.

4. Discussion

In order to estimate the effect of the design on the dry band dimensions, the total dry-band extension of the selected designs for a given ESDD and fog spray rate were plotted on the same graph, as shown in Figure 10. The graph shows a selection of calculated dry band extensions for all four insulator designs with an applied pollutant on the surface equal to ESDD 0.64 mg/cm^2, and a fog spray rate equal to 3 L/h. It is clear that the conventional design exhibits dry-band formation much earlier than textured insulators. As early as after 1 min, the conventional (CONV) samples start to develop dry bands which continue to extend with increasing applied voltage. This trend is significantly delayed on textured (TT) samples, where the total dry-band length starts to be significant only after 4 min. The curves before flashover shows different trends. In fact, the dry-band dimensions on the TT6 design increases sharply, whereas for TT4 and TTS4 a delayed increase and a lower maximum width is observed before the flashover event. If the analysis was limited to only flashover voltage levels, the test results would not have highlighted any other details such as the clearly visible differences in this graph [10]. This suggests that the new procedure can offer a valuable tool to analyse and compare insulator performance under pollution.

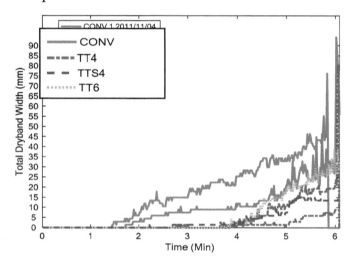

Figure 10. Comparison of the dry-band extension on all the trunks for the five designs, with ESDD = 0.64 mg/cm^2, and a fog spray rate of 3 L/h.

5. Conclusions

The paper presented a novel procedure to assess the formation, and to identify the location, of dry bands based on the automatic identification of selected areas and the calculation of the associated average temperature. In particular, the division of sub zones and left and right areas increase the detection capabilities from the IR recordings.

The proposed methodology has been applied on artificially polluted insulator surfaces under a ramp-shape voltage test in clean-fog chamber and the results of the analysis show that the two features are clearly a function of the insulator surface design and the pollution level applied to the insulator surface (ESDD). The automatic identification of the insulator boundaries allows an accurate estimation of the average local temperature, disregarding any area related to the surrounding environment. In addition, the choice of selected area on the trunks permits to take into account the presence of the sheds that can perturb the wetting action of the fog. The evaluation of local average temperature on symmetrical areas along the vertical axis permits to confirm the correct uniform application of a pollution layer and to warn about possible localized defects on the surface.

Pollution level is one of the most important parameters in insulator design selection. Any increased discharge activity caused by pollution on a specific insulator design affects the life expectancy, which is a key parameter that network operators have to asses in the adoption of a new design. Since one of the major causes of degradation is caused by continuous discharges on the polymeric insulator surfaces, this proposed new procedure based on the analysis of IR video recordings and on the spatial and time characterization of dry bands may provide an indication of the selection of the most appropriate surface design to maximize insulator-life extension.

The results show it is possible to estimate the location, extension and development-over-time of dry bands, and these features offer good indications to select the appropriate design for dry-band control.

Author Contributions: Conceptualization, M.A. and A.M.H.; writing—original draft preparation, M.A.; writing—review and editing, M.A. and A.M.H.; visualization, N.B.

Appendix A

Appendix A shows an example of ramp test performed on insulator design CONV, with applied pollution of ESDD 0.42 mg/cm^2 and a fog rate 3 L/hr. Figures A1 and A2 shows the average temperature of the main trunk and top and lower trunk areas during the whole test respectively. Figure A3 shows the total dry-band length computed for the same selected test. Figure A4 shows the dry-band development on each trunk as a maximum temperature profile on a 3D plot versus time and vertical axis position.

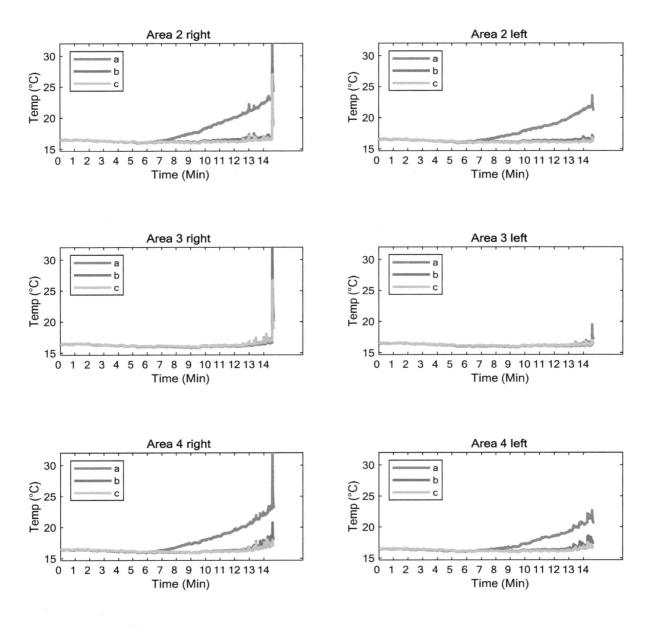

Figure A1. Average temperature of the main trunk areas during the whole test (left and right zones). Design CONV, ESDD = 0.42 mg/cm^2, and a fog rate of 3 L/hr.

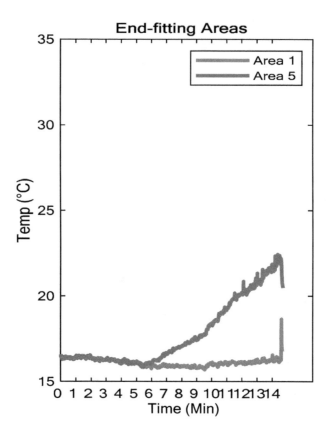

Figure A2. Average temperature of top and lower trunk areas during the whole test (left and right zones). Design CONV, ESDD = 0.42 mg/cm^2, and a fog rate of 3 L/hr.

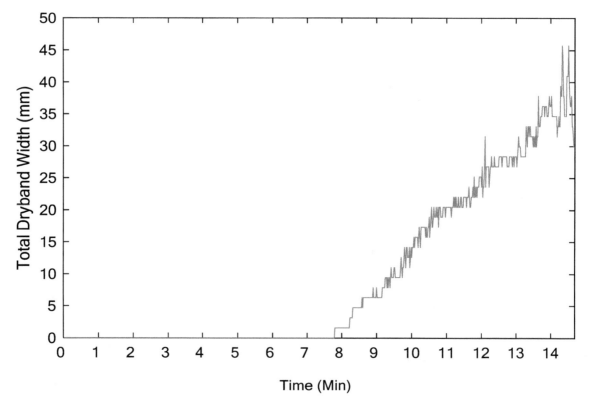

Figure A3. Total dry-band length. Ramp-test: Design CONV, ESDD = 0.42 mg/cm^2, and a fog rate of 3 L/hr.

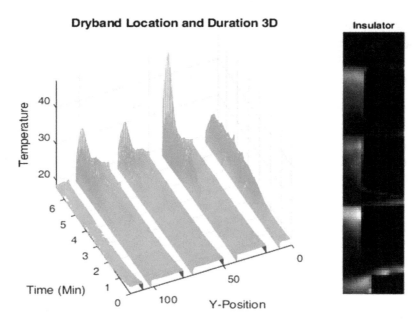

Figure A4. 3D representation of dry-band development on each trunk. Ramp-test: Design CONV, ESDD = 0.42 mg/cm^2, and a fog rate of 3 L/hr.

References

1. Gorur, R.S.; Cherney, E.A.; Hackam, R.; Orbeck, T. The Electrical Performance of Polymeric Insulating Materials Under Accelerated Aging in a Fog Chamber. *IEEE Trans. Power Deliv.* **1988**, *3*, 1157–1163. [CrossRef]
2. Moreno, V.M.; Gorur, R.S. Effect of long-term corona on non-ceramic outdoor insulator housing materials. *IEEE Trans. Dielectr. Electr. Insul.* **2001**, *8*, 117–128. [CrossRef]
3. Blackmore, P.; Birtwhistle, D. Urface discharges on polymeric insulator shed surfaces. *IEEE Trans Dielectr. Electr. Insul.* **1997**, *4*, 210–217. [CrossRef]
4. Moreno, V.M.; Gorur, R.S. Impact of corona on the long-term performance of nonceramic insulators. *IEEE Trans. Dielectr. Electr. Insul.* **2003**, *10*, 80–95. [CrossRef]
5. Meng, D.; Zhang, B.-Y.; Chen, J.; Lee, S.-C.; Jong-Yun Lim, J.-Y. Tracking and erosion properties evaluation of polymeric insulating materials. In Proceedings of the 2016 IEEE International Conference on High Voltage Engineering and Application (ICHVE), Chengdu, China, 19–22 September 2016; pp. 1–4.
6. Vita, V.; Ekonomou, L.; Chatzarakis, G.E. Design of artificial neural network models for the estimation of distribution system voltage insulators' contamination. In Proceedings of the 12th WSEAS International Conference on Mathematical Methods, Computational Techniques and Intelligent Systems (MAMECTIS '10), Kantaoui, Sousse, Tunisia, 3–6 May 2010; pp. 227–231.
7. Pappas, S.S.; Ekonomou, L.; Heraklion, N. Comparison of adaptive techniques for the prediction of the equivalent salt deposit density of medium voltage insulators. *WSEAS Trans. Power Syst.* **2017**, *12*, 220–224.
8. Nekahi, A.; McMeekin, S.G.; Farzaneh, M. Influence of Dry Band Width and Location on Flashover Characteristics of Silicone Rubber Insulators. In Proceedings of the 2016 Electrical Insulation Conference (EIC), Montréal, QC, Canada, 19–22 June 2016.
9. Nekahi, A.; McMeekin, S.G.; Farzaneh, M. Measurement of surface resistance of silicone rubber sheets under polluted and dry band conditions. *Electr. Eng.* **2017**. [CrossRef]
10. Dhahbi-Megriche, N.; Slama, M.E.A.; Beroual, A. Influence of dry bands on polluted insulator performance. In Proceedings of the 2017 International Conference on Engineering & MIS (ICEMIS), Monastir, Tunisia, 8–10 May 2017; pp. 1–4. [CrossRef]
11. Arshad; Mughal, M.; Nekahi, A.; Khan, M.; Umer, F. Influence of Single and Multiple Dry Bands on Critical Flashover Voltage of Silicone Rubber Outdoor Insulators: Simulation and Experimental Study. *Energies* **2018**, *11*, 1335. [CrossRef]
12. Albano, M.; Haddad, A.; Griffiths, H.; Waters, R.T. Dry-band characterisation using visual and IR data analysis. In Proceedings of the International Conference on High Voltage Engineering and Application—ICHVE2014, Poznan, Poland, 8–10 September 2014. [CrossRef]

13. Albano, M.; Waters, R.T.; Charalampidis, P.; Griffiths, H.; Haddad, A. Infrared Analysis of Dry-band Flashover of Silicone Rubber Insulators. *IEEE Trans. Dielectr. Electr. Insul.* **2016**, *23*, 304–310. [CrossRef]

14. Albano, M.; Haddad, A.; Bungay, N. Is the dry-band characteristic a function of pollution and insulator design? In Proceedings of the 2018 IEEE International Conference on High Voltage Engineering and Application (ICHVE), Athens, Greece, 10–13 September 2018; pp. 1–4. [CrossRef]

15. Charalampidis, P.; Albano, M.; Griffiths, H.; Haddad, A.M.; Waters, R.T. Silicone Rubber Insulators for Polluted Environments Part 1: Enhanced Artificial Pollution Tests. *IEEE Trans. Dielectr. Electr. Insul.* **2014**, *21*, 740–748. [CrossRef]

16. Albano, M.; Charalampidis, P.; Griffiths, H.; Haddad, A.M.; Waters, R.T. Silicone Rubber Insulators for Polluted Environments Part 2: Textured Insulators. *IEEE Trans. Dielectr. Electr. Insul.* **2014**, *21*, 749–757. [CrossRef]

3

Application of Machine Learning in Transformer Health Index Prediction

Alhaytham Alqudsi [1] and Ayman El-Hag [2,*]

[1] Mechanical Engineering Department, École de technologie supérieure, Montréal, QC H3C 1K3, Canada
[2] Electrical and Computer Engineering, University of Waterloo, Waterloo, ON N2L 3G1, Canada
* Correspondence: ahalhaj@uwaterloo.ca.

Abstract: The presented paper aims to establish a strong basis for utilizing machine learning (ML) towards the prediction of the overall insulation health condition of medium voltage distribution transformers based on their oil test results. To validate the presented approach, the ML algorithms were tested on two databases of more than 1000 medium voltage transformer oil samples of ratings in the order of tens of MVA. The oil test results were acquired from in-service transformers (during oil sampling time) of two different utility companies in the gulf region. The illustrated procedure aimed to mimic a realistic scenario of how the utility would benefit from the use of different ML tools towards understanding the insulation health index of their transformers. This objective was achieved using two procedural steps. In the first step, three different data training and testing scenarios were used with several pattern recognition tools for classifying the transformer health condition based on the full set of input test features. In the second step, the same pattern recognition tools were used along with the three training/testing scenarios for a reduced number of test features. Also, a previously developed reduced model was the basis to reduce the needed number of tests for transformer health index calculations. It was found that reducing the number of tests did not influence the accuracy of the ML prediction models, which is considered as a significant advantage in terms of transformer asset management (TAM) cost reduction.

Keywords: feature selection; insulation health index; machine learning; oil/paper insulation; transformer asset management

1. Introduction and Background

One of the major parameters that define the operation and planning of an electrical utility is the transformer asset health condition. Based on their health condition, electrical utility engineers can predict the transformer useful remnant lifetime. Such an understanding can benefit utility companies to prepare a proper financial plan to estimate the future cost of maintenance and replacement for the transformer units. Significant research has been conducted to help utility companies in cutting their asset maintenance costs. This area of research is commonly referred to as the transformer asset management (TAM) practice [1]. TAM, as explained in [1–3], defines a strategic set of future maintenance and replacement activities for the utility transformer asset based on diagnostic testing methods of the transformer health condition. The ultimate objective of TAM is to ensure the power system reliability within an economic platform. Abu-Elanien in [2] defines the diagnostic testing methods in its two-part forms of condition monitoring (CM) and condition assessment (CA). CM refers to all the electrical, chemical, and physical tests that are used collectively towards CA tools that determine the transformer health condition. Azmi et al. states that TAM practices are at their best when they are comprised of both the CA and financial information [3]. Having knowledge of the transformer history (loading and failure history), associated risk index (based on the load it feeds), current health condition, and all related financial costs (maintenance, operation, and failure) would result in an economic risk management plan with adequate subsequent decisions.

1.1. Insulation Health Index Computation

According to the literature, the transformer health condition is governed by its oil-paper insulation condition [1,4,5]. According to [1], analyzing the transformer oil samples would be more advantageous than testing other transformer components (turns ratio, winding resistance, leakage reactance, etc.) for fault detection and life expectancy. With the health condition of all transformer components being taken into consideration, the overall health condition can be defined using the health index (HI). The health index, as explained by Jahromi et al. [6,7], is a single index factor which combines operating observations, field inspections, and laboratory tests to aid in the TAM cycle. The insulation condition is a vital part of the HI computation that could suffice when limited data is available for the transformer service record and design. Understanding the insulation condition would require thorough transformer oil sample analysis through electrical, chemical, and physical laboratory tests.

All conducted oil insulation laboratory tests fall into one of three major test categories, which are namely the dissolved gases (DGA), oil quality (OQA), and furan (FFA) tests [6,7]. DGA tests are typically conducted for the detection of transformer internal faults (electrical and/or thermal). The dissolved gases include hydrogen (H_2), methane (CH_4), ethane (C_2H_6), ethene (C_2H_4), ethyne (C_2H_2), carbon monoxide (CO), and carbon dioxide (CO_2) [8]. For the second category of tests, OQA, the oil quality is determined by testing the oil breakdown voltage (BDV), acidity, water content, interfacial tension (IFT), dielectric dissipation factor (DDF), and color [9]. Finally, the measurement of FFA determines the extent of paper insulation degradation through determining the furfuraldehyde (or commonly known as furan) content in the transformer oil. Furan is a chemical compound that dissolves in the insulation oil upon the breakdown of the cellulose chain of the paper material [10]. The furan test is a strong indicator of transformer paper insulation ageing. Collectively, laboratory tests on dissolved gases, oil, and paper quality would be used to compute the HI value based on a given formula that is developed by experts in the TAM field. Examples of such different formulas can be found in [6–13], with the method illustrated in [6,7] being solely used for computing the HI for the majority of publications.

1.2. Novel Methods for HI Computation

The approach of using artificial intelligence tools, such as machine learning (ML) technologies and fuzzy logic for the HI computation, was well studied in several publications. In Ref. [14], the assessment of the HI was done using a neuro-fuzzy approach. The aim of the study was to test the performance of the five-layer based neuro-fuzzy model in computing the HI as per the scoring method dictated in [6]. The inputs to the model were the oil test features that have been taken from in-service transformers records. Though limited in the number of available data, the author was able to show a prediction accuracy of more than 50% using the developed neuro-fuzzy model. In other works, such as [15], a general regression neural network (GRNN) was developed for the HI of four-class based condition assessment (namely, very poor, poor, fair, and good). For this particular work, the transformer health was graded based on international oil assessment standards (such as that of [9]). Six key inputs, including the oil total dissolved combustible gas, furan content, dielectric strength, acidity, water content, and dissipation factor, were utilized. An 83% success was reported in predicting the transformer health condition. Zeinoddini-Meymand et al. in [16] used an artificial neural network and neuro-fuzzy models to include the basic oil assessment test features and additional economical parameters that have not been accounted for in many publications. These economical parameters include the transformer percentage economical life-time and ageing acceleration factor. According to [16], the inclusion of such parameters in the HI condition-class problem resulted in an excellent assessment performance, which correlated with that of the field experts.

In the work of [17], probabilistic Markov chain models were used in predicting the future performance of the transformer asset based on their HI computation for a defined span of time. Transition probabilities have been derived (using a non-linear optimization technique) to be the core element of the Markov chain model, which in turn was used to predict the future HI of the transformer asset. The reported results indicate satisfactory prediction performances for a number of tested transformers. An interesting approach was presented by Tee et al. in [18] for determining the transformer health condition using principle component analysis (PCA) and analytical hierarchy process (AHP). Data of the transformer oil quality tests and age have been used in both the aforementioned techniques to rank the transformer asset based on the insulation health condition. The ranking obtained by both techniques showed a comparable performance to that of an expert-based empirical formula assessment for the same transformer asset.

In Ref. [19], a fuzzy-based support vector machine (SVM) was used in HI-condition assessment. The HI condition of a given transformer was determined based on a number of factors, including industry standards and utility expert judgments. The SVM model showed a classification rate of 87.8%. A fuzzy-logic based model was also incorporated in [20], where the HI class condition (three classes) was predicted using the technical oil test features as inputs. In other attempts, such as [21], a general study was conducted using different conventional feature selection methods on oil test features for predicting the HI condition with different ML techniques.

With reference to all the reported works in this paper, a promising future is foreseen for the use of ML in the TAM field. As was indicated earlier, predicting the HI of the transformer asset will substantially impact the financial strategy of the utility company in asset maintenance plans. The objective of this work was to extend our previous research of [22] and establish a platform for using a wide range of ML tools in understanding the transformer health condition.

1.3. Organization of the Presented Work

In the following sections of this paper, the transformer databases used for this study will be introduced. The methodology of computing the HI value and accordingly classifying the health condition for a given sample in the oil databases are illustrated. Thereafter, the different ML tools used in the pattern recognition/classification problem of this study are introduced. The stepwise regression feature selection tool will also be introduced. Accordingly, two major steps will be done in achieving the objectives of the presented work:

- Full-feature modelling: The pattern recognition model will be trained and tested based on the complete number of available test features (which is 10 in this study). Eight different pattern recognition methods will be used with three different training and testing scenarios based on the two different oil databases that were acquired in this study.

- Reduced-feature modelling: Based on the reported work of [22], stepwise regression was used as a feature selection tool for predicting the HI value of a given oil sample. Accordingly, it was concluded which oil test features are of the highest statistical significance in computing the HI value of a given transformer. Only the indicated oil test features from [22] will be used from the two databases in a reduced-feature modelling step. Again, eight different pattern recognition methods will be used along with the same three training and testing scenarios used in the full-feature modelling procedure.

2. Materials and Methods

2.1. Transformer Oil Samples

As mentioned earlier, transformer oil sample databases were acquired from two utility companies in the gulf region. For confidentiality purposes, the two companies will be referred to as *Util1* and *Util2*. In total, 730 transformer oils samples were obtained from *Util1*, while 327 transformer oil samples were obtained from *Util2*. Transformers from both databases are medium voltage distribution transformers.

Util1 transformers are 66/11 kV, 12.5 to 40 MVA, while those of *Util2* are 33/11 kV and 15 MVA. For every transformer in both databases, 10 different oil test results are available, namely: H_2, CH_4, C_2H_6, C_2H_4, C_2H_2, BDV, IFT, water content, acidity, and furan.

2.2. Structuring the HI Database

The transformer databases have a total of 1057 data samples combined. The prime objective of this paper is to be able to estimate the transformer insulation health condition using ML with either the entire oil feature set or a partial part of it. The published work in [6,7] is considered as the base method for computing the HI. In this method, all input test features are assigned a score value based on a predefined scale. The scored test feature is then multiplied or scaled by a predefined weight factor and is quantifiably added to the other scaled test features. With the inclusion of other arithmatic operations, three quantitative factors related to the DGA, OQA, and FFA tests are calculated (denoted by β). The β factors are discrete values that range from 1 to 5 (or 1 to 4 for β_{OQA}), with the ascending order of the values indicating a deteriorating condition of the transformer health. The three factors will then be multiplied by their associated weights (denoted by α) to eventually produce the HI value using Equation (1) as illustrated in [6]:

$$HI = \frac{(\beta_{DGA} \times \alpha_{DGA}) + (\beta_{OQA} \times \alpha_{OQA}) + (\beta_{FFA} \times \alpha_{FFA})}{4 \times (\alpha_{DGA} + \alpha_{OQA} + \alpha_{FFA})} \times 100\%. \tag{1}$$

The HI value ranges from 0% to 100%, with 100% being a transformer in the healthiest possible condition. Once the HI is computed, the data sample is classified into one of three health condition classes, which are: Bad (B), fair (F), and good (G). Table 1 illustrates the HI value range for a given class and the number transformers from each utility company in that class.

To have a better visualization of the 10 input variables and how they are related to the health condition of the transformer data, a 3-D plot of the data using the β factors from Equation (1) is depicted in Figure 1. As can be seen in Figure 1, there is a clear and distinct difference between the three classes of data that is influenced by the three β factors.

Table 1. Health index value range of health condition classes and the corresponding number of samples.

Data Group	Good (G), HI > 85%	Fair (F), 50% < HI ≤ 85%	Bad (B), HI ≤ 50%
Util1	496	206	28
Util2	238	84	5
Total from *Util1* and *Util2*	734	290	33

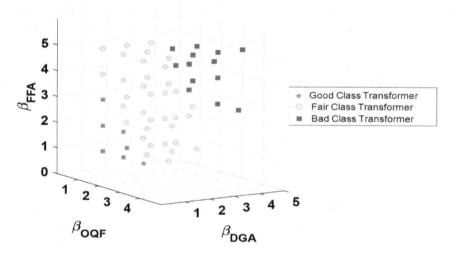

Figure 1. A 3D representation of the input variables through β_{DGA}, β_{OQA}, and β_{FFA}.

2.3. The Machine Learning Methodology

Transformer condition assessment requires the analysis of substantial data, whose instances represent investigated transformers and whose features represent variables measured to predict the transformer HI condition. A pattern recognition or classification model (classifier) can be trained on the dataset so that learning algorithms can operate faster and more effectively; in other words, costs are reduced, and learning accuracy is improved [23]. The methodology presented in this paper is based on feature selection and pattern classification for assessing the HI of power transformers. ML helps in gaining insights into the properties of data dependencies, and the significance of individual attributes in the dataset. Classification is an ML technique used to assign labels (classes) to unlabeled input instances based on discriminant features. Class labels in this study are the three health condition classes. Feature selection techniques determine the important features to include in the classification process of a particular data collection. Ten features are available for this study as was illustrated earlier. The entire process of applying these ML techniques to predict the class of unseen data consists of the typical phases of training and testing.

Consider a binary classification problem with positive (P) and negative (N) classes (i.e., two class problem). As a generalization for such multi-class classification problems, the overall classification accuracy measure is assessed in terms of the quantity of truly and falsely classified samples. Accordingly, a confusion matrix is constructed for recording the frequency of truly positive (TP), truly negative (TN), falsely positive (FP), and falsely negative (FN) samples. Table 2 shows a confusion matrix for binary classification problems, in which the class is either P or N.

Table 2. Confusion matrix of binary classification problems.

		Classified as	
		P	N
Really is	P	TP	FN
	N	FP	TN

As illustrated in [24], the overall classification accuracy measure is calculated by:

$$Accuracy\ Rate = \frac{TP + TN}{TP + FP + TN + FN}. \qquad (2)$$

To validate the classification model, the k-fold cross-validation technique is used. Wherein, the dataset is randomized then equally subdivided into k subsets, of which $k - 1$ subsets are used for training and the remaining subset is retained for testing to validate the resulting model. Then, a different subset is used as a test set and the remaining $k - 1$ are used for training; thus, a second model is built. The process is repeated k times (folds) until k models are built. The final estimation is based on the average of the k results. A 10-fold cross-validation is usually used [25]. This method has the advantage of using all data for both the testing and validation. Moreover, it reduces the standard deviation with random seeds, as compared to methods that split the dataset into two sets, a training set and a testing set.

Feature selection is the process of selecting a subset of relevant, high quality, and non-redundant features for building learning models [26], with improved accuracy [27]. Quite often, datasets contain features with different qualities, which can influence the performance of the entire learning framework. For instance, noisy features can decrease the classifiers' performance. Moreover, some features are redundant and are highly correlated, i.e., they do not give additional information. Considering all available information, provided by all kinds of features, would make it hard for the classifier to discover the real distinguishing characteristics, and would cause it to be overly specified (over fitted) on the examples it is trained with, hence reducing its generalization power drastically. Therefore, it is important to select the appropriate features to base classification on them [28,29]. As explained in the earlier publication of [22] and detailed in [30], stepwise regression deals with studying the statistical significance of a number of test features as they relate to the variable of interest. In other words, stepwise regression will be used to determine which of the 10 oil test features will be adequately sufficient to predict the HI value with the absence of the least significant ones. In the stepwise regression process, a final multilinear regression model is developed for predicting the variable of interest by adding or removing test features in a stepwise manner. The process starts with a single feature present in the regression model. A feature is then added to assess the incremental performance of model in computing the variable of interest (the HI value in this case). In each step, the F-statistic of the added feature in the model is computed. The F-statistic is found by:

$$F_j = \frac{SS_R\left(\gamma_j | \gamma_0, \gamma_1, \gamma_2 \ldots, \gamma_{j-1}\right)}{MS_E} \tag{3}$$

where γ is the regression coefficient of the associated feature in the multilinear regression model. F_j is the F-statistic value with the inclusion of the jth term in the regression model given the other existing test features in the model. SS_R is the regression sum of squares of the data sets' computed model output as compared to data sets' actual values, and MS_E is the mean square of error of the model with all its existing and currently tested features. During each step, a p-value for the F-statistic of the added feature is determined and tested against the null hypothesis. The null hypothesis rejects the idea that the feature in question is statistically significant to the variable of interest. Thus, when performing the stepwise feature selection in the forward manner and the p-value of the added term to the model is found to be below the pre-defined entrance tolerance, the null hypothesis is rejected and the term is added to the model. Once all the forward stepwise process is done, the stepwise process starts to move in the backward manner. If a given test feature that exists in the model has its p-value for the F-statistic above the exit tolerance, the null hypothesis is confirmed and that feature is removed from the final model.

2.4. Classifiers Used in the Study

In this paper, WEKA version 3.8.2 was used as the platform for the different classifiers. It is a collection of machine learning algorithms for data mining tasks developed at the University of Waikato, New Zealand. A brief description of the different machine learning algorithms used by WEKA in this study is presented below.

- *Random Forest (RForest):* It is a form of the nearest neighbor predictor that starts with a standard ML technique called a decision tree [31,32]. RForest is a meta-estimator that fits a number of decision tree classifiers on various sub-samples of the dataset and uses averaging to improve the predictive accuracy and control over-fitting. RForest is a fast classifier that can process many classification trees [22].

- *Decision Tree (J48):* J48 is the Java implementation of the C4.5 decision tree algorithm in the WEKA data mining software [32]. The algorithm builds decision trees by calculating the information gain of the attributes, and then uses the attribute that has the highest normalized information gain to split the instances into subsets of the same class label (called a leaf node of the tree). The algorithm repeats the same process with all split subsets as children of a node. Finally, the tree is pruned by removing the branches that do not help in classification.

- *Support Vector Machines (SVMs):* SVM classifies instances by constructing a set of hyperplanes to separate the class categories [33]. SVMs belong to the general category of kernel methods [34,35], which depend on the data only through dot products. To ensure that the hyperplane is as wide as possible between classes, the kernel function computes a projection product in some possibly high dimensional feature space. SVMs have the advantages of being less computationally intense than other classification algorithms, a good performance in high-dimensional spaces, and efficiency in handling nonlinear classification using the kernel trick that indirectly transforms the input space into another high dimensional feature space.

- *Artificial Neural Networks (ANNs):* ANNs are bio-inspired methods of data processing that enable computers to learn similarly to human brains [24]. ANNs are typically structured in layers made up of a number of interconnected nodes. Patterns are presented to the network via the input layer, which communicates to one or more hidden layers where the actual processing is done via a system of weighted connections. The hidden layers then link to an output layer where the answer (health index level in our case) is the output. In the learning phase, weights are changed until the best ANN model that fits the input data is built [24].

- *k-Nearest Neighbor (kNN):* The kNN algorithm is a supervised learning technique that has been used in many ML applications. It classifies objects based on the closest training examples in the feature space. The idea behind kNN is to find a predefined number of training samples closest in distance to a given query instance and predict the label of the query instance from them [24]. kNN is similar to a decision tree algorithm in terms of classification, but instead of finding a tree, it finds a path around the graph. It is also faster than decision trees.

- *OneR:* One rule is made for each attribute in the predictor variable in the dataset in which a given class is assigned to the value of that attribute [36]. The rule is created by counting the frequency of target classes that appear for a given attribute in the predictor variable. The most frequent target class is assigned for that attribute and the error for using that one rule for the entire data set is computed. The attribute of the predictor variable with the least error is considered as the final one rule. For the problem in this study, the attributes of each predictor variable are continuous numerical values and thus they must be discretized to create the one rule. The discretization process is thoroughly explained in [36].

- *Multinomial Logistic Regression (MLR):* A linear regression model attempts to fit a linear equation that maps the predictor variables to an estimated response variable [37]. Logistic regression or binary logistic regression, on the other hand, is more of a two-class classification-based model that is considered as a generalized linear model. The algorithm aims to define linear decision boundaries between the different classes [38]. The linear logistic regression model maps the input feature vector into a probability value via the sigmoid logistic function that is set during the training process. Based on the obtained probability, a decision is made of whether the given sample belongs to a particular class or not. For multiple classes, the MLR model is developed, which is a set of binary logistic regression models of a given class against all other classes. The classification of the sample is based on the maximum computed probability amongst the different logistic regression models [39].

- *Naïve Bayes (NB):* The NB classifier is a probabilistic classifier that is based on Bayes theorem [40]. This classifier is based on the assumption that the predictor variables are conditionally independent given the class of the data sample in question. In other words, the posterior probability of the sample being in a particular class given the predictor variables is computed using Bayes theorem [40]. For that, the likelihood of the predictor variable given the class, predictor prior probability, and class prior probability are determined. The samples are classified based on the outcome of the maximum posterior probability computed amongst all the different classes. This type of classifier is easily modelled and is typically suitable for large datasets.

3. Results

As illustrated earlier, two subsequent procedural steps were followed to achieve the main objective of this paper. In the first step, eight different classifiers were modelled for classifying the transformer health condition as being B, F, or G with three different training/testing scenarios involving the two utility databases (*Util1* and *Util2*). The full number of 10 features (as obtained from *Util1* and *Util2*) will be used to model the classifiers, and thus such classifiers will be named as the full-feature classifiers. In the following step, predetermined stepwise regression features from the reported results of [22] were used as the only features in modelling the eight classifiers with the same three training/testing scenarios of the full-feature model. Such classifiers will be named as the reduced-feature classifiers.

An assessment of the performance of both the full-feature and reduced feature classifiers was done by means of the accuracy rate obtained as per Equation (2). The mean accuracy rate (MAR) is shown in the presented results, which is basically the average value obtained for the accuracy rate for 10 trials.

3.1. Full-Feature Classifier Modelling

Eight different types of classifiers were used, which are NB, MLR, ANN, SVM, kNN, OneR, J48, and RF. Different training and testing scenarios were designed for the study. For a training/testing scenario, *Tr-Util1, Ts-Util1*, the classifier would be trained on data from *Util1* and tested using the unused data from *Util1*. Similarly is the case with a training/testing scenario, *Tr-Util2, Ts-Util2*.

In order to validate the generalized nature of the classifiers, a training/testing scenario, *Tr-Util1, Ts-Util2*, would have the classifier being trained on data from *Util1* and tested on different data from *Util2*.

To have a better understanding of how the results are obtained, consider the following example. When applying the training/testing scenario *Tr-Util1, Ts-Util1* with the RF classifier, the confusion matrix will indicate the frequency of truly and falsely classified data samples. Table 3 shows the confusion matrix obtained for the *Tr-Util1, Ts-Util1* scenario using the RF classifier. As can be seen in Table 3, 99.2% (492 of 496) for the G class of transformers were correctly classified. Similarly, 94.2% and 68% of the data samples were correctly classified as the F and B class, respectively. The accuracy rate was calculated using Equation (2) as:

$$Accuracy\ Rate = \frac{TG+TF+TB}{TG+TF+TB+FG+FF+FB} = \frac{492+194+19}{492+194+19+13+3+9} \times 100\% = 96.58\%, \quad (4)$$

where TG, TF, and TB are the truly classified data samples in the G, F, and B classes, respectively, and FG, FF, and FB are the falsely classified data samples in the G, F, and B classes, respectively. The same simulation was done 10 times and the MAR was noted. Table 4 shows the summary of the obtained MAR results for eight classifier types with the three training/testing scenarios.

Table 3. Confusion matrix for the *Tr-Util1, Ts-Util1* scenario using the Random Forest classifier.

		Classified as		
		G	F	B
Really is	G	492	4	0
	F	9	194	3
	B	0	9	19

Table 4. Summary of full-feature mean accuracy rate results for the eight classifier types with the three training/testing scenarios.

Training/Testing Scenario	NB	MLR	ANN	SVM	kNN	OneR	J48	RF
Tr-Util1, Ts-Util1	92.6%	95.5%	94.9%	92.6%	93.0%	86.7%	95.6%	96.6%
Tr-Util2, Ts-Util2	93.3%	94.8%	92.7%	86.9%	94.5%	85%	98.2%	96.6%
Tr-Util1, Ts-Util2	90.2%	95.4%	95.4%	85.6%	87.8%	85.9%	95.1%	93.6%

3.2. Reduced-Feature Classifier Modelling

In the published work of [22], it was concluded that the four test features of furan, IFT, C_2H_6, and C_2H_2 are the concise test features of the highest statistical significance in the regression problem of the HI value. The approach presented in this paper differs than that of [22] such that the ML approach deals with the prediction of the health condition class rather than the HI value (thus a classification problem rather than regression). Thus, the indicated four test features in [22] were used in the reduced-feature classifier modelling. Similar to the approach followed in the full-feature classifier models, eight classifiers with three different training/testing scenarios were used. Table 5 shows the summary of the obtained results for the MAR.

Table 5. Summary of reduced-feature MAR results.

Training/Testing Scenario	NB	MLR	ANN	SVM	kNN	OneR	J48	RF
Tr-Util1, Ts-Util1	94.4%	95.3%	95.1%	92.1%	95.6%	86.7%	95.3%	96.6%
Tr-Util2, Ts-Util2	90.8%	93.3%	91.1%	77.7%	93.6%	86.9%	97.2%	96.9%
Tr-Util1, Ts-Util2	92.4	90.5%	93.3%	86.2%	92.4%	85.9%	92.4%	93.6%

4. Discussions of the Results

Tables 4 and 5 summarize the results obtained for more than 640 simulations combined. This section of the paper will highlight the most important results that would significantly support the objective of using ML for the transformer health index problem.

4.1. Full-Feature Classifier Results

With reference to Table 4, the following points should be noted:

- The overall MAR results obtained for the full-feature classifier models were above 85%. It was observed that the highest classification error was observed for the B class samples. Though many data samples that actually belong to the B class were misclassified, they were always misclassified as being in the F class rather than the G class. This sort of error is of minimal risk in the sense that the classifiers would never give a misleading information of the transformer being in an excellent health condition while truly being in the worst health condition. Table 6 shows examples of the confusion matrices obtained in which a significant number of the B samples were misclassified as being F samples. The misclassification in most cases is attributed to the fact that a limited number of transformer oil samples of the B class are available for training (only 33 samples in total as can be observed in Table 1).

- For the remaining health condition classes, the accuracy rate was high and acceptable for most simulations. Most of the classifiers performed well in distinguishing between an F sample and a G sample. This is mainly attributed to the fact that a significant number of samples are available from both classes for training data.

- One of the extremely important training/testing scenarios would be that of *Tr-Util1, Ts-Util2*. With reference to Table 4, it was observed that most of the classifiers for this particular scenario did not perform as well as in the other training/testing scenarios. This inferior performance is mainly attributed to the fact that the classifier training was done with data from one utility company and testing was done with completely unseen data from another utility company. Still, the MAR results obtained are excellent given the previously unseen testing data. The MLR and ANN classifiers were the best classifiers, which resulted in an MAR of 95.4%. These excellent results support the generalized nature of the proposed approach in such a way that a utility company can use pre-modelled classifiers for their own transformer oil samples.

Table 6. Confusion matrix for (**a**) the *Tr-Util1, Ts-Util1* scenario using the kNN classifier, (**b**) *Tr-Util1 and Util2, Ts-Util1 and Util2* scenario using the ANN.

		(a)		
		Classified as		
		G	**F**	**B**
Really is	**G**	483	13	0
	F	22	181	3
	B	0	13	15

Table 6. *Cont.*

		Classified as		
(b)				
		G	**F**	**B**
Really is	**G**	719	15	0
	F	20	262	8
	B	0	17	16

4.2. Feature-Reduced Classifier Results

The selected features are basically the furan (which is an indicator of the paper insulation condition), the IFT (which is an indicator for the oil quality), and C_2H_2 and C_2H_6 that reveal the presence of any faults inside the transformer. Also, the ML can detect any correlation between the tests and hence remove highly correlated tests. With reference to Table 5, the obtained MAR results using most of the feature-reduced classifiers were above 90%. Similar to the full classifier models, the J48 and RF classifiers had the best performance amongst the eight reduced classifier models. These results give a significant conclusion that the feature selection technique of stepwise regression was very successful. The impact of these results would help utility companies reduce the cost of performing TAM practices by reducing the number of test samples required for computing the health index of the transformer asset. However, it is important to note that the frequency of oil sampling and hence overall TAM planning will not be improved by the results of the presented paper.

To illustrate the correlation of the health index with the selected features, Figure 2 shows the plots obtained for the furan, water, IFT, and C_2H_2 content against the health index. Although, furan, IFT, and C_2H_2 were selected by the stepwise regression, the water content was not selected [22]. Water clearly had the least correlation with the health index value, which would explain why it was not selected in either stepwise regression. On the other hand, the other three selected features showed a strong correlation with the health index.

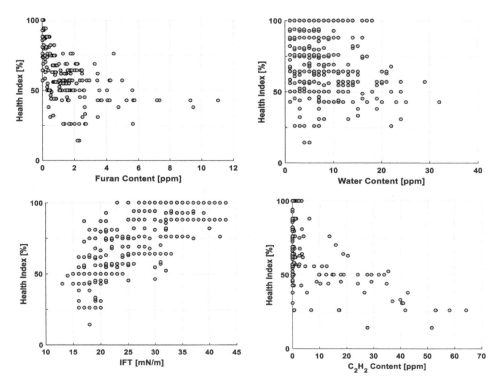

Figure 2. Plots indicating the relationship between a particular predictor variable and the health index value.

An additional study was conducted to observe if the age of the transformer was correlated with the obtained health index value. The age factor was not included in the study due to the fact that the transformer age data was only available for the transformer asset of *Util1*. Figure 3 shows the relationship between the transformer energization date and the health index value. It is apparent that there is no strong correlation between the transformer age and health index value. Merely, the transformer age may not be a good health index indicator without the inclusion of other factors, like the manufacturer, design, loading, and any refurbishing history. This observation is in agreement with that of [18], which basically indicates that a number of factors can influence the transformer conditions based on the service record and fault history, which differs from one transformer to another.

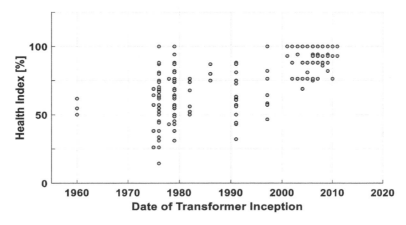

Figure 3. Date of transformer inception versus health index value.

5. Conclusions

The presented work was developed to validate the approach of using ML (through pattern classification tools) for the health index problem. The proposed approach was supported by the significant number of 1000+ transformers that were used for this study from different utility companies. The performance of the classifiers was proven as successful in both the full-feature and reduced-feature classifiers. The average results for all the MAR values were well beyond 85%. The worst results were shown with the OneR classifiers given the fact that these classifiers were trained on one feature. The use of stepwise regression as the feature selection tool for the health index problem was proven successful through the obtained results. The overall conclusion of the reported and discussed work significantly encourages the use of ML-feature selection methodology in the TAM industry for understanding the health condition of the transformer asset.

Author Contributions: Conceptualization, A.A. and A.E.-H.; methodology, A.A.; software, A.E.-H.; validation, A.A.; formal analysis, A.A.; investigation, A.E.-H.; resources A.A.; data curation, A.E.-H.; writing—original draft preparation, A.A.; writing—review and editing, A.E.-H.; visualization, A.A.; supervision, A.E.-H.; project administration, A.A.

References

1.   Zhang, X.; Gockenbach, E. Asset-Management of Transformers Based on Condition Monitoring and Standard Diagnosis. *IEEE Electr. Insul. Mag.* **2018**, *24*, 26–40.

2.   Abu-Elanien, A.E.B.; Salama, M.M.A. Asset management techniques for transformers. *Electr. Power Syst. Res.* **2010**, *80*, 456–464. [CrossRef]

3.   Azmi, A.; Jasni, J.; Azis, N.; Ab Kadir, M.Z.A. Evolution of transformer health index in the form of mathematical equation. *Renew. Sustain. Energy Rev.* **2017**, *76*, 687–700. [CrossRef]

4.   Abu-Siada, A.; Islam, S. A new approach to identify power transformer criticality and asset management decision based on dissolved gas-in-oil analysis. *IEEE Trans. Dielectr. Electr. Insul.* **2012**, *19*, 1007–1012. [CrossRef]

5.   Jalbert, J.; Gilbert, R.; Denos, Y.; Gervais, P. Methanol: A Novel Approach to Power Transformer Asset Management. *IEEE Trans. Power Deliv.* **2012**, *27*, 514–520. [CrossRef]

6.   Jahromi, A.; Piercy, R.; Cress, S.; Service, J.; Fan, W. An approach to power transformer asset management using health index. *IEEE Electr. Insul. Mag.* **2009**, *25*, 20–34. [CrossRef]

7.   Naderian, A.; Cress, S.; Piercy, R.; Wang, F.; Service, J. An Approach to Determine the Health Index of Power Transformers. In Proceedings of the Conference Record of the 2008 IEEE International Symposium on Electrical Insulation, Vancouver, BC, Canada, 9–12 June 2008.

8.   Tenbohlen, S.; Coenen, S.; Djamali, M.; Müller, A.; Samimi, M.H.; Siegel, M. Diagnostic Measurements for Power Transformers. *Energies* **2016**, *9*, 347. [CrossRef]

9.   Institute of Electrical and Electronics Engineers (IEEE). *IEEE Guide for Acceptance and Maintenance of Insulating Oil in Equipment, IEEE Std. C57.106-2006*; IEEE: Piscataway Township, NJ, USA, 2006.

10. Kachler, A.J.; Hohlein, I. Aging of Cellulose at Transformer Service Temperatures. Part 1: Influence of Type of Oil and Air on the Degree of Polymerization of Pressboard, Dissolved Gases, and Furanic Compounds in Oil. *IEEE Electr. Insul. Mag.* **2005**, *21*, 15–21. [CrossRef]

11. Hernanda, I.G.N.S.; Mulyana, A.C.; Asfani, D.A.; Negara, I.M.Y.; Fahmi, D. Application of health index method for transformer condition assessment. In Proceedings of the TENCON 2014-2014 IEEE Region 10 Conference, Bangkok, Thailand, 22–25 October 2014.

12. Singh, A.; Swanson, A.G. Development of a plant health and risk index for distribution power transformers in South Africa. *SAIEE Afr. Res. J.* **2018**, *109*, 159–170. [CrossRef]

13. Ortiz, F.; Fernandez, I.; Ortiz, A.; Renedo, C.J.; Delgado, F.; Fernandez, C. Health indexes for power transformers: A case study. *IEEE Electr. Insul. Mag.* **2016**, *32*, 7–17. [CrossRef]

14. Kadim, E.J.; Azis, N.; Jasni, J.; Ahmad, S.A.; Talib, M.A. Transformers Health Index Assessment Based on Neural-Fuzzy Network. *Energies* **2018**, *11*, 710. [CrossRef]

15. Islam, M.M.; Lee, G.; Hettiwatte, S.N. Application of a general regression neural network for health index calculation of power transformers. *Int. J. Electr. Power Energy Syst.* **2017**, *93*, 308–315. [CrossRef]

16. Zeinoddini-Meymand, H.; Vahidi, B. Health index calculation for power transformers using technical and economical parameters. *IET Sci. Meas. Technol.* **2016**, *10*, 823–830. [CrossRef]

17. Yahaya, M.S.; Azis, N.; Mohd Selva, A.; Ab Kadir, M.Z.A.; Jasni, J.; Kadim, E.J.; Hairi, M.H.; Yang Ghazali, Y.Z. A Maintenance Cost Study of Transformers Based on Markov Model Utilizing Frequency of Transition Approach. *Energies* **2018**, *11*, 2006. [CrossRef]

18. Tee, S.; Liu, Q.; Wang, Z. Insulation condition ranking of transformers through principal component analysis and analytic hierarchy process. *IET Gener. Transm. Distrib.* **2017**, *11*, 110–117. [CrossRef]

19. Ashkezari, A.D.; Ma, H.; Saha, T.K.; Ekanayake, C. Application of Fuzzy Support Vector Machine for Determining the Health Index of the Insulation System of In-Service Power Transformers. *IEEE Trans. Dielectr. Electr. Insul.* **2013**, *20*, 965–973. [CrossRef]

20. Abu-Elanien, A.E.B.; Salama, M.M.A.; Ibrahim, M. Calculation of a Health Index for Oil-Immersed Transformers Rated Under 69 kV Using Fuzzy Logic. *IEEE Trans. Power Deliv.* **2012**, *27*, 2029–2036. [CrossRef]

21. Benhmed, K.; Mooman, A.; Younes, A.; Shaban, K.; El-Hag, A. Feature Selection for Effective Health Index Diagnoses of Power Transformers. *IEEE Trans. Power Deliv.* **2018**, *33*, 3223–3226. [CrossRef]

22. Alqudsi, A.; El-Hag, A. Assessing the power transformer insulation health condition using a feature-reduced predictor model. *IEEE Trans. Dielectr. Electr. Insul.* **2018**, *25*, 853–862. [CrossRef]

23. Geng, X.; Liu, T.Y.; Qin, T.; Li, H. Feature selection for ranking. In Proceedings of the 30th Annual International ACM SIGIR Conference on Research and Development in Information Retrieval. ACM, New York, NY, USA, 1 January 2007.

24. Witten, I.H.; Frank, E.; Hall, M.A. *Data Mining: Practical Machine Learning Tools and Techniques*, 3th ed.; Elsevier: Amsterdam, The Netherlands, 2016; ISBN 978-0-12-374856-0.

25. Efron, B.; Tibshirani, R. Improvements on Cross-Validation: The 0.632+ Bootstrap Method. *J. Am. Stat. Assoc.* **1997**, *92*, 548–560.

26. Guyon, I.; Elissee, A. An introduction to variable and feature selection. *J. Mach. Learn. Res.* **2003**, *3*, 1157–1182.

27. Wang, H.; Khoshgoftaar, T.M.; Gao, K.; Seliya, N. High-dimensional software engineering data and feature selection. In Proceedings of the 21th IEEE International Conference on Tools with Artificial Intelligence, Newark, NJ, USA, 2–4 November 2009.

28. Mierswa, I.; Wurst, M.; Klinkenberg, R.; Scholz, M.; Euler, T. YALE: Rapid Prototyping for Complex Data Mining Tasks. In Proceedings of the 12th ACM SIGKDD International Conference on Knowledge Discovery and Data Mining (KDD-06), Philadelphia, PA, USA, 20–23 August 2006.

29. Liu, H.; Motoda, H. *Feature Selection for Knowledge Discovery and Data Mining*; Kluwer Academic Publishers: Dordrecht, The Netherland, 2000.

30. Montgomery, D.; Runger, G. *Applied Statistics and Probability for Engineers*; Wiley: Hoboken, NJ, USA, 1994.

31. Rokach, L.; Maimon, O. *Data Mining with Decision Trees: Theory and Applications*; World Scientific Publishing Co., Inc.: River Edge, NJ, USA, 2008.

32. Hall, M.A. Correlation-Based Feature Subset Selection for Machine Learning. Ph.D. Thesis, University of Waikato, Hamilton, New Zealand, 1998.

33. Alpaydin, E. *Introduction to Machine Learning*; MIT Press: Cambridge, MA, USA, 2010; p. 5.

34. Breiman, L. Random Forests. *Mach. Learn.* **2001**, *45*, 5–32. [CrossRef]

35. Quinlan, R. *C4.5: Programs for Machine Learning*; Morgan Kaufmann Publishers: San Mateo, CA, USA, 2014.

36. Holte, R.C. Very Simple Classification Rules Perform Well on Most Commonly Used Datasets. *Mach. Learn.* **1993**, *11*, 63–90. [CrossRef]

37. Department of Statistics & Data Science at Yale University: Online course on Multiple Linear Regression. Available online: http://www.stat.yale.edu/Courses/1997-98/101/linmult.htm (accessed on 20 May 2019).

38. Gudivada, V.N.; Irfan, M.T.; Fathi, E.; Rao, D.L. Cognitive Analytics: Going Beyond Big Data Analytics and Machine Learning. In *Handbook of Statistics*; Gudivada, V.N., Raghavan, V.V., Govindaraju, V., Rao, C.R., Eds.; Elsevier: Amsterdam, The Netherlands, 2016; Volume 35, pp. 169–205.

39. PennState Elberly College of Science-Analysis of Discrete Data: Polytomous (Multinomial) Logistic Regression. Available online: https://newonlinecourses.science.psu.edu/stat504/node/172/ (accessed on 25 May 2019).

40. Mitchell, T.M. Generative and Discriminative Classifiers: Naive Bayes And Logistic Regression. In *Machine Learning*; Mitchell, T.M., Ed.; McGraw Hill: New York, NY, USA, 2015.

4

Computation of Transient Profiles along Nonuniform Transmission Lines Including Time-Varying and Nonlinear Elements using the Numerical Laplace Transform

Rodrigo Nuricumbo-Guillén [1,*], **Fermín P. Espino Cortés** [1], **Pablo Gómez** [2] and **Carlos Tejada Martínez** [1]

[1]   Departamento de Ingeniería Eléctrica SEPI-ESIME ZAC, Instituto Politécnico Nacional, Mexico City 07738, Mexico

[2]   Electrical and Computer Engineering Department, Western Michigan University, Kalamazoo, MI 49008, USA

*   Correspondence: rodrigo.ng.85@gmail.com

†   This paper is an extended version of our paper presented in the 2018 IEEE International Conference on High Voltage Engineering and Application (ICHVE), Athens, Greece, 10–13 September 2018.

Abstract: Electromagnetic transients are responsible for overvoltages and overcurrents that can have a negative impact on the insulating elements of the electrical transmission system. In order to reduce the damage caused by these phenomena, it is essential to accurately simulate the effect of transients along transmission lines. Nonuniformities of transmission line parameters can affect the magnitude of voltage transients, thus it is important to include such nonuniformities correctly. In this paper, a frequency domain method to compute transient voltage and current profiles along nonuniform multiconductor transmission lines is described, including the effect of time-varying and nonlinear elements. The model described here utilizes the cascade connection of chain matrices in order to take into consideration the nonuniformities along the line. This technique incorporates the change of parameters along the line by subdividing the transmission line into several line segments, where each one can have different electrical parameters. The proposed method can include the effect of time-dependent elements by means of the principle of superposition. The numerical Laplace transform is applied to the frequency-domain solution in order to transform it to the corresponding time-domain response. The results obtained with the proposed method were validated by means of comparisons with results computed with ATP (Alternative Transients Program) simulations, presenting a high level of agreement.

Keywords: electromagnetic transients; nonuniform transmission line; numerical Laplace transform; time-dependent elements; transmission line modeling

---

1. Introduction

Electromagnetic transients can produce overvoltages and overcurrents that can have a negative impact on electric power systems. In order to reduce the potential deterioration or damage due to this condition, accurate transient simulations are needed [1]. Typically, the transient analysis of electrical systems is performed by means of two-port models of uniform transmission lines, thus the voltage/current measurements are available at certain nodes of the network. However, the maximum transient overvoltages and overcurrents may appear at interior points of the transmission line [2]. In such cases, the traditional simulation methods may not be well suited to correctly analyze this kind of phenomena. Additionally, nonuniformities can be present along the transmission line in the form of parameter variations such as the height of the line or the properties of the terrain; if these nonuniformities are too prominent, the distribution and magnitude of the voltage and currents can drastically be affected along the transmission line in comparison with uniform transmission lines (lines with space-independent per-unit-length parameters) [3,4].

There has been a considerable interest in developing methods to accurately model nonuniform transmission lines during electromagnetic transients. Previous works have presented line models applying different techniques such as the numerical Laplace transform, the method of characteristics, rational approximations, among others, with good results [5–12]. However, in general, these methods are only able to provide voltage and current information at the ends of the line, which in some cases may not be enough to correctly analyze the transient behavior of a transmission line [2], such as for insulation design or protection purposes.

A frequency domain method for the computation of transient voltage and current profiles along transmission lines was reported in [13,14], and later extended to include time-varying and nonlinear elements [15]. However, this method cannot be applied to nonuniform transmission lines. In [16], Laplace-domain and time-domain methods were tested in the computation of transient profiles on a nonuniform electronic system. It was found that the line model that used the inverse numerical Laplace transform (INLT) provided the most accurate results. However, the method described in [16] can only be applied to time-invariant linear systems.

Expanding upon the aforementioned publications, the main contribution of the present paper is the complete description and verification of a method for the computation of transient voltage and current profiles along nonuniform transmission lines. This method utilizes a modeling approach defined in the frequency domain and based on the cascaded connection of chain matrices. Furthermore, using the superposition technique, the proposed method can include time-dependent and nonlinear elements (such as switching devices and surge arresters) in the computation of transient profiles along nonuniform lines, something that has not been conducted in any previously reported research. Since the line model is defined in the frequency domain, it can take into account the frequency dependence of electrical parameters in a straightforward manner, providing more accurate results in comparison with existing methods defined in the time domain.

The method presented here makes use of the inverse numerical Laplace transform [17,18] to convert the computed frequency domain solution to a time-domain transient response. This method has strong potential for application in fault location and insulation coordination, with particular accuracy benefits for lines with prominent nonuniformities, such as river crossings, hilly terrains, and other substantial sagging conditions, which are commonly encountered in large countries such as China, Canada, India, Russia, and Brazil. For example, very challenging river crossings are found in Brazil for overhead lines constructed to connect the power generation in the Amazon Basin to the main load centers of the country. These river crossings are in the order of 2 km leading to very tall towers and wide line spans [19]. Accurate fault location under these circumstances requires an appropriate consideration of wave propagation along nonuniform lines, which can be achieved with the method described here. Additionally, the proposed method can be expanded to the modeling of other power system nonuniform elements, such as transmission towers [20] and rotating machines [21].

In order to validate the accuracy of the proposed method, the results obtained are compared with those obtained from simulations performed with ATP (Alternative Transients Program) [22]. In the ATP simulations, the J. Marti line model [23] was used, and the transmission line was subdivided into several line segments to allow the connection of measuring probes at internal points of the line, as well the inclusion of nonuniformities.

The computation of transient profiles along nonuniform transmission lines including nonlinear and time-varying conditions has not been previously reported, providing an original contribution to the current state of the art of the topic.

2. Nonuniform Transmission Line Model for the Transients Profiles Computation

This section describes the transmission line model used to compute the transient profiles as well as the technique used to include time-varying and nonlinear elements. Additionally, a brief explanation for the implementation of the INLT algorithm is presented.

2.1. Nonuniform Transmission Line Model

This work introduces the nonuniformities along transmission lines by means of the technique of cascade connection of chain matrices, as it has been previously shown to be an effective technique for the simulation of electromagnetic transients when nonuniform transmission lines are considered [11,24].

Initially, the uniform transmission line of Figure 1 is considered. This line can be represented by a two-port model known as transfer or **ABCD** matrix model:

$$\begin{bmatrix} V_L(s) \\ I_L(s) \end{bmatrix} = \begin{bmatrix} A & B \\ C & D \end{bmatrix} \begin{bmatrix} V_0(s) \\ I_0(s) \end{bmatrix} \tag{1}$$

where

$$\begin{aligned} A &= \cosh(\gamma L) \\ B &= -Z_0 \sinh(\gamma L) \\ C &= Y_0 \sinh(\gamma L) \\ D &= -\cosh(\gamma L) \\ \gamma &= \sqrt{ZY} \end{aligned} \tag{2}$$

In Equation (2) Z_0, Y_0, Z, Y, and L are the line's characteristic impedance, characteristic admittance, series impedance, shunt admittance and length, respectively. Additionally, the Laplace variable is defined as $s = c + j\omega$, where ω is given by $2\pi f$.

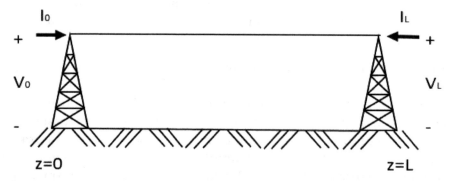

Figure 1. Uniform transmission line representation. V_0 and V_L are the voltages of the sending and receiving node, and I_0 and I_L are the injected currents at the sending and receiving node, respectively, and L is the line's length.

By changing the direction of I_L in Figure 1, Equation (1) is modified in the following manner:

$$\begin{bmatrix} V_L(s) \\ I_L(s) \end{bmatrix} = \begin{bmatrix} A & B \\ -C & -D \end{bmatrix} \begin{bmatrix} V_0(s) \\ I_0(s) \end{bmatrix} \tag{3}$$

or in a compact form:

$$\begin{bmatrix} V_L(s) \\ I_L(s) \end{bmatrix} = \boldsymbol{\Phi} \begin{bmatrix} V_0(s) \\ I_0(s) \end{bmatrix} \tag{4}$$

Matrix $\boldsymbol{\Phi}$ in Equation (4) is the chain matrix of the transmission line. Due to the fact that the currents at both ends of the line have the same direction, multiple transmission lines can be cascade-connected by using their corresponding chain matrices, as it can be seen in Figure 2.

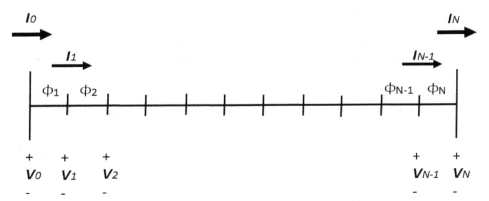

Figure 2. Cascaded connection of chain matrices, where Φ_n, V_n, and I_n are the chain matrix, sending voltage, and sending current of the n-th cascade-connected line, respectively.

Figure 2 can also be interpreted as one transmission line subdivided into several smaller line segments, where each segment is represented by a unique chain matrix with its own independent electrical properties. Using this approach, it is possible to build a transmission line model that includes nonuniformities. With this consideration in mind, it can be deduced from Figure 2 that the voltage and current at the beginning of each line segment and the chain matrix of the same segment can be used to compute the voltage and current of the next segment:

$$
\begin{aligned}
\begin{bmatrix} V_1(s) \\ I_1(s) \end{bmatrix} &= \Phi_1 \begin{bmatrix} V_0(s) \\ I_0(s) \end{bmatrix} \\
\begin{bmatrix} V_2(s) \\ I_2(s) \end{bmatrix} = \Phi_2 \begin{bmatrix} V_1(s) \\ I_1(s) \end{bmatrix} &= \Phi_2 \Phi_1 \begin{bmatrix} V_0(s) \\ I_0(s) \end{bmatrix}
\end{aligned}
\tag{5}
$$

or in a general way:

$$
\begin{bmatrix} V_N(s) \\ I_N(s) \end{bmatrix} = \Phi_N \Phi_{N-1} \cdots \Phi_3 \Phi_2 \Phi_1 \begin{bmatrix} V_0(s) \\ I_0(s) \end{bmatrix}
\tag{6}
$$

Equation (6) can be used to compute the voltage and current profiles along a nonuniform transmission line. The transient profiles are computed in a sequential manner, obtaining the voltage and current at the end of the first line segment from the chain matrix and from the voltages and currents at the beginning of the same segment; this process is repeated until the voltage and current along the whole line have been computed. It can also be observed in Equation (6) that V_0 and I_0 are required as initial values of the algorithm; in order to compute such values, a two-port admittance representation of the complete transmission line from the chain matrix Φ_{FL} is defined as:

$$
\Phi_{FL} = \Phi_N \Phi_{N-1} \cdots \Phi_3 \Phi_2 \Phi_1 = \begin{bmatrix} \Phi_{FL11} & \Phi_{FL12} \\ \Phi_{FL21} & \Phi_{FL22} \end{bmatrix}
\tag{7}
$$

$$
\begin{bmatrix} I_0(s) \\ I_L(s) \end{bmatrix} = \begin{bmatrix} Y_{ss} & -Y_{sr} \\ -Y_{rs} & Y_{rr} \end{bmatrix} \begin{bmatrix} V_0(s) \\ V_L(s) \end{bmatrix}
\tag{8}
$$

where

$$
\begin{aligned}
Y_{ss} &= -\Phi_{FL12}^{-1} \Phi_{FL11} \\
Y_{sr} &= -\Phi_{FL12}^{-1} \\
Y_{rs} &= \Phi_{FL21} - \Phi_{FL22} \Phi_{FL12}^{-1} \Phi_{FL11} \\
Y_{rr} &= -\Phi_{FL22} \Phi_{FL12}^{-1}
\end{aligned}
\tag{9}
$$

$V_0(s)$ is obtained by solving (8) for the voltages vector. $I_0(s)$ is computed as follows:

$$
I_0(s) = Y_{ss} V_0(s) - Y_{rr} V_L(s)
\tag{10}
$$

2.2. Modeling of Time-Varying Elements

It can be difficult to include time-varying conditions, such as switching maneuvers, when methods defined in the frequency domain are used to simulate electromagnetic transients. However, the principle of superposition has demonstrated to be an efficient method to include such conditions in the frequency domain [25], as explained below.

The closing of a switch can be computed using the circuits presented in Figure 3, which represent the state of the circuit before and after the closing maneuver. V_E, V_S, and V_R are the voltage at the source side, at the line's sending node, and at the line's receiving node, respectively. The nodal voltage vector $V_{NL}(s)$ is formed by the three subvectors in Figure 3 as shown below:

$$V_{NL}(s) = \begin{bmatrix} V_E(s) & V_0(s) & V_L(s) \end{bmatrix}^T \tag{11}$$

and can be obtained from the following expression [11]:

$$V_{NL}(s) = Ybus_0^{-1}I_{N0} + Ybus_1^{-1}I_{N1} \tag{12}$$

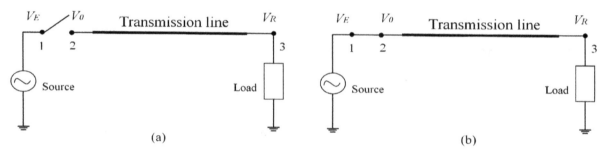

Figure 3. (a) Transmission line before the closing operation of a switch connected at the sending node. (b) Transmission line after the closing operation of a switch connected at the sending node.

In Equation (12), $Ybus_0$ is the nodal admittance matrix before the switch closing (Figure 3a), $Ybus_1$ is the admittance matrix modified by the closing operation (Figure 3b), I_{N0} contains the initially injected currents, and I_{N1} is the injection current vector due to the switch operation. An example of the construction of the nodal admittance matrices $Ybus_0$ and $Ybus_1$ for the transmission line in Figure 3 is presented below:

$$Ybus_{0,1} \begin{bmatrix} Y_{11} & -Y_{12} & -Y_{13} \\ -Y_{21} & Y_{22} & -Y_{23} \\ -Y_{31} & -Y_{32} & Y_{33} \end{bmatrix} \tag{13}$$

If $Ybus_0$ (Figure 3a) is to be built using (13), the elements Y_{11}, Y_{12}, Y_{21}, and Y_{22} would assume the following values: $Y_{12} = Y_{21} = Y_{switcho}$, where $Y_{switcho}$ is the open switch's admittance between its poles, ideally $Y_{switcho} = 0$. $Y_{11} = Y_s + Y_{swtcho}$, with Y_s being the admittance connected at the sending node, and $Y_{22} = Y_{LL} + Y_{switcho}$ where Y_{LL} represents the self-admittance of the transmission line. The matrix $Ybus_1$ (Figure 3b) can be built in a similar way, but replacing $Y_{switcho}$ by $Y_{switchc}$, that is, the admittance of the switch when closed.

The procedure presented above can be extended to any number of changes in the circuit topology using the following expression:

$$V_{NL}(s) = Ybus_0^{-1}I_{N0} + \sum_{i=1}^{m} Ybus_i^{-1}I_{Ni} \tag{14}$$

where $Ybus_i$ and I_{Ni} are the modified admittance matrix and the injection current vector corresponding to the n-th switch operation, respectively. A comprehensive explanation of this procedure can be found in [11].

$\mathbf{V}_0(s)$ in (11) is the voltage at the sending node of the line needed in (6) to compute the transient voltage and current profiles. Meanwhile, the current at the beginning of the line is computed with (10) using $\mathbf{V}_0(s)$ and $V_L(s)$ from (11).

2.3. Inclusion of Nonlinear Elements

It has been demonstrated that the principle of superposition can be used to overcome the difficulties of the inclusion of nonlinear components in the frequency domain [25]. This is achieved by means of a sequence of switching operations.

In order to include a nonlinear element in a simulation performed using a method based in the frequency domain, first, its nonlinear characteristic curve must be approximated by a piece-wise curve, as illustrated in Figure 4.

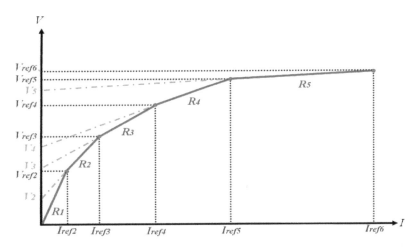

Figure 4. Five-segment piece-wise approximation of a nonlinear v-i characteristic curve [11].

The element R_n of the n-th linear element from Figure 4 and the voltage V_n are computed as follows:

$$R_n = \frac{V_{refn+1} - V_{refn}}{I_{refn+1} - I_{refn}} \tag{15}$$

$$V_n = \left(-R_{n+1}I_{refn}\right) + V_{refn+1} \tag{16}$$

By approximating the v-i characteristic of a nonlinear element as shown above, such an element can be represented as a network of N parallel-connected branches, as shown in Figure 5.

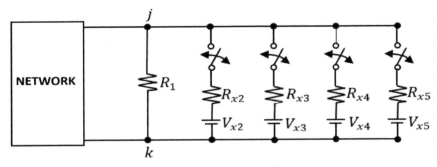

Figure 5. Circuital representation of the piecewise approximation for the characteristic curve of a nonlinear element presented in Figure 5 [11].

In the circuit presented in Figure 5, the switch in the n-th branch operates according to the reference voltage; it closes when the voltage across nodes j and k goes above V_{refn} and it opens when the voltage drops below V_{refn}. It is important to mention that the switches must operate in a successive manner, meaning that a switch can only close when the one in the preceding branch is closed, and a switch can only open when the one in the following branch is open.

When the switch connected to the n-th branch is closed, the equivalent Thevenin resistance of the circuit in Figure 5 must be equal to the value of R_n, that is, the slope of the n-th linear segment in the approximation of Figure 4:

$$R_n = \frac{R_{n-1}R_{xn}}{R_{n-1} + R_{xn}} \tag{17}$$

Finally, by solving (17) for R_{xn}, the value of such resistance can be computed:

$$R_{xn} = \frac{R_{n-1}R_n}{R_{n-1} - R_n} \tag{18}$$

V_{xn} is computed as:

$$V_{xn} = \frac{R_{n-1}V_n - V_{n-1}R_n}{R_{n-1} - R_n} \tag{19}$$

With this approach, it is possible to compute the transient profiles along nonuniform transmission lines with nonlinear and/or time-varying elements using (6), calculating the voltage and current at the sending node with (10) and (14), and implementing the nonlinear element model presented in this section.

2.4. Inverse Numerical Laplace Transform

As the last step, the INLT is applied to the computed transient profiles in the frequency domain with (6); this is done in order to transform the results to the time domain. This method has proven to be very accurate for the study of electromagnetic transients [11,24,25]. The implementation of the INLT algorithm is described below in a brief manner; references [17,18] can be consulted for a thorough explanation. Considering a time function $f(t)$ as a real and causal function, the inverse Laplace transform can be written as:

$$f(t) = Re\left\{ \frac{e^{ct}}{2\pi} \int_0^\infty F(s)e^{j\omega t}d\omega \right\} \tag{20}$$

In this work, the numerical evaluation of (20) is done considering an odd sampling in the frequency spectrum (using a spacing of $2\Delta\omega$), and normal time steps Δt in the time domain. With these considerations in mind, the following definitions for the discrete functions in the time and frequency domain for N equally spaced samples are made:

$$f_n = f(n\Delta t) F_m = F(c + j(2m + 1)\Delta\omega) \tag{21}$$

where $n, m = 0, 1, 2, \ldots, N-1$ and c hat reduces the aliasing errors of the algorithm; in this work, c is defined as $c = 2\Delta\omega$.

Additionally, it is necessary to establish finite integration limits for the numerical evaluation of (20), that is, the maximum frequency Ω and the observation time T. The observation time is computed from:

$$T = \frac{\pi}{\Delta\omega} \tag{22}$$

and the following relations can be established:

$$\Delta t = \frac{T}{N}$$
$$\Delta\omega = \frac{\Omega}{2N} = \frac{\pi}{T} \tag{23}$$

Finally, by implementing the odd sampling presented in (21) and including a window function σ_m, the Laplace transform in (20) can be numerically approximated by:

$$f_n = Re\left\{C_n\left[\sum_{m=0}^{N-1} F_m\sigma_m exp\left(\frac{j2\pi m}{N}\right)\right]\right\} \text{for } n, m = 0, 1, 2, \ldots, N-1 \qquad (24)$$

where:

$$C_n = \frac{2\Delta\omega}{\pi}exp\left(cn\Delta t + \frac{j\pi n}{N}\right) \qquad (25)$$

In Equation (24), the term inside the brackets corresponds to the fast Fourier transform algorithm; this allows computing time savings if N is equal to an integer power of two.

3. Test Cases

Three test cases are presented in order to validate the proposed method. The frequency dependence of the line's electrical parameters was taken into account by means of the application of the complex image method to introduce the earth-return impedance, as well as the complex penetration depth to include the skin effect in the line conductors [26]. The accuracy of the method was validated through comparisons with measurements from simulations performed with ATP, where the frequency-dependent J. Marti line model was discretized in several smaller segments to allow the introduction of nonuniformities and the connection of measuring probes at interior points. The length of the simulated lines in the test cases was kept short in order to reduce the implementation burden in ATP; with the proposed method, the length and discretization of the line did not represent a problem.

3.1. Highly Nonuniform Transmission Line

A highly nonuniform three-phase transmission line was considered. The nonuniformity was included by the sagging of the conductors as shown in Figure 6. The line was excited by an ideal unit step voltage source (1 p.u.) connected at phase A of the sending node, while all of the other nodes were left open.

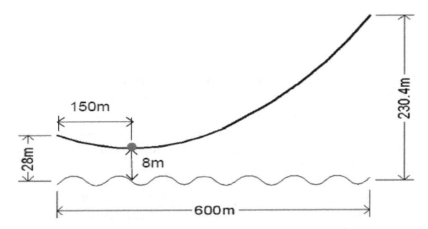

Figure 6. Three-phase non-uniform transmission line used in the example in Section 3.1. A second-degree polynomial equation was utilized to approximate the conductors' height, as presented in [23]. There was a 10 m separation between the conductors.

Figure 7 presents the transient voltage profile along phase A computed with the proposed method; in order to include the nonuniformities along the transmission line, it was subdivided into 80 chain matrices. This figure illustrates the propagation of the traveling waves along the line and how these waves were reflected when they reached the receiving end. It can also be observed in the transient profiles that there were periods where negative voltages appeared at some points along the line, which is not a commonly observed phenomenon when similar simulations with uniform lines are performed.

Additionally, a comparison at the middle of the line between the transient voltage waveforms obtained with the proposed method and results from ATP simulations is presented (Figure 8). The simulations performed with ATP required a time step 10 times smaller than the proposed method to achieve similar results.

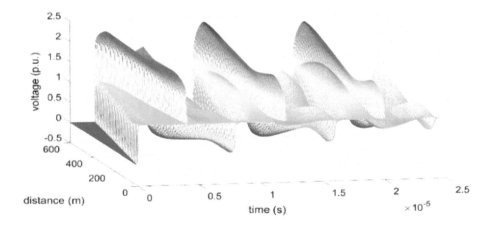

Figure 7. Computed voltage profile along phase A (example 3.1).

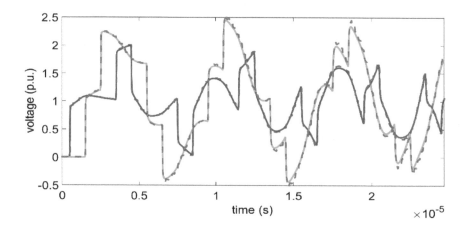

Figure 8. Comparison of transient voltage waveforms at two specific distances between the computed results with the proposed method (solid lines) and results obtained with ATP simulations (dashed lines). The solid blue and dashed red lines correspond to voltage measurements at 150 m from the sending node; the solid green and the dashed orange lines were obtained at 450 m. The oscillations observed in the results from the ATP simulations are attributed to the error accumulation due to the discretization of the transmission line [2].

This example was simulated with the proposed method considering 20 and 40 chain matrices. Figure 9 presents a comparison of voltage measurements at the middle point of the transmission line from simulations considering 20, 40, and 80 chain matrices. From this comparison, it can be observed that the curves from the three simulations, in general, have the same shape (curves for 40 and 80 chain matrices are overlapped). However, the curve corresponding to the simulation performed with 20 chain matrices presents some oscillations in comparison with the other two. As it can be seen with the curves corresponding to 40 and 80 chain matrices, with an increase in the number of chain matrices used in the simulations, the curves become smoother; however, a further increase in the number of chain matrices used is barely noticeable. This indicates that although the method's accuracy is dependent on the number of chain matrices considered, it does not require a large number of chain matrices to achieve good results.

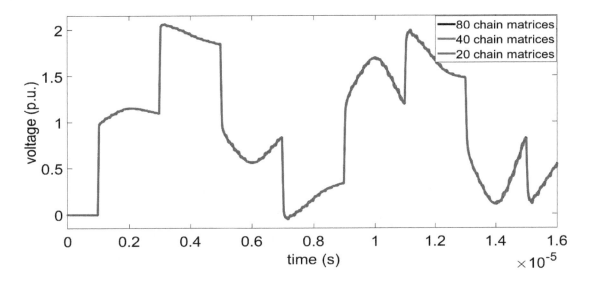

Figure 9. Comparison of transient voltage waveforms at the middle point of the line computed with different numbers of chain matrices. The curve computed with 20 chain matrices is not as smooth as the ones computed with 40 and 80.

3.2. Sequential Energization

A 5 km transmission line was used in this case. The nonuniformity was introduced by means of the line's sag: there was a transmission tower every 500 m, the conductors' height was maximum at the towers (20 m), and the height was minimum at the midspan between towers (15 m). The consideration for the height's variation as well as the horizontal separation between conductors was the same as in the previous example. The line was connected to an AC voltage source (1 p.u.) at the sending node through a three-phase switch. The switch's poles operated in a sequential manner with closing times of 3, 6, and 9 ms in an ABC sequence. The line's receiving node was left open.

The transient voltage profiles of phases A and B are presented in Figures 10 and 11, respectively. The comparison with ATP simulations is shown in Figure 12.

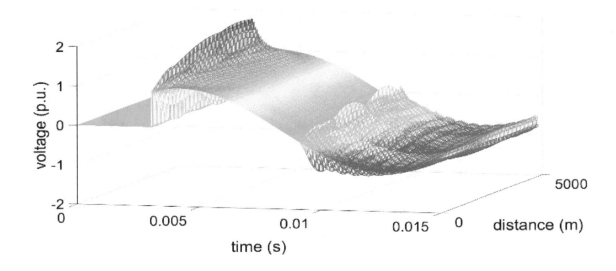

Figure 10. Voltage profile computed along phase A (example 3.2).

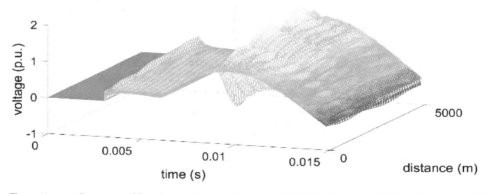

Figure 11. Transient voltage profile along phase B in example 3.2. The complete voltage profile induced in phase B due to the energization of phase A can be observed. This type of detailed information cannot be easily achieved using ATP-like simulation software.

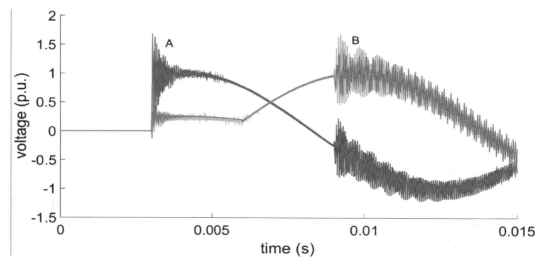

Figure 12. Comparison between the transient voltage computed with the proposed method (blue and green lines) and those obtained with ATP simulations (red and orange lines) in example 3.2. The comparisons are made at the middle of the transmission line (2500 m from the sending node) in phases A and B. A good level of agreement can be observed, but the ATP simulations needed a time step 50 times smaller than the proposed method in order to achieve such results.

3.3. Surge Arrester Operation

This example presents the operation of surge arresters during a direct lightning strike; due to their *v-i* characteristic, the arresters were modeled as nonlinear elements as described in Section 2.3. The same line configuration presented in Section 3.1 was considered. The line was excited by a direct lightning strike (1.2/50 μs), the impact point was at the phase A of the sending node, and the line was impedance-matched at both ends in order to avoid reflections. The injected current to the line was approximated by a double exponential current source defined as:

$$i(t) = I_0\left(e^{-\frac{t}{\tau_1}} - e^{-\frac{t}{\tau_2}}\right) \tag{26}$$

where $\tau_1 = 68.199$ μs, $\tau_2 = 0.405$ μs, and $I_0 = 10.37$ kA.

Surge arresters were connected to the three phases at the receiving node. The nonlinear characteristic of the arresters was simulated by means of a five-segment piecewise-linear approximation. Table 1 presents the *v-i* coordinates of this representation.

Table 1. Nonlinear v-i characteristic of the arresters [27].

Vref (kV)	Iref (kA)
0	0
484	0.1760
616	0.3226
748	0.7626
836	1.6426
880	12.6426

First, a simulation was performed without surge arresters connected to the line. Figure 13 presents the transient voltage profile along phase A computed in this simulation. On the other hand, Figure 14 shows the voltage profile along phase A when the arresters were connected at the receiving node of the line. By comparing Figures 13 and 14, the influence that the surge arresters' operation had on the magnitude of the voltages along the line is easily observed. The voltage along the line in Figure 14 was considerably lower in comparison with the transient profile in Figure 13. Additionally, a comparison is presented with results obtained from ATP simulations (Figure 15). In a similar way to the example in Section 3.1, the time step required by ATP was 10 times smaller than the proposed method in order to obtain similar results.

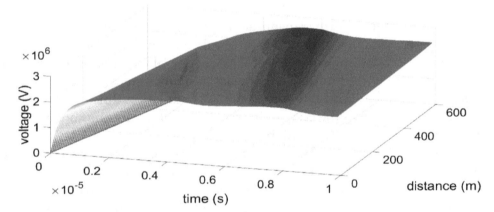

Figure 13. Computed transient voltage profile along phase A when there were no surge arresters connected at the receiving end of the line (example 3.3).

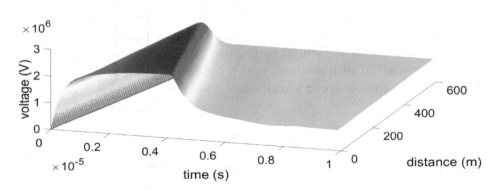

Figure 14. Computed transient voltage profile along phase A with surge arresters connected at the receiving node (example 3.3). The voltage magnitude significantly decreased along the transmission line compared to the transient profile presented in Figure 13.

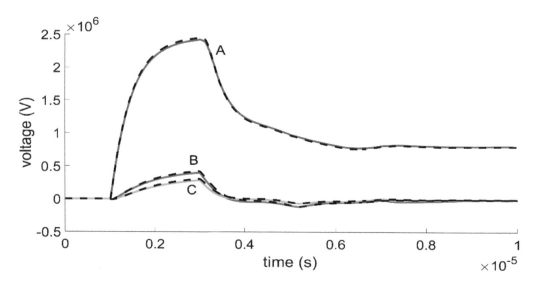

Figure 15. Voltage comparison between the results from the proposed method (solid lines) and those obtained from ATP simulations (dashed lines) for example 3.3. The comparisons were made at the middle point of the line (300 m from the sending node) for phases A, B, and C.

4. Discussion

Transient simulation of nonuniform transmission lines using traditional software (such as ATP in this paper) is a challenging task. This difficulty is due to the fact that the nonuniformities are typically approximated by cascade-connecting several small line segments. Such approximation requires the use of a very small time step in the simulation (at least five times smaller than the traveling time of the line segments [6]), which may translate into excessive simulation times and the saturation of the software's available memory, as was the case in this work. Additionally, the connection of several line segments can introduce errors in the simulation results [2].

In contrast, the proposed method is able to accurately compute transient profiles along nonuniform transmission lines with substantially larger time steps in comparison to ATP, resulting in a better memory usage as expected from the findings reported in [16] with regards to the use of the INLT algorithm. As it can be observed in Figures 7, 10, 11 and 14, the main advantage of the proposed method is the fact that it allows visualizing the voltage and current transient behavior along the line and not only at its ends, which can be advantageous when designing lines with a high level of nonuniformities and cannot be easily done using traditional simulation software.

The precision of the proposed method was validated by comparisons with the results obtained with ATP simulations. In general, there was a very good level of agreement between the results from both methods, as it can be seen by the mean relative difference presented in Tables 2–4. There is a slight difference in the comparison of results presented in Figure 12 corresponding to the simulation of sequential pole closure. This is attributed to the previously mentioned ATP limitations and the difference in the time step used in each method, which can result in deviations in the operation time of the switch model, leading to the observed variations.

Table 2. Mean relative difference of transient voltage in example 3.1.

Measuring Point	Mean Relative Difference (%)		
	Phase A	Phase B	Phase C
150 m	0.0708	0.1837	0.2893
450 m	0.3839	0.3351	0.6832

Table 3. Mean relative difference of transient voltage in example 3.2.

Measuring Point	Mean Relative Difference (%)		
	Phase A	Phase B	Phase C
2.5 km	0.3889	0.9861	1.2761

Table 4. Mean relative difference of transient voltage in example 3.3.

Measuring Point	Mean Relative Difference (%)		
	Phase A	Phase B	Phase C
300 m	0.0374	0.7738	1.9183

5. Conclusions

This paper describes a frequency domain method to compute transient voltage and current profiles along nonuniform multiconductor transmission lines, where the nonuniformities along the line are introduced in the model by means of the cascaded connection of chain matrices. The method can incorporate nonlinear and time-dependent elements by using the superposition principle. The profiles obtained provide useful information to locate possible overvoltages at interior points along the transmission line, in contrast to traditional methods that only provide information at the line's ends. This information can be used as a helpful instrument in the insulation coordination design of transmission lines, as well as an educational tool in electrical engineering courses.

The results computed with the described method were compared with time-domain simulations using the well-known software ATP. In all of the comparisons, a high level of agreement was observed, demonstrating that the proposed method has a high level of accuracy.

It is worth mentioning that, although the results from both methods were very similar, the ATP simulations required substantially more time samples (10 to 50 times more samples) to achieve such results, which can result in a large computational burden. Additionally, in order to obtain measurements at interior points of the transmission line and to include nonuniformities along the transmission line in the ATP simulations, it is necessary to subdivide the line into many line segments, which can be a time-consuming process and can lead to error accumulation. Because of these issues, the proposed frequency domain method is a superior alternative to analyze the transient behavior of a transmission line when the voltage at interior points of the line is of interest.

The relevance of the proposed method lies in its strong potential for application in the accurate fault location and insulation coordination of transmission systems with prominent nonuniformities, which are commonly encountered in many power systems around the world.

Author Contributions: Conceptualization, R.N.-G. and F.P.E.C.; investigation, P.G. and F.P.E.C.; methodology, R.N.-G.; funding acquisition, P.G. and C.T.M.; software, R.N.-G.; validation, R.N.-G. and F.P.E.C.; visualization, C.T.M.; writing—original draft, R.N.-G.; writing—review & editing, R.N.-G., F.P.E.C. and P.G.

References

1. Martínez-Velasco, J.A. Parameter determination for electromagnetic transient analysis in power systems. In *Power System Transients: Parameter Determination*, 1st ed.; CRC Press: Boca Raton, FL, USA, 2010; pp. 1–16.

2. Marti, L.; Dommel, H.W. Calculation of voltage profiles along transmission lines. *IEEE Trans. Power Deliv.* **1997**, *12*, 993–998. [CrossRef]

3. Pinto, A.J.G.; Costa, E.C.M.; Kurokawa, S.; Monteiro, J.H.A.; de Franco, J.L.; Pissolato, J. Analysis of the electrical characteristics and surge protection of EHV transmission lines supported by tall towers. *Int. J. Electr. Power Energy Syst.* **2014**, *57*, 358–365. [CrossRef]

4. Pinto, A.J.G.; Costa, E.C.M.; Kurokawa, S.; Pissolato, J. Analysis of the electrical characteristics of an alternative solution for the Brazilian-Amazon transmission system. *Electr. Power Comp. Syst.* **2011**, *39*, 1424–1436. [CrossRef]

5. Martínez, D.; Moreno, P.; Loo-Yau, R. A new model for non-uniform transmission lines. In Proceedings of the 12th International Conference on Electrical Engineering, Computing Science and Automatic Control (CCE), Mexico City, Mexico, 28–30 October 2015.

6. Lima, A.C.S.; Moura, R.A.R.; Gustavsen, B.; Schroeder, M.A.O. Modelling of non-uniform lines using rational approximation and mode revealing transformation. *IET Gener. Transm. Distrib.* **2017**, *11*, 2050–2055. [CrossRef]

7. Saied, M.M.; Al-Fuhaid, A.S.; El-Shandwily, M.E. S-domain analysis of electromagnetic transients on non-uniform lines. *IEEE Trans. Power Deliv.* **1990**, *5*, 2072–2081. [CrossRef]

8. Oufi, E.A.; Al-Fuhaid, A.S.; Saied, M.M. Transient analysis of loss-less single phase nonuniform transmission lines. *IEEE Trans. Power Deliv.* **1994**, *9*, 1694–1700. [CrossRef]

9. Correia de Barros, M.T.; Almeida, M.E. Computation of electromagnetic transients on nonuniform transmission lines. *IEEE Trans. Power Deliv.* **1996**, *11*, 1082–1087. [CrossRef]

10. Semlyen, A. Some frequency domain aspects of wave propagation on nonuniform lines. *IEEE Trans. Power Deliv.* **2003**, *18*, 315–322. [CrossRef]

11. Gómez, P.; Escamilla, J.C. Frequency domain modeling of nonuniform multiconductor lines excited by indirect lightning. *Int. J. Electr. Power Energy Syst.* **2013**, *45*, 420–426. [CrossRef]

12. Nuricumbo-Guillén, R.; Espino-Cortés, F.P.; Gómez, P.; Tejada-Martínez, C. Computation of transient profiles along non-uniform trasmission lines using the numerical Laplace transform. In Proceedings of the IEEE International Conference on High Voltage Engineering and Applications, Athens, Greece, 10–13 September 2018.

13. Nuricumbo-Guillén, R.; Gómez, P.; Espino-Cortés, F.P.; Uribe, F.A. Accurate computation of transient profiles along multiconductor transmission lines by means of the numerical Laplace transform. *IEEE Trans. Power Deliv.* **2014**, *29*, 2385–2395. [CrossRef]

14. Gómez, P.; Vergara, L.; Nuricumbo-Guillén, R.; Espino-Cortés, F.P. Two dimensional definition of the numerical Laplace transform for fast computation of transient profiles along power transmission lines. *IEEE Trans. Power Deliv.* **2016**, *31*, 412–414. [CrossRef]

15. Nuricumbo-Guillén, R.; Vergara, L.; Gómez, P.; Espino-Cortés, F.P. Laplace-based computation of transient profiles along transmission lines including time-varying and non-linear elements. *Int. J. Electr. Power Energy Syst.* **2019**, *106*, 138–145. [CrossRef]

16. Brancik, L. Time and Laplace-domain methods for MTL transient sensitivity analysis. *Int. J. Comput. Math. Electr. Electron. Eng.* **2011**, *30*, 1205–1223. [CrossRef]

17. Gómez, P.; Uribe, F.A. The numerical Laplace transform: An accurate tool for analyzing electromagnetic transients on power system devices. *Int. J. Electr. Power Energy Syst.* **2009**, *31*, 116–123. [CrossRef]

18. Moreno, P.; Ramírez, A. Implementation of the numerical Laplace transform: A review. *IEEE Trans. Power Deliv.* **2008**, *23*, 2599–2609. [CrossRef]

19. Souza, L.; Lima, A.C.S.; Carneiro, S., Jr. Modeling Overhead Transmission Line with Large Asymmetrical Spans. In Proceedings of the 2011 International Conference on Power Systems Transients, Delft, The Netherlands, 14–17 June 2011.

20. Guo, X.; Fu, Y.; Yu, J.; Xu, Z. A non-uniform transmission line model of the ±1100 kV UHV tower. *Energies* **2019**, *12*, 445. [CrossRef]

21. Hussain, M.K.; Gómez, P. Optimal filter tuning to minimize the transient overvoltages on machine windings fed by PWM inverters. In Proceedings of the 2017 North American Power Symposium (NAPS), Morgantown, VA, USA, 17–19 September 2017.

22. Dommel, H.W. *Electromagnetic Transient Program Theory Book*; Bonneville Power Administration: Portland, OR, USA, 1986.

23. Marti, J. Accurate modelling of frequency-dependent transmission lines in electromagnetic transient simulations. *IEEE Trans Power App. Syst.* **1982**, *PAS-101*, 147–157. [CrossRef]

24. Gómez, P.; Moreno, P.; Naredo, J.L. Frequency domain transient analysis of nonuniform lines with incident field excitation. *IEEE Trans. Power Deliv.* **2005**, *20*, 2273–2280. [CrossRef]

25. Moreno, P.; Gómez, P.; Naredo, J.L.; Guardado, J.L. Frequency domain transient analysis of electrical networks including non-linear conditions. *Int. J. Electr. Power Energy Syst.* **2005**, *27*, 139–146. [CrossRef]

26. Gary, C. Approche complète de la propagation multifilaire en haute fréquence par utilisation des matrices complexes. *EDF Bull. Dir. des Études et Rech.* **1976**, *3*, 4.
27. Gómez, P.; Moreno, P.; Naredo, J.L.; Guardado, L. Frequency domain transient analysis of transmission networks including non-linear conditions. In Proceedings of the 2003 IEEE Bologna Power Tech Conference, Bologna, Italy, 23–26 June 2003.

5

Surface Discharges and Flashover Modelling of Solid Insulators in Gases

Mohammed El Amine Slama [1,*], Abderrahmane Beroual [2] and Abderrahmane (Manu) Haddad [1]

[1] Advanced High Voltage Engineering Centre, School of Engineering. Cardiff University, Queen's Buildings The Parade, Cardiff, Wales CF24 3AA, UK; haddad@cardiff.ac.uk

[2] Laboratoire Ampère, University of Lyon, 36 Avenue Guy de Collongues, 69130 Ecully, France; abderrahmane.beroual@ec-lyon.fr

* Correspondence: slamame@cardiff.ac.uk

Abstract: The aim of this paper is the presentation of an analytical model of insulator flashover and its application for air at atmospheric pressure and pressurized SF_6 (Sulfur Hexafluoride). After a review of the main existing models in air and compressed gases, a relationship of flashover voltage based on an electrical equivalent circuit and the thermal properties of the discharge is developed. The model includes the discharge resistance, the insulator impedance and the gas interface impedance. The application of this model to a cylindrical resin-epoxy insulator in air medium and SF_6 gas with different pressures gives results close to the experimental measurements.

Keywords: surface discharge; flashover; gas; modelling; pressure; thermal properties

1. Introduction

In order to optimize the insulation level for high-voltage components (air insulated substations (AIS), gas insulated substations (GIS) and gas insulated lines (GIL), breakers, overhead lines ...), a special attention is given to creeping or surface discharges because of the thermal effects and the faults that they can produce by sparking or flashover. Then, the knowledge of the parameters characterizing this kind of discharge is essential to understand the complexity of the mechanisms involved in their development. Thus, it is fundamental to acquire such information to enable building a mathematical model that can help in optimizing the insulation efficiency.

This paper aims to carry out a review of existing models of creeping discharges and to propose an analytical approach for the calculation of flashover voltage of solid insulators in gases under lightning voltage stress.

2. Review of Surface Discharges and Flashover Models in Gases

From the insulation viewpoint, the triple junction (metal-gas-solid) constitutes the weakest point in high-voltage equipment. Indeed, when the electric field reaches a critical value, partial discharges (PDs) can be initiated in the vicinity of this region. The increase of the voltage leads these PDs to develop and to transform into surface discharges (creeping discharges) that propagate over the insulator up to flashover [1–4]. In the case of GIS and GIL, the worst case is when insulators (spacer, post-type insulator) are contaminated by metallic particles on their surfaces [5,6].

The physical mechanisms responsible for the surface discharge propagation are still not well known because of the complexity of the phenomena and the interaction of different factors, such as the interaction between the discharges, nature of gas and the proprieties of the solid insulating material, gas pressure, surface charges and pollution (metallic particle), geometrical parameters (insulator shape, electrodes form ...), etc. Fundamental studies have been conducted to understand the inception and propagation of creeping discharges in various gases [7–14]. It appears from the reported results that the phenomena start with corona discharges that evolves into ramified streamers. When the streamer discharge reaches a certain length, a leader channel with streamers at its head appears.

The creeping discharge propagation dynamics in SF_6 (sulfur hexafluoride) has been investigated by many researchers [7–11]. Okubo et al. [8] reported that the creeping discharge has the same dynamics as in air (Figure 1). Tenbohlen and Schröder [9] analysed the surface discharge under lightning impulse (LI) voltage with different electrical charges deposition on the insulator surface. Figure 1 illustrates the current waveform from the inception to flashover with different electrical charges on the insulator surface. From Figure 1, some similarities with discharge current propagating in air [13] can be noted: the current increases with the leader elongation until the discharge reaches the critical length. Then, the final jump occurs causing the full flashover.

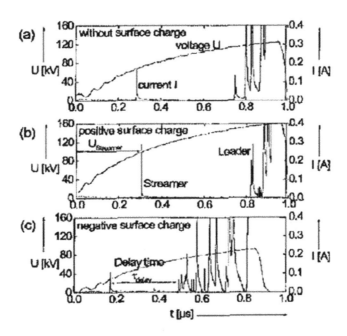

Figure 1. Instantaneous current and voltage during flashover at the surface of cylindrical epoxy insulator according to Reference [9].

Hayakawa et al. [7] analysed the mechanism of impulse creeping discharge propagation on charged PMMA (poly-methyl methacrylate) surface. Their results showed that the discharge propagation is influenced by the charged surface and can be explained by the streamer propagation and streamer-to-leader transition based on the precursor mechanism. On the other hand, according to Okubo et al. [8], Beroual [3] and Beroual et al. [10,11], the creeping discharge propagation depends on the specific capacitance of the solid insulator. The permittivity, the conductivity and the geometry of the insulator affect the propagation of the surface discharge [10–12].

Modelling and calculation of flashover voltage is not an easy task because of the interaction of different parameters, such as gas pressure and its chemical constitution, physicochemical properties of the solid insulator, nature and distribution of the surface charges, etc. Different models have been proposed in order to compute the inception voltage of creeping discharges and flashover voltage of insulator in air at atmospheric pressure [2,3,13,14]. Figure 2 depicts the different evolution steps of creeping discharge on the insulator.

According to Reference [2], the corona inception voltage depends on the equivalent capacitance of the system. It can be calculated with the following relationship [2]:

$$U_{inception} = \frac{A}{C^a} \tag{1}$$

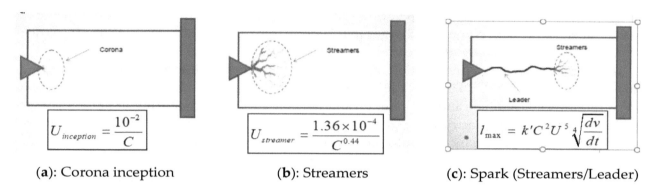

(a): Corona inception   (b): Streamers   (c): Spark (Streamers/Leader)

Figure 2. Steps of creeping discharges according to Reference [2].

The second step is the appearance of streamers (Figure 2a). The streamers voltage inception is given by [2]:

$$U_{streamer} = \frac{B}{C^b} \tag{2}$$

According to Toepler [3], the maximum (critical) length of the discharge that leads to flashover is:

$$l_{max} = k \cdot C^2 \cdot U^5 \cdot \sqrt[4]{\frac{du}{dt}} \tag{3}$$

Then, if the voltage is increased, the discharge will be irreversible and propagates until flashover. In this case, the flashover voltage U_{fov} can be calculated as well:

$$U_{fov} = \frac{D}{C^d} \tag{4}$$

where:

- C is the equivalent capacitance,
- A, B and D are parameters that depend on the geometry and the material of insulator, the kind of the discharge and the experimental conditions (gas, pressure, temperature, humidity, electrodes shape, voltage waveform ...), respectively. Terms a, b and d are empirical parameters the values of which vary in the range 0.2–0.44.

These models are empirical and involve only the capacitance of the insulator.

In the case of SF$_6$, Laghari [15] proposed a relationship of flashover voltage based on the efficiency coefficient that represents the ratio of the flashover voltage for uniform electrical gradient distribution to the voltage breakdown of the same gap without an insulator with the same configuration of insulator as well:

$$V_{fov} = \frac{12.4}{\ln(V_b)} \cdot \frac{k_1}{k_2} \cdot \frac{\ln(\varepsilon_r)}{\varepsilon_r} \cdot V_b \tag{5}$$

where,

$$V_b = const + \left(\frac{E}{p}\right)_{cri} \cdot p \cdot d \tag{6}$$

V_b is the breakdown voltage calculated according to Paschen law. k_1 and k_2 are parameters that depend on the roughness and the contact nature between the insulator and the electrodes. ε_r is the permittivity of the insulator.

Hama et al. [16] proposed a semi-empirical relationship of flashover voltage based on the mechanism leader/precursor:

$$V_{fov} = \frac{X_{Leader}}{D_{pol} V_{Leader}} + V_{Leader} \tag{7}$$

where X_{Leader} and V_{Leader} are the length and the voltage of the leader discharge respectively, and D_{pol} is a coefficient that is dependent on the polarity of the applied voltage, the reduced critical electrical gradient and the shape of the electrodes, with:

$$D_{pol} = const \times \left(\frac{E}{p}\right)_{cri} \times \phi(Ry_{electrodes}) \tag{8}$$

The application of this model shows results close to the experimental measurements, but it is limited to the shape of the used insulators and the experimental conditions.

In the following, we recall the main principles of an analytical static model based on the electrical equivalent circuit and thermal discharge temperature we previously developed [1,13].

3. Principal of Circuit Model

Surface discharges are like spark (streamer/leader) discharges, i.e., a hot leader column and a streamers zone at its head [8,9,13,17]. Based on Figure 3, the voltage along the discharge can be written as follows:

$$V_d = V_l + V_s = x_l E_l + x_s E_s = x_d\, r_d\, I \tag{9}$$

where V_d, V_l and V_s are the voltages of the discharge, the leader channel and the streamers, respectively. x_l, x_s, E_l and E_s are respectively the length and the electrical gradient of the leader channel and the streamers. x_d, r_d, and I are respectively the discharge length, the discharge resistance and the current.

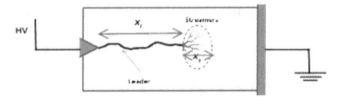

Figure 3. Illustration of leader column and streamers head of a discharge at the surface of an insulator.

The discharge resistance can be deduced from Equation (9):

$$r_d = \frac{x_l E_l + x_s E_s}{x_d I} \tag{10}$$

where $x_d = x_s + x_l$.

According to Equation (10), creeping discharge can be considered as a resistance and it can be assumed that the discharge channel is a uniform cylinder.

Many researchers published photos of surface discharges indicating that there are two regions: the main luminous discharge (leader + streamer head) and less luminous branches, as illustrated in Figures 4 and 5 [10,18]. So, the presence of those less luminous discharges can be represented as a resistor in parallel to the insulator surface. On the other hand, several research investigations demonstrate the existence of a dark current in high-pressurized gases that contribute to increase the insulator conductivity [19]. These currents contribute to the appearance of the second region (called luminous plasma), as depicted in Figures 4 and 5.

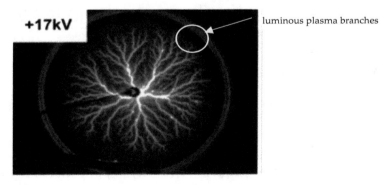

Figure 4. Surface discharge at the surface of insulator in SF6 with 3 bars under LI+ according to Reference [10].

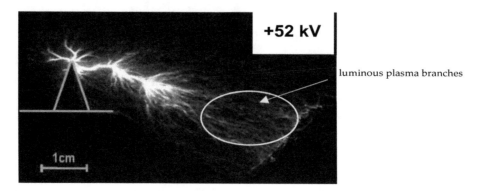

luminous plasma branches

Figure 5. Surface discharge at the surface of a coated electrode in SF6 with 1 bar under LI+ according to Reference [18].

3.1. Parameters of the Circuit

The proposed model is constituted by an equivalent electrical circuit representing the electrical discharge, in series with the unbridged gap. The gap (the distance between the head of discharge and the opposite electrode) consists of a gas layer and of the solid dielectric at the interface (Figure 6). The gas layer is assumed to be equal to the diameter of the discharge channel. This model was developed elsewhere [13] in the case of air at atmospheric pressure and represents the instant when the discharge reaches a maximum length (called critical length) before the final jump [13]. In the following, the same approach [13] will be adopted with the assumption that the LI (lightning impulse) voltage waveform can be considered as a quart-cycle of sine signal with a frequency about 0.3 MHz.

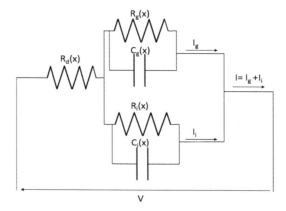

Figure 6. Insulator cylindrical model with a discharge channel and the corresponding equivalent electrical circuit.

The electrical Equation describing this circuit is:

$$V = R_d(x)I + \left[Z_g(x)//Z_i(x)\right]I \tag{11}$$

where,

$$Z_g(X) = R_g(x)//C_g(x) = \frac{R_g(x)}{1+j\omega R_g(x)C_g(x)} = \frac{r_g(L-x)}{1+(r_g c_g \omega)^2} - j\frac{r_g^2 c_g \omega(L-x)^2}{1+(r_g c_g \omega)^2}$$

$$Z_i(X) = R_i(x)//C_i(x) = \frac{R_i(x)}{1+j\omega R_i(x)C_i(x)} = \frac{r_i(L-x)}{1+(r_i c_i \omega)^2} - j\frac{r_i^2 c_i \omega(L-x)^2}{1+(r_i c_i \omega)^2} \tag{12}$$

and

$R_d(x) = r_d x = x\frac{\rho_d}{s_d};$

$C_i(X) = \frac{c_i}{L-x} = \varepsilon_i \frac{s_i}{L-x}; C_g(X) = \frac{c_g}{L-x} = \varepsilon_g \frac{s_g}{L-x}; R_i(x) = r_i(L-x) = \rho_i \frac{L-x}{s_i}; R_g(x) = r_g(L-x) = \rho_g \frac{L-x}{s_g}$

r_d is the linear resistance of the discharge channel. r_i, r_g, c_i, c_g, ε_I, ε_g, ρ_i, and ρ_g, are respectively the linear resistance, capacitance, the permittivity and the resistivity, respectively of the solid insulator and the unbridged gap. s_d is the cross-section of the discharge channel, s_i and s_g are respectively the cross-sections of the solid insulator and the layer of the unbridged gap. ω is the pulsation ($\omega = 2\pi f$, f being the frequency).

Then, the equivalent impedance of the system will be:

$$Z_{eq}(x) = r_d x + \frac{r_g r_i}{\alpha_g \alpha_i} G_1 (L - x) + j \frac{r_g r_i}{\alpha_g \alpha_i} G_2 (L - x) \tag{13}$$

Let us put:

$$\begin{aligned} \tau_g &= \rho_g \varepsilon_g \omega = r_g c_g \omega \\ \tau_i &= \rho_i \varepsilon_i \omega = r_i c_i \omega \end{aligned} \tag{14}$$

Product $\tau_i^2 \gg 1$ and $\tau_g^2 \gg 1$, then:

$$\begin{cases} \alpha_i \approx \tau_i^2 \\ \alpha_g \approx \tau_g^2 \end{cases} \tag{15}$$

The terms G_1 and G_2 are:

$$\begin{aligned} G_1 &= \frac{z_1}{z_3 + z_4} \\ G_2 &= \frac{z_2}{z_3 + z_4} \end{aligned} \tag{16}$$

where,

$$\begin{cases} z_1 = \left(\frac{r_g}{\tau_g^2} + \frac{r_i}{\tau_i^2}\right)(1 - \tau_g \tau_i) + \left(\frac{r_g}{\tau_g} + \frac{r_i}{\tau_i}\right)(\tau_g + \tau_i) \\ z_2 = \left(\frac{r_g}{\tau_g} + \frac{r_i}{\tau_i}\right)(1 - \tau_i) - \left(\frac{r_g}{\tau_g^2} - \frac{r_i}{\tau_i^2}\right)(\tau_g + \tau_i) \\ z_3 = \left(\frac{r_g}{\tau_g^2} + \frac{r_i}{\tau_i^2}\right)^2 \\ z_4 = \left(\frac{r_g}{\tau_g} + \frac{r_i}{\tau_i}\right)^2 \end{cases} \tag{17}$$

The square of the modulus of the equivalent impedance is:

$$|Z_{eq}|^2 = \gamma x^2 + 2Lx\left[r_d\left(r_d - \frac{r_g r_i}{\tau_g^2 \tau_i^2} G_1\right) - \gamma\right] \tag{18}$$

where,

$$\gamma = \left(r_d - \frac{r_g r_i}{\tau_g^2 \tau_i^2} G_1\right)^2 + \left(\frac{r_g r_i}{\tau_g^2 \tau_i^2} G_2\right)^2 \tag{19}$$

According to Reference [20], when the discharge length increases, the equivalent impedance decreases:

$$\frac{d|Z_{eq}|^2}{dx} \leq 0 \tag{20}$$

By differentiating Equation (18) with respect to x, we get:

$$\frac{d|Z_{eq}|^2}{dx} = 2\gamma x + 2L\left[r_d\left(r_d - \frac{r_g r_i}{\tau_g^2 \tau_i^2} G_1\right) - \gamma\right] \leq 0 \tag{21}$$

Then,

$$\frac{x}{L} - 1 \leq \left[\frac{r_d}{\gamma \tau_g^2 \tau_i^2}\left(r_g r_i G_1 - \tau_g^2 \tau_i^2 r_d\right)\right] \tag{22}$$

Flashover of the solid dielectric occurs when Equation (22) is equal to zero, i.e., when the discharge length is equal to the total creeping (leakage) distance. This Equation can be considered as "the flashover condition". Therefore, the maximum (or critical) length of the discharge corresponding to flashover is:

$$x_{cri} = \frac{L}{\gamma \alpha_g \alpha_i}\left[\gamma \tau_g^2 \tau_i^2 - r_d\left(r_d - \tau_g^2 \tau_i^2 G_1\right)\right] = L \cdot n \tag{23}$$

where,

$$n = \frac{1}{\gamma \tau_g^2 \tau_i^2}\left[\gamma \tau_g^2 \tau_i^2 - r_d(r_d - r_g r_i G_1)\right] \tag{24}$$

where $0 < n < 1$.

The worst case can be derived from Equation (12), it corresponds to:

$$\frac{r_d}{\gamma \tau_g^2 \tau_i^2}\left(r_g r_i G_1 - \tau_g^2 \tau_i^2 r_d\right) \geq 0 \tag{25}$$

The term $\frac{r_d}{\gamma \tau_g^2 \tau_i^2}$ is always positive, then:

$$r_g r_i G_1 \geq \tau_g^2 \tau_i^2 r_d \tag{26}$$

Equation (26) can be written as:

$$\frac{\tau_g^2 \tau_i^2}{G_1} \cdot \frac{r_d}{r_g r_i} = K \leq 1 \tag{27}$$

where,

$$0 < K \leq 1 \tag{28}$$

or:

$$r_d \leq K \cdot G_1 \cdot \frac{r_g r_i}{\tau_g^2 \tau_i^2} \tag{29}$$

Condition (28) indicates that the discharge propagates when the ratio K is less than or equal to 1. This corresponds to the propagation criterion in which the discharge length is sufficient for causing the final jump, provoking flashover [13].

On the other hand, the power loss per unit length p_d in the discharge channel is:

$$p_d = r_d I^2 \tag{30}$$

By combining Equations (30) and (26), it yields:

$$I = \sqrt{\frac{p_d}{r_d}} \tag{31}$$

The square of the modulus of the voltage—Equation (11) is:

$$|V|^2 = |I|^2 \cdot |Z_{eq}|^2 \tag{32}$$

By substituting Equations (23), (24) and (30) in Equation (18), it yields:

$$|Z_{eq}|^2 = \beta L^2 \left(\frac{r_g r_i}{\tau_g^2 \tau_i^2}\right)^2 \tag{33}$$

with:

$$\beta = K^2 G_1{}^2 n^2 + \left(G_1{}^2 + G_2{}^2\right)(1-n)^2 + 2K^2 G_1 n(1-n) \tag{34}$$

By substituting Equations (31) and (33) in Equation (32), the Equation of flashover voltage will be deduced as:

$$V_{FOV} = \frac{L}{\tau_g \tau_i} \sqrt{p_d \cdot \frac{r_g r_i}{r_d} \cdot \beta}. \tag{35}$$

3.2. Thermal Conductivity and Discharge Resistance

According to the solution proposed by Frank-Kamenetski [21,22], the energy dissipated by thermal conduction within the discharge channel is:

$$P_d = 16\pi \lambda_d \frac{K_B}{W_i} T^2 \tag{36}$$

By combining Equations (36) and (30), the final Equation of flashover voltage will be:

$$V_{FOV} = 4\frac{L}{\tau_g\tau_i}T\sqrt{\frac{\pi K_B}{W_i} \cdot \lambda_d \cdot \frac{r_g r_i}{r_d} \cdot \beta}$$

(37)

In the case of air at atmospheric pressure, the thermal conductivity is calculated according to the following Equation [23]:

$$\lambda(\theta) = \frac{\lambda_a}{1+\frac{A_a(1-v_a)}{v_a}}$$

(38)

where λ_a, v_a and A_a are the thermal conductivity, volume fraction and kinetic gas coefficient for air, respectively.

Also, the discharge resistance in air at atmospheric pressure is given by [24]:

$$r_d(T) = r_{0d}\exp\left(\frac{W_i}{2K_BT}\right).$$

(39)

where r_{0d} is a constant in the range of operating temperatures of the discharge. W_i represents the first ionization energy of the different species constituting the discharge channel and K_B is the Boltzmann constant.

In the case of SF$_6$, both discharge resistance and discharge thermal conductivity are functions simultaneously of gas pressure and plasma temperature [25,26].

$$\lambda_d = \Gamma(T,p)$$
$$\sigma_d = \Sigma(T,p)$$

(40)

According to Pinnekamp and Niemeyer [27], and Niemeyer et al. [28], the temperature of the leader discharge is between 2400 K and 2800 K. On the other hand, based on the transport parameters data of SF$_6$ published in the literature [25,26], the thermal conductivity was plotted as a function of gas pressure (Figure 7) and the discharge resistance against gas pressure (Figure 8) for a range of temperatures between 2500 K and 3500 K. From these figures, numerical empirical formulae of the discharge thermal conductivity and discharge resistance against pressure for a given temperature was deduced:

$$\sigma_d = A.p^{-m}$$

(41)

$$\lambda_d = a_0 + \ldots + a_n p^n$$

(42)

where p is the gas pressure and a and A are empirical parameters.

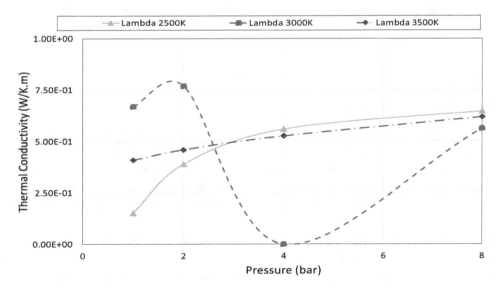

Figure 7. Discharge thermal conductance of discharge versus variation with pressure with for different temperatures.

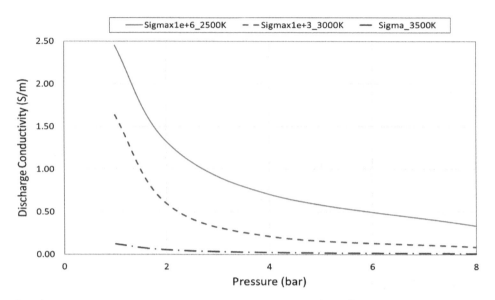

Figure 8. Discharge resistance variation versus pressure with different temperatures at 2500 and 3000 K, respectively.

4. Application

To validate the proposed model, first the model was applied for the calculation of the flashover voltage of cylindrical epoxy insulators in air at atmospheric pressure. The second application will be for the same kind of insulator in SF$_6$ gas medium. The computed flashover voltages are compared with the experimental data reported by other researchers, as in References [12,13,15,29]. Table 1 gives the characteristics of the used insulator in the computations.

Table 1. Characteristics of used insulators from literature used in modelling.

Insulator	Material	Diameter	Length	Reference	Gas	Pressure
1	Epoxy	25	60	[13]	Air	Atmospheric
2	Epoxy	25	60	[12]	SF$_6$	Variable
3	Epoxy	25	45	[29]	SF$_6$	Variable
4	Epoxy	30	10	[15]	SF$_6$	Variable

The lightning impulse voltage frequency is calculated based on the following Equation [30]:

$$f = 0.35 / T_R \qquad (43)$$

T_R is the rising time of the voltage front equal to 1.2 μs.

4.1. Air at Normal Atmospheric Conditions

Figure 9 illustrates the results of the application of the proposed model in air at atmospheric pressure. The model is compared with the experimental data of Reference [13], a previous model developed earlier [1] and Toepler's model. The temperature of discharge was taken between 1800 K and 2000 K, which corresponds to a leader phase on the insulator surface [13]. The resistance of air ranges from 10^{23} to 10^{25} Ω/cm, its dielectric constant being equal to 1. The effect of humidity and roughness are neglected.

By comparing flashover voltage given by Equation (36) and the other models, we can remark that the computed values are close to the measured ones and follow the same trend. According to this result, we can deduced that the impedance of the interface between the head of the discharge and the opposite electrode plays affects the result (Figure 9). It contributes to the breakdown process before the final jump of the discharge (flashover), as described in Reference [12]. The maximum deviation is 18.2% and the average deviation is less than 5%.

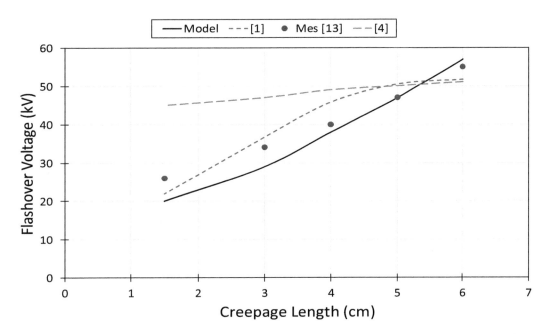

Figure 9. Comparison between calculated and measured flashover voltage versus creepage distance in air at atmospheric pressure for insulator 1.

4.2. SF₆ at Variable Pressure

In the case of SF_6, we use Equations (41) and (42), to compute the flashover voltage and its dependency on the gas pressure and temperature. The temperature of the discharge was taken between 2500 K and 3500 K.

A specific consideration for the resistance of the gas at the interface is required in the case of pressurised SF_6. In fact, experimental results concerning flashover of solid insulators on pressurised gases suggest that the discharge tends to stick to the insulator surface when the gas pressure increases [8,12]. On the other hand, according to Figures 4 and 5, the gap between the discharge's head and the ground electrode appears like an ionized cylinder. Knowing that the attachment of the pressurised gas also increases with pressure, it can be deduced that the resistance of the interface between the discharge head and the grounding electrode depends on the gas pressure as well.

Based on the data reported in the literature [25,26], the resistance of the interface can be represented as a cylindrical plasma with a temperature between 1000 K and 1500 K. In this range of temperature, the plasma resistivity increases with the gas pressure, as depicted in Figure 10. As can be observed in this figure, the assumption of a plasma with a temperature varying between 1200 K and 1400 K is a good approximation, since the resistivity is increasing with pressure for all temperatures. The dielectric constant being equal to 1 and the effects of surface charge accumulation and humidity are not considered.

Figure 11 illustrates the comparison of the calculated flashover voltage with the data of Slama et al. [12] for insulator 2 of Table 1. It can be observed that the calculated flashover voltages are close to the measured values, indicating that flashover voltage tends to be stable with the pressure increase. The maximum deviation is 8.2% and the average deviation is around 4.5%.

A comparison of the calculated flashover voltage with the data of Reference [29] obtained with insulator 3, is depicted in Figure 12. In this work, Moukengué and Feser [29] present results of flashover voltages as a function of gas pressure for different tests: one for a single impulse shot and the second for five impulse shots. It is noted that the calculated flashover voltages are close to the experimental measured values. The maximum deviation is 8.2% and the average deviation is around 6.5%.

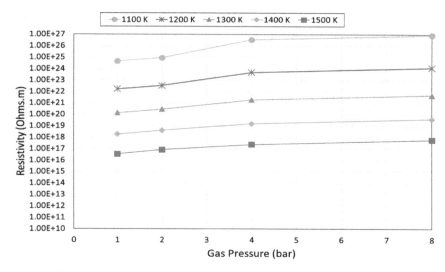

Figure 10. Resistivity of the SF_6 plasma at non-thermal regime versus gas pressure et different temperatures.

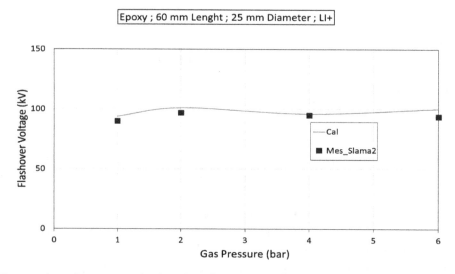

Figure 11. Comparison between calculated and measured flashover voltage versus gas pressure for insulator 2 with 60 mm length and 25 mm diameter.

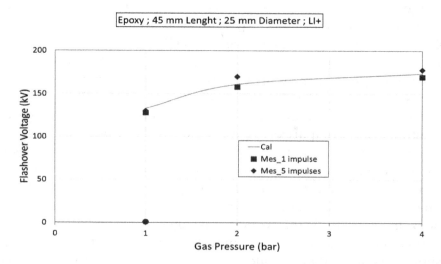

Figure 12. Comparison between calculated and measured flashover voltage versus gas pressure for a cylindrical epoxy insulator 4 with 45 mm length and 25 mm diameter.

Figure 13 shows the comparison of the results using the developed model and the data of Reference [15] with insulator 4. Again, it is observed that the calculated flashover voltages are close to the experimental ones and the maximum deviation is 10% and the average deviation is less than 4%.

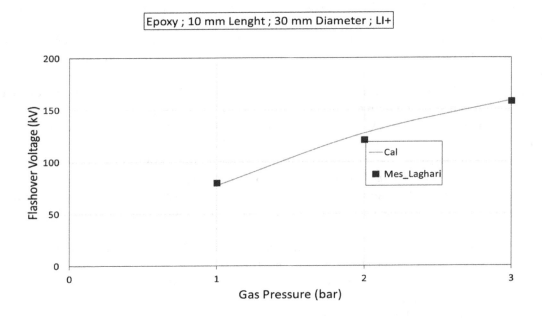

Figure 13. Comparison between calculated and measured flashover voltage versus gas pressure for a cylindrical epoxy insulator with 10 mm length and 30 mm diameter.

5. Conclusions

In this paper, a model was developed for surface discharges and flashover voltage in air at atmospheric pressure and compressed SF_6. The proposed analytical model is based on the equivalent electrical circuit representing the discharge along the insulator surface and the thermal properties of the discharge by assuming that the area between the discharge head and the ground electrode as a cylindrical plasma.

The proposed model was first applied for air at atmospheric pressure to validate it against existing models and data. It was noticed that the simulated results are very close to the experimental ones. Thus, the impedance of the interface between the head of the discharge and the opposite electrode significantly affects the result. It contributes to the breakdown process before the final jump of the final discharge that is flashover. In the case of SF_6, the application of this model to various configurations taken from the literature shows that the computed flashover voltage magnitudes are close to the measured values and exhibit similar trends.

The proposed model constitutes a first step for developing a tool for flashover prediction in ambient air and for the design of the solid insulation in GIS and GIL filled with SF_6.

Author Contributions: Conceptualization, M.E.A.S., A.B. and A.M.H.; methodology, M.E.A.S., A.B. and A.M.H.; validation, M.E.A.S., A.B. and A.M.H.; formal analysis, M.E.A.S., A.B. and A.M.H.; investigation, M.E.A.S.; writing—original draft preparation, M.E.A.S.; writing—review and editing, M.E.A.S., A.B. and A.M.H.; visualization, M.E.A.S., A.B. and A.M.H. All authors have read and agreed to the published version of the manuscript.

Acknowledgments: The authors would like to thank Sylvain Nichele, Alain Girodet and Paul Vinson from SuperGrid Institute (France) for their help.

References

1. Slama, M.E.A.; Beroual, A.; Girodet, A.; Vinson, P. Characterization and Mathematical Modelling of Surface Solid Insulator in Air. In Proceedings of the International Symposium on High Voltage Engineering, Plzen, Czech Republic, 24–28 August 2015.

2. Al-Arainy, A.B.; Malik, N. *Experiments in High Voltage Engineering King, Saud University*; Academic Publishing and Press: Riyadh, Saudi Arabia, 2014.

3. Beroual, A. Creeping Discharges at Liquid/solid and Gas/Solid Interfaces: Analogies and Involving Mechanisms. In Proceedings of the IEEE 20th International Conference on Dielectric Liquids (ICDL), Roma, Italy, 23–27 June 2019.

4. Toepler, M. Über die physikalischen Grundgesetze der in der Isolatorentechnik auftretenden elektrischen Gleiterscheinungen. *Archiv Elektrotechnik* **1921**, *10*, 157–185. [CrossRef]

5. Cookson, A.H. Review of high-voltage gas breakdown and insulators in compressed gas. *IET* **1981**, *128*, 303–312. [CrossRef]

6. Sudarshan, T.S.; Dougal, R. Mechanisms of surface flashover along solid dielectrics in compressed gases: A Review. *IEEE Trans. Electr. Insul.* **1986**, *21*, 727–746. [CrossRef]

7. Hayakawa, N.; Ishida, Y.; Nishiguchi, H.; Hikita, M.; Okubo, H. Mechanism of impulse creepage discharge propagation on charged dielectric surface in SF6 gas. *IEEJ Trans. Power Energy* **1996**, *116-B*, 600–606. [CrossRef]

8. Okubo, H.; Kanegami, M.; Hikita, M.; Kito, Y. Creepage Discharge Propagation in Air and SF6 Gas Influenced by Surface Charge on Solid Dielectrics. *IEEE Trans. Dielectr. Electr. Insul.* **2004**, *1*, 204–304.

9. Tenbohlen, S.; Schröder, G. Discharge Development over Surfaces in SF6. *IEEE Trans. Dielectr. Electr. Insul.* **2000**, *7*, 241–246. [CrossRef]

10. Beroual, A.; Coulibaly, M.-L.; Aitken, O.; Girodet, A. Investigation on creeping discharges propagating over epoxy resin and glass insulators in presence of different gases and mixtures. *Eur. Phys. J. Appl. Phys.* **2011**, *56*, 30802–30809. [CrossRef]

11. Beroual, A.; Coulibaly, M.-L.; Aitken, O.; Girodet, A. Effect of micro-fillers in PTFE insulators on the characteristics of surface discharges in presence of SF6, CO_2 and SF6 –CO_2 mixture. *IET Gener. Transm. Distrib.* **2012**, *6*, 951–957. [CrossRef]

12. Slama, M.E.A.; Beroual, A.; Girodet, A.; Vinson, P. Barrier effect on Surface Breakdown of Epoxy Solid Dielectric in SF6 with Various Pressures. In Proceedings of the Conference on Electrical Insulation and Dielectrics Phenomena, Toronto, ON, Canada, 23 October 2016. CEIDP 2016.

13. Slama, M.E.A.; Beroual, A.; Girodet, A.; Vinson, P. Creeping Discharge and Flashover of Solid Dielectric in Air at Atmospheric Pressure: Experiment and Modelling. *IEEE Trans. Dielectr. Electr. Insul.* **2016**, *23*, 2949–2956. [CrossRef]

14. Douar, M.A.; Beroual, A.; Souche, X. Creeping discharges features propagating in air at atmospheric pressure on various materials under positive lightning impulse voltage—Part 2: Modelling and computation of discharges' parameters. *IET Gener. Transm. Distrib.* **2018**, *6*, 1429–1437. [CrossRef]

15. Laghari, R. Spacer Flashover in Compressed Gases. *IEEE Trans. Electr. Insul.* **1985**, *EI-20*, 83–92. [CrossRef]

16. Hama, H.; Inami, K.; Yoshimura, M.; Nakanishi, K. Estimation of Breakdown Voltage of Surface Flashover Initiated from Triple Junction in SF6 Gas. *IEEJ* **1996**, *116*. [CrossRef]

17. Waters, R.T.; Haddad, A.; Griffiths, H.; Harid, N.; Sarkar, P. Partial-arc and Spark Models of the Flashover of Lightly Polluted Insulators. *IEEE Trans. Dielectr. Electr. Insul.* **2010**, *17*, 417–424. [CrossRef]

18. Kessler, J.E.M. Isoliervermögen Hybrider Isoliersysteme in Gasisolierten Metallgekapselten Schaltanlagen (GIS). Ph.D. Thesis, Technische Universität München, Munich, Germany, 2011.

19. Zavattoni, L. Conduction phenomena through gas and insulating solids in HVDC GIS, and consequences on electric field distribution. Ph.D. Thesis, University of Grenoble, Grenoble, France, 2014.

20. Dhahbi-Megriche, N.; Beroual, A.; Krähenbühl, L. A New Proposal Model for Polluted Insulators Flashover. *J. Phys. D Appl. Phys.* **1997**, *30*, 889–894. [CrossRef]

21. Fridman, A.; Nester, S.; Kennedy, L.A.; Saveliev, A.; Mutaf-Yardimci, O. Gliding arc discharge. *Prog. Energy Combust. Sci.* **1999**, *25*, 211–231. [CrossRef]

22. Fridman, A. *Plasma Chemistry*; Cambridge University Press: Cambridge, UK, 2008.

23. McElhannon, W.; McLaughlim, E. Thermal Conductivity of Simple Dense Fluid Mixtures. In Proceedings of the Fourteenth International Conference on Thermal Conductivity, Storrs, CT, USA, 2–4 June 1975; Klemens, P.G., Clen, T.K., Eds.;

24. Slama, M.E.A.; Beroual, A.; Hadi, H. Analytical Computation of Discharge Characteristic Constants and Critical Parameters of Flashover of Polluted Insulators. *IEEE Trans. Dielectr. Electr. Insul.* **2010**, *17*, 1764–1771. [CrossRef]

25. Zhong, L.; Rong, M.; Wang, X.; Wu, J.; Han, G.; Lu, Y.; Yang, A.; Wu, Y. Compositions, thermodynamic properties, and transport coefficients of high temperature $C_5F_{10}O$ mixed with CO_2 and O_2 as substitutes for SF_6 to reduce global warming potential. *AIP Adv.* **2017**, *7*, 075003. [CrossRef]

26. Assael, M.J.; Koini, I.A.; Antoniadis, K.D.; Huber, M.L.; Abdulagatov, I.M.; Perkins, R.A. Reference Correlation of the Thermal Conductivity of Sulfur Hexafluoride from the Triple Point to 1000 K and up to 150 MPa. *J. Phys. Chem. Ref. Data* **2012**, *41*, 023104. [CrossRef]

27. Niemeyer, L.; Pinnekamp, F. Leader discharges in SF6. *J. Phys. D: Appl. Phys.* **1983**, *16*, 1031–1045. [CrossRef]

28. Wiegart, N.; Niemeyer, L.; Pinnekamp, F.; Boeck, W.; Kindersberger, J.; Morrow, R.; Zaengl, W.; Zwicky, I.; Gallimberti, M.; Boggs, S.A. Inhomogeneous Field Breakdown in GIS—The Prediction of Breakdown Probabilities and Voltages: Parts I, II, and III. *IEEE Trans. Power Deliv.* **1988**, *3*, 923–946. [CrossRef]

29. Moukengué lmano, A.; Feser, K. Flashover behavior of conducting particle on the spacer surface in compressed N_2, 90%N_2+10%SF_6 and SF_6 under lightning impulse stress. In Proceedings of the International Symposium on Electrical Insulation (ISEI 2000), Anaheim, CA, USA, 2–5 April 2000.

30. Ianovici et, M.; Morf, J.-J. *Compatibilité Électromagnétique*; Presses Polytechniques et Universitaires Romandes: Lausanne, Suisse, 1979.

6

Accumulation Behaviors of Different Particles and Effects on the Breakdown Properties of Mineral Oil under DC Voltage

Min Dan [1,2], Jian Hao [1,*], Ruijin Liao [1], Lin Cheng [3,4], Jie Zhang [3,4] and Fei Li [3,4]

[1] State Key Laboratory of Power Transmission Equipment & System Security and New Technology, Chongqing University, Chongqing 400044, China; danmin@cqu.edu.cn (M.D.); rjliao@cqu.edu.cn (R.L.)
[2] State Grid Chongqing Nanan Power Supply Company, Chongqing 401223, China
[3] Najing NARI Group Corporation, State Grid Electric Power Research Institute, Nanjing 211000, China; chenglin@sgepri.sgcc.com.cn (L.C.); zhangjie3@sgepri.sgcc.com.cn (J.Z.); lifei6@sgepri.sgcc.com.cn (F.L.)
[4] Wuhan NARI Co. Ltd., State Grid Electric Power Research Institute, Wuhan 430077, China
* Correspondence: haojian2016@cqu.edu.cn.

Abstract: Particles in transformer oil are harmful to the operation of transformers, which can lead to the occurrence of partial discharge and even breakdown. More and more researchers are becoming interested in investigating the effects of particles on the performance of insulation oil. In this paper, a simulation method is provided to explore the motion mechanism and accumulation characteristics of different particles. This is utilized to explain the effects of particle properties on the breakdown strength of mineral oil. Experiments on particle accumulation under DC voltage as well as DC breakdown were carried out. The simulation results are in agreement with the experimental results. Having a DC electrical field with a sufficient accumulation time and initial concentration are advantageous for particle accumulation. Properties of impurities determine the bridge shape, conductivity characteristics, and variation law of DC breakdown voltages. Metal particles and mixed particles play more significant roles in the increase of current and electrical field distortion. It is noteworthy that cellulose particles along with metal particles cannot have superposition influences on changing conductivity characteristics and the electrical field distortion of mineral oil. The range of electrical field distortion is enlarged as the particle concentration increases. Changes in the electrical field distribution and an increase in conductivity collectively affect the DC breakdown strength of mineral oil.

Keywords: mineral oil; different particles; accumulation behavior; breakdown voltage; DC voltage

1. Introduction

In order to meet the urgent demand for energy delivery, China has vigorously developed large-capacity and long-distance ultra-high voltage transmission technology over the last 10 years. As a result, the market for large transformers has enlarged, which has strengthened the requirement for large volume and high-quality transformers. Mineral oil-paper insulation is widely used in large transformers, and this determines the safety, stability, and insulation performance of transformers. Transformer damage is mainly caused by insulation problems, among which particle pollution of insulating oil is a significant factor [1].

A large body of literature has expounded sources of particles in power equipment. For power transformers, metal and non-metal particles are the main components of solid particles. Metal particles mainly include copper and iron particles, while non-metallic particles are mainly made up of carbon particles and cellulose particles. Cellulose impurities are the main part of non-metallic impurities, accounting for more than 90% of these impurities [2–4]. Owing to the process of production, assembly, transportation, and incomplete oil filtering, impurities may be left in oil [5–9]. Meanwhile, impurities may be caused by the aging of oil–paper insulation, partial discharge, and part wearing due to long-term operation [5–9]. A few particles move to the joint position between different components with the oil flow. Others gather in the region where the flow rate is slower to form a path connecting conductors, which brings a severe challenge to the stable operation of transformers as well as leading to partial discharge or breakdown [10–12].

The CIGRE Working Group 12.17 conducted a statistical analysis of the fault causes of twenty-two transformers and forty bushings (765 kV) and pointed out that the source and hazard of particles must be paid attention to for transformers 400 kV and above [6]. In recent years, in view of the influence of particles on the insulation performance of oil, scholars at home and abroad have carried out research on the movement characteristics of particles in mineral oil as well as the influence of impurity bridges on insulation performance. Mahmud S. et al. studied the bridging phenomena of cellulose particles in mineral oil based on spherical electrodes and needle-plate electrodes under different DC and AC voltages. It was found that DC voltage was the main factor in bridge formation [13–17]. Yuan Li et al. studied the generation and development of cellulose bridges and their influences on partial discharge under DC voltage [18]. In [19,20], a millimeter-diameter metal particle's movement properties and its movement were studied using a theoretical model. In [21,22], the motion trajectory of a metal particle in flowing oil under AC voltage was provided. The movement process of a millimeter-diameter metal particle between electrodes has been investigated under various forms of voltage, and it has been stated that the voltage form and particle concentration have important impacts on the motion characteristics of metal particles [23,24]. The movement properties of millimeter-diameter metal particles under AC voltage have been investigated by numerous researchers. Nevertheless, studies of the motion characteristics of smaller metal particles and their DC breakdown features are lacking. The bridging processes of non-metallic and metal impurities has scarcely been researched. In addition, most simulation models used to analyze the impact of particles on the breakdown performance of insulating oil are still being devoted to the study of electrical field distortion of insulating oil caused by large spherical particles, and comprehensive analyses on the influence of current characteristics along with changes in the electrical field owing to particle accumulation in insulation oil are lacking.

In this paper, experiments on the motion characteristics of cellulose particles, copper particles, and mixed particles are carried out under DC voltage. In addition, DC breakdown voltages of mineral oil contaminated by different concentration particles are measured. A simulation model of particle accumulation is built through Comsol software. The computation model can be utilized to explain the differences in the motion and accumulation properties of different particles in mineral oil. Moreover, the variation in the DC breakdown strength of mineral oil containing different particles with different concentration is analyzed.

2. Simulation and Experiment

2.1. Motion Model of Metal Particles

Assuming that insulating oil cannot be compressed, suspended metal particles in oil withstand gravity and buoyancy in the vertical direction. In the horizontal direction, forces include the dielectrophoretic force (F_{DEP1}) [25], the viscous drag force (F_{drag1}) [25], and the Coulomb force (F_{C1}) [25]. Based on the microstructure of metal particles in Figure 1a, the dielectrophoretic force of spherical metal particles with the radius r can be calculated:

$$F_{DEP1} = 2\pi\varepsilon_m r^3 \frac{\delta_p - \delta_m}{\delta_p + 2\delta_m}\nabla|E|^2 \tag{1}$$

where ε_m represents the permittivity of insulation oil, and σ_m and σ_p stand for the conductivity of insulation liquid and the metal particle respectively.

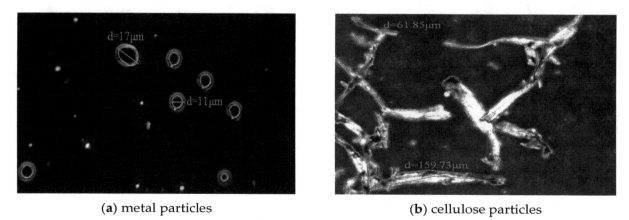

(a) metal particles   (b) cellulose particles

Figure 1. The microstructures of metal particles (**a**) and cellulose particles (**b**).

Viscous drag force (F_{drag1}) acts on the metal particle, as shown in Formula (2). Because of the nonlinearity of the viscous drag force, the correction factor (C_d) related to the Reynolds coefficient (Re) should be used to correct the viscous drag force.

$$F_{drag1} = 6\pi\eta r v_{p1} C_d \tag{2}$$

$$C_d = \begin{cases} 0.17 + 0.72\ln(\text{Re}) & \text{Re} \geq 2 \\ 1 & \text{Re} < 2 \end{cases} \tag{3}$$

$$\text{Re} = \frac{2r\rho_m |v_{p1}|}{\eta} \tag{4}$$

where η and ρ_m are the viscosity and density of the insulation liquid, and v_{p1} represents the velocity of the metal particle.

When the metal particle with the radius of r collides with one electrode, a certain amount of charge (q_1) is obtained, as calculated by Formula (5). Then, plugging Formula (5) into (6) yields the following expression:

$$q_1 = \frac{2}{3}\pi^3 \varepsilon_m \varepsilon_0 r^2 E \tag{5}$$

$$F_{C1} = 0.83 q_1 E = 0.553\pi^3 E^2 r^2 \varepsilon_m \varepsilon_0 \tag{6}$$

Owing to the dynamic model of metal particles represented by Equations (1) to (6), the velocity of the metal particle can be calculated by Formula (7):

$$v_{p1} = \frac{r^2 \varepsilon_m (\delta_p - \delta_m)}{3\eta(\delta_p + 2\delta_m)}\nabla E^2 \tag{7}$$

2.2. Motion Model of Cellulose Particles

As a result of the difference in micromorphology between metal particles and cellulose particles, their kinetic models are different. The micromorphology of cellulose particles is illustrated in Figure 1b. It is assumed that cellulose particles are regarded as ellipsoids to simplify the computational investigation. The kinetic model of cellulose particles is as follows:

$$F_{DEP2} = \frac{\pi\varepsilon_m \varepsilon_0}{12} d^2 l \left[\frac{\alpha}{\alpha - 1} - f(\beta)\right]^{-1} \nabla E^2 \tag{8}$$

$$\alpha = \frac{\varepsilon_p}{\varepsilon_m} \tag{9}$$

$$f(\beta) = \xi[(1 - \xi^2)\coth^{-1}\xi + \xi] \tag{10}$$

$$\xi = \beta(\beta^2 - 1)^{-1/2} \tag{11}$$

$$\beta = l/d \tag{12}$$

$$F_{drag2} = 3\pi\eta d v_{p2} g(\beta) \tag{13}$$

$$g(\beta) = \frac{8}{3}\left[\frac{-2\beta}{\beta^2 - 1} + \frac{2\beta^2 - 1}{(\beta^2 - 1)^{3/2}} \ln\frac{\beta + (\beta^2 - 1)^{1/2}}{\beta - (\beta^2 - 1)^{1/2}}\right]^{-1} \tag{14}$$

$$F_{C2} = 0.553\pi^3 E^2 r^2 \varepsilon_m \varepsilon_0 \tag{15}$$

where ε_m and ε_p represent the permittivity of insulation oil and cellulose impurities, and d and l stand for the diameter and length of the cellulose particle. Using Formulas (8)–(15), the velocity of the cellulose particle v_{p2} can be attained using

$$v_{p2} = \lim_{\beta\to\infty} \frac{\varepsilon_m \varepsilon_0}{24\eta} l^2 \frac{\ln 2\beta - 0.5}{\ln 2\beta - 1} \nabla E^2 \tag{16}$$

2.3. Accumulation Model of Particles

Based on the results of the impurity accumulation experiments, the velocity of the particles (v_p) is closely related to the concentration (c). There is a certain particle concentration (c_{crit}) that prevents particle movement, because a large number of particles are bound to form bridges [26]. Thus, the relationship between the velocity and concentration of particles is as follows [26]:

$$v'_p = \begin{cases} 0 & c \geq c_{crit} \\ v_p - v_p \frac{c}{c_{crit}} & c < c_{crit} \end{cases} \tag{17}$$

Fick's diffusion law, which contains the velocity of particles, is utilized to calculate the dynamic behavior of particle concentration. Because of no chemical reaction takes place, the reaction rate expression for species R_i is equal to 0. The flux vector (or molar flux) N is associated with the Fick equation and is used under boundary conditions and for flux computation. u_m (10^{-7} s*mol/kg) is the charge mobility, where c is the concentration of impurities, and D (10^{-11} m^2/s) is seen as the diffusion parameter:

$$\frac{\partial c_i}{\partial t} + \nabla \bullet N_i = R_i$$
$$N_i = -D_i \nabla c_i - z_i u_{m,i} F c_i \nabla V \tag{18}$$

The different physics interface involving only the scalar electric potential can be interpreted in terms of the charge relaxation process. The basic equation is Ohm's law:

$$J = \delta E + J_e$$
$$\delta = \delta_p^{2.3c} * \delta_{oil}^{1-2.3c} \tag{19}$$

where J_e is an externally generated current density, and $J_e = 0$. σ is the collective conductivity of insulation oil and impurities, which is derived according to the Looyenga Formula (31). The static form of the equation of current continuity then reads

$$\nabla \bullet J = 0 = -\nabla \bullet (\delta \nabla V - J_e) = -\nabla \bullet (\delta \nabla V) = 0 \tag{20}$$

Then, according to the fundamental equation of electrostatic field, the electric field distribution can be illustrated by

$$E = -\nabla V$$
$$\nabla^2 V = 0 \tag{21}$$

Based on motion models and accumulation model of particles, the results of the accumulation concentration together with the electrical field distribution variation for cellulose particles, metal particles, as well as mixed particles can be obtained.

2.4. Particle Accumulation and Oil Breakdown Measurement

A standard oil cup containing a sphere–sphere electrode made of copper material with a diameter of 13 mm was used in the experiments. In accordance with IEC 60156, an electrode distance of 2.5 mm was used in the DC breakdown experiments. To study the accumulation characteristics of the three kinds of particles, 7.5 mm and 12.5 kV were utilized. The oil cup was situated under a digital camera which recorded the process of particle accumulation. The current was measured by a Keithley electrometer (6517B). The experimental setup for the DC breakdown voltage testing is shown in Figure 2. HCDJC-100kV/5kVA was utilized to provide high DC voltage. The signal together with data was controlled and attained by the computer system.

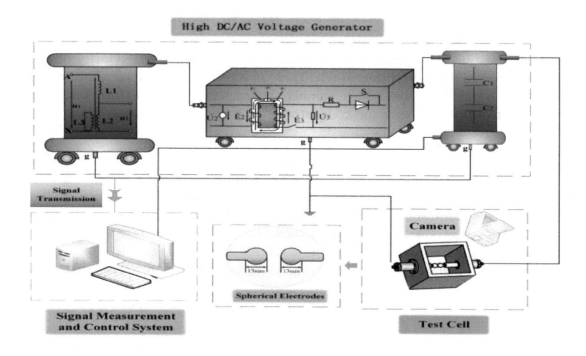

Figure 2. The experimental platform for the particle accumulation and DC breakdown test.

Three kinds of impurities were used in all experiments: cellulose particles, metal particles, and their mixture. Cellulose particles were produced by rubbing new insulation paper through metal files of different sizes. In this paper, cellulose particles with sizes of 63–150 μm and 150–250 μm were prepared. Spherical copper particles with 15 μm diameters came from Beijing Hongyu New Materials Co., Ltd. (Beijing, China). The contamination levels of particles can be seen in Table 1. Before the experiments, all samples were dried for 48 h using a vacuum box at 90 °C and 133 Pa. The parameters of clean mineral oil can be seen in Table 2. For the sake of simplicity, abbreviations are used in Table 3 to stand for different samples. After homogeneous mixing of particles and mineral oil, the samples were sealed in the vacuum box for 24 h.

Each particle accumulation experiment lasted 1500 s. To ensure even particle distribution, a stirrer was used to stir samples. Moreover, every sample was repeatedly broken down eight to ten times. The average value was treated as the final DC breakdown voltage for this sample. All tests were performed at room temperature.

Table 1. Parameters of mineral oil.

	Contamination Level				
Cellulose particles	0.001%	0.003%	0.006%	0.009%	0.012%
Metal particles	0.1 g/L	0.3 g/L	0.6 g/L	1 g/L	1.5 g/L
Mixed particles	0.003% + 0.1 g/L	0.003% + 0.3 g/L	0.003% + 0.6 g/L	0.003% + 1 g/L	—
	0.012% + 0.1 g/L	0.012% + 0.3 g/L	0.012% + 0.6 g/L	0.012% + 1g/L	—

Table 2. Parameters of mineral oil.

Parameters	Mineral Oil
Density in g/cm^3	0.89
Dynamic viscosity in mm^2/s (20 °C)	25.70
Permittivity (20 °C, 50 Hz)	2.20
Volume resistivity in Ω·m (20°)	4.68×10^{13}

Table 3. Abbreviations of samples analyzed in the experiments.

Samples	Sample Composition
DMCP	Dry mineral oil + cellulose particles
DMMP	Dry mineral oil + metal particles
DMCM	Dry mineral oil + mixed particles of cellulose and metal particles
MO	Pure mineral oil

3. Experimental Results and Discussion

3.1. Particle Accumulation Simulation Results

The parameters of the particle accumulation simulation model are described in Table 4. Three-dimensional simulation diagrams of the accumulation of cellulose particles, metal particles, and mixed particles in insulation oil at 10 s and 600 s are shown in Figure 3. With an increase in computational time, impurity accumulation between electrodes was more evident. Samples containing cellulose particles formed initial filamentous bridges faster than those only containing metal particles at the start stage of voltage application. At 600 s, the concentration of mixed particles was the largest, followed by cellulose particles and metal particles. The concentrations of cellulose particles and copper particles were, respectively, 1.2 and 3.7 times less than that of mixed particles.

Table 4. Parameters of the particle accumulation model.

Parameters	Cellulose Particles	Metal Particles
Density, g/cm^3 (20 °C)	1.2	8.6
Permittivity (20 °C, 50 Hz)	4.4	10^5
Size, μm	63–150	15
Volume resistivity, Ω·m (20 °C)	2×10^6	2×10^{-8}
Initial concentration (c_0), mol/m^3	0.0005	0.0005
DC voltage, kV	11.25	11.25
Distance, mm	7.5	7.5

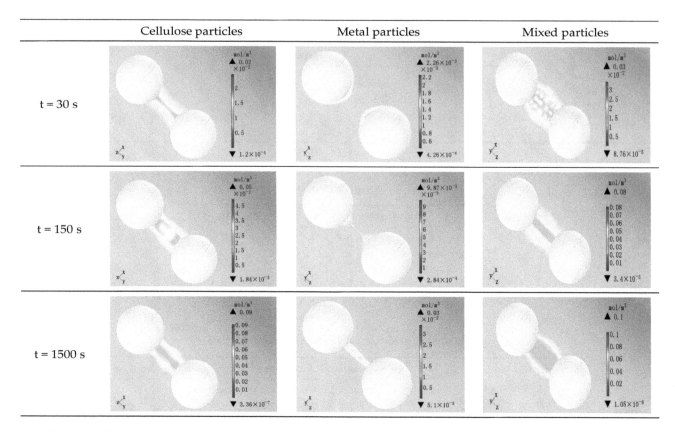

Figure 3. Simulation results of different kinds of particle accumulation in mineral oil at different times.

3.2. *Experimental Results of Particle Accumulation.*

Figure 4 illustrates the dynamic behavior of particle accumulation under DC voltage, in which the concentration levels of cellulose particles, metal particles, and mixed particles were respectively 0.009% by weight, 0.1 g/L, and the sum of both. The distance and DC voltage between the spherical electrodes were 7.5 mm and 11.25 kV, respectively. The left electrode was a cathode, and the right one was an anode.

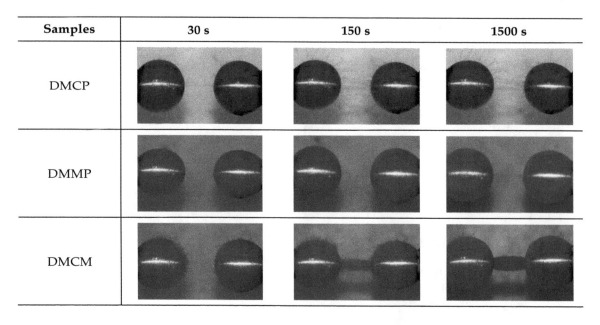

Figure 4. Particle accumulation status under the same DC voltage in mineral oil at different time.

It is obvious that impurity bridges form for the three kinds of particles, as shown in Figure 4. The non-charged particles move toward the higher electrical field area under the attraction of dielectrophoresis, i.e., the central position of the electrode. The particles are charged on the side of the adsorbed electrode. When the Coulomb force of the particles are enough to resist the dielectrophoresis force, the particle moves uniformly to the opposite electrode under combined forces including the Coulomb force, the viscous drag force, and dielectrophoresis force. A few of the particles move back and forth between the electrodes according to the above rule. It is noteworthy that large particles are more likely to move rapidly, because the electrical field force and dielectrophoresis force of the particle are proportional to the square of a particle's radius, and the viscous drag force is proportional to the particle size. However, as a result of lacking enough charge to resist the dielectrophoresis attraction in a certain time period, the rest of particles adhere to electrodes which form a tip on the surface of electrodes. The tip may enhance the local electrical field, attract surrounding particles to move there, and continuously lengthen the particle chain. The initial filamentous bridge is not completely parallel to the electrical field line, which results in parallel movement of the particle bridge under the action of force in the horizontal direction and merges with other small bridges; thus, its thickness increases.

As shown in Figure 4 (sample DMCP) a thin fiber bridge was observed after 30 s in mineral oil containing cellulose particles. After that, the thickness of cellulose bridges increased gradually. The complete bridge formed until approximately 600 s. Then, notable changes in the cellulose bridge were not seen. For mineral oil contaminated by metal particles, most metal particles sunk down, owing to their larger densities. A filamentous bridge formed at about 500 s (Figure 4, sample DMMP), which was so unstable that it migrated back and forth between electrodes, which may be attributed to the impact of the strong electrical field. The ultimate metal bridge appeared until 1200 s. As for mixed particles (Figure 4, sample DMCP), the complete bridge formed more quickly than those for single impurities. Furthermore, it was observed that the complete bridge was densest when the transverse force among thin bridges made up of mixed particles was larger. It can be inferred that the simulation model depicting the changing course of particles aggregation is in agreement with the actual process. Thus, the simulation model can accurately describe the impact of impurities on motion as well as on the accumulation characteristics of particles. Different attributes of particles determined the difference in the trajectory. The trajectories of cellulose particles and metal particles under DC electrical field observed from Figure 4 is shown in Figure 5.

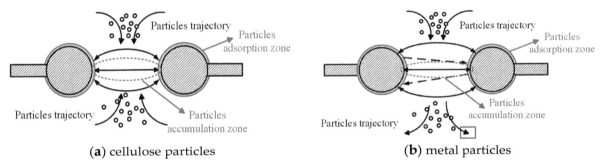

(a) cellulose particles   (b) metal particles

Figure 5. Motion trajectory of cellulose particles (a) and metal particles (b) under DC voltage in mineral oil.

3.3. Effect of Particles Accumulation on Oil DC Breakdown Voltage

The Weibull distribution model was utilized to analyze the DC breakdown voltages of samples. Its mathematical expression is shown by Equation (22), where t stands for the DC breakdown strength; α stands for the scale parameter, which describes the characteristic breakdown strength of oil; and β represents the shape parameter, which reflects the changing rate of breakdown probability with an increase in DC voltage. The data samples used in this study were complete, so the empirical distribution function $F_n(t_i)$ could be calculated by Formula (23), where i represents the order of test samples, and n is the number of samples. Figure 6 is the Weibull probability distribution plot of different samples under the DC breakdown voltage. With an increase in the concentration of particles, the Weibull curve moved to the left. Namely, the average breakdown values of samples reduced as the particle concentration increased. The relationship between the average breakdown voltages of mineral oil containing cellulose particles, metal particles, and mixed particles and the particles concentration are shown in Figure 7. The DC breakdown voltages of oil samples contaminated with metal particles and mixed particles were found to be lower than those of mineral oil containing cellulose particles. In summary, metal impurities and mixed particles have more significant impacts on the DC breakdown characteristics of mineral oil:

$$F(t; \alpha, \beta) = 1 - \exp[-(\frac{t}{\alpha})^{\beta}] \tag{22}$$

$$F_n(t_i) = \frac{i - 0.375}{n + 0.25} \times 100\% \tag{23}$$

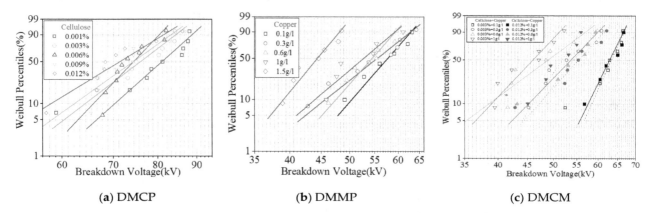

Figure 6. Weibull probability distribution plot of DC breakdown voltage for different samples DMCP (**a**), DMMP (**b**) and DMCM (**c**).

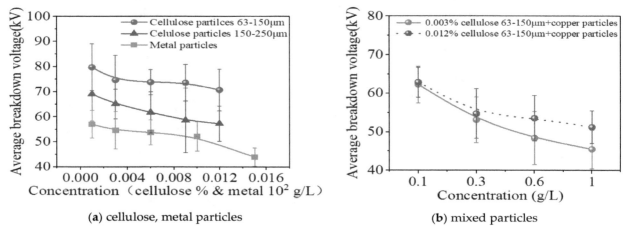

Figure 7. Average DC breakdown voltage of mineral oil with cellulose or metal particles (**a**), and mixed particles (**b**).

3.4. Difference Analysis of the Effects of Different Particles on the Oil DC Breakdown Voltage

First of all, impurities have a prominent effect on the conductivity of mineral oil. The changing current properties of mineral oil are displayed in Figure 8. Upon the application of DC voltage, instantaneous polarization current appears in both clean oil and contaminated oil. For clean oil, the polarization current decreases to a conduction current. Yet, due to the effect of particles' back and forward motion between electrodes, causing charging and discharging together with formation of bridges, the current in contaminated oil increases little by little until the saturation level is reached. The increasing degree of current saturation in contaminated oil compared to clean oil occurs in decreasing order for mixed particles, metal particles, and cellulose particles. The saturated current in mineral oil containing mixed particles is the largest, almost seven times larger than that of clean oil, followed by copper particles (5.5 times) and cellulose particles (2.8 times). As a result, metal particles or mixed particles more easily cause insulation oil to partially discharge and even break down. Since the sum of the saturated current of cellulose particles and metal particles is not equal to the saturated current of mixed particles, cellulose particles coupled with metal particles cannot have a superposition effect on the conductivity of oil. There is no doubt that the bridge formation of particles under a DC electrical field can significantly improve the conductivity of mineral oil, which is one of main reasons for the decrease in the breakdown strength of oil. Furthermore, there is a corresponding relation between current and particle aggregation state, which is shown in Figure 9.

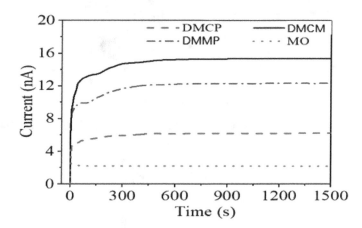

Figure 8. Changes in the current in mineral oil containing different particles under the same DC voltage.

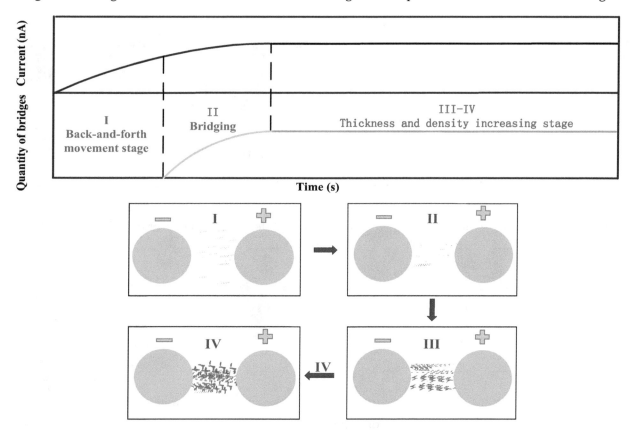

Figure 9. Relationship between oil conductivity current and particle aggregation.

Then, taking advantage of the effective simulation model shown in Section 3.1, relationships among concentrations, particles properties, and the variation in electrical field strength were analyzed in order to explain the DC breakdown results. Figures 10 and 11 display three dimensional particle accumulation patterns as well as the DC electric field distribution of insulation oil under three different initial concentrations at 600 s. It can be observed that firstly, accumulation degrees increase as the initial concentration increases. Secondly, the range of electrical field distortion caused by particle accumulation is also gradually enlarged.

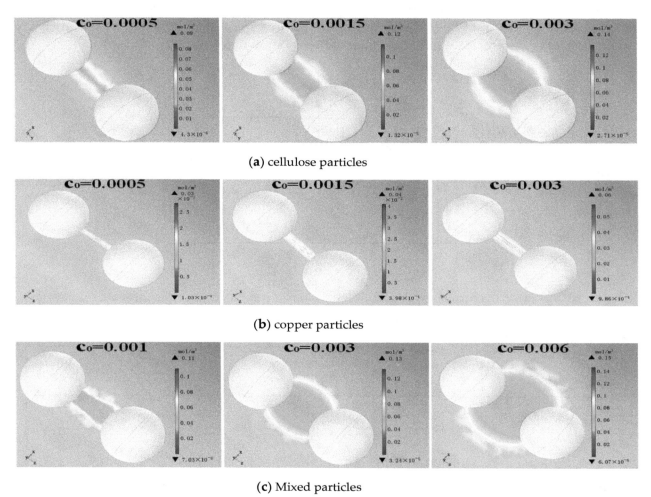

Figure 10. Particle accumulation patterns under three different initial concentrations for cellulose particles (**a**), copper particles (**b**) and mixed particles (**c**) at 600 s.

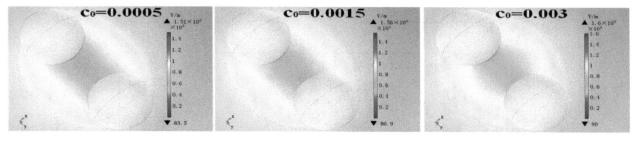

(**a**) cellulose particles

Figure 11. *Cont.*

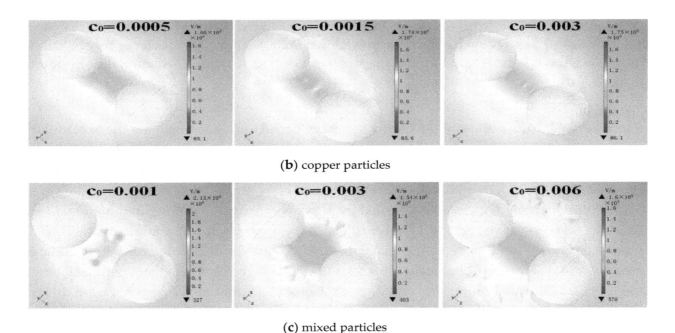

(b) copper particles

(c) mixed particles

Figure 11. DC electric field distribution of mineral oil contained cellulose particles (a), copper particles (b) and mixed particles (c) under three different initial concentrations at 600 s.

Figure 12 shows the influence of the particle concentration on the maximum electrical field strength (E_{max}) of contaminated mineral oil at 600 s. It can be seen that metal particles almost play a more prominent part in electrical field distortion than cellulose particles. Additionally, the saturated current of metal particles is larger than that for cellulose impurities, so the breakdown voltage of mineral oil polluted by metal particles is smaller than that contaminated by cellulose particles. Figures 10 and 11 also imply that cellulose particles as well as metal particles cannot have a superposition effect on the electric field distortion of insulation oil. For mineral oil contaminated by cellulose or metal particles, E_{max} increases as the particle concentration increases, which indicates that electrical field distortion combined with the conductivity variation leads to degradation of the insulation strength of mineral oil. As for mineral oil containing mixed particles, though electrical field distortion weakens with an increasing concentration, conductivity obviously increases because of the increasing accumulation degree. Hence, changes in the electrical field distribution together with an increase in conductivity collectively affects the DC breakdown characteristics of mineral oil.

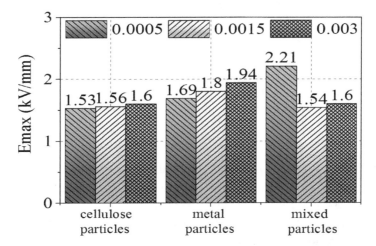

Figure 12. E_{max} of mineral oil and natural ester containing different particles under three different initial concentrations at 600 s.

The effect of pressure on breakdown voltage of insulation oil has been investigated, which indicates that the increasing of pressure can increase breakdown voltage of oil [27]. However, few researchers have studied the effects of temperature or pressure effect on particles' accumulation properties. This needs further study in the future.

4. Conclusions

In this paper, simulations and experiments were carried out to investigate the accumulation characteristics of cellulose, copper, and mixed particles in mineral oil under DC voltage. The DC breakdown voltages of mineral oil with different particle concentrations were measured. The conclusions obtained from this study are as follows.

The simulation model was able to reveal the process of particle bridging in mineral oil under DC voltage. The simulation results showed that as the experiment duration increased, particle accumulation became more evident. The accumulation concentration of mixed particles was the largest, followed by cellulose particles and metal particles, which is in agreement with the experimental results. In the particle accumulation experiments, it was also obvious that metal particles have difficulty forming stable bridges, while bridges of mixed particles were the densest and thickest among the tested compounds. Moreover, the larger the initial concentration of particles, the more obvious the accumulation phenomenon as well as the electrical field distortion.

The properties of particles determine the accumulation shapes, conductivity characteristics, and variation law of breakdown voltages. The increasing degrees of saturated current in contaminated oil compared to clean oil in decreasing order were found to be mixed particles, metal particles, and cellulose particles. Moreover, metal particles were also shown to play a more prominent part in electrical field distortion than cellulose particles. Therefore, the breakdown voltage of mineral oil contaminated by cellulose particles was larger than that of mineral oil containing copper particles. Nevertheless, it is noteworthy that cellulose particles along with metal particles cannot have a superposition effect on the conductivity characteristics and electrical field distortion of insulation oil.

The test and computational results indicated that changes in the electrical field distribution together with an increase in conductivity collectively affect the breakdown strength of mineral oil. The bridge formation of particles under DC electrical field was shown to significantly improve the conductivity of oil, which is one of main reasons for the decrease in breakdown strength of mineral oil containing particles. The simulation results of the particle accumulation model showed that as the duration of the experiment increased, the range of electrical field distortion caused by particles accumulation gradually expanded, which is also one of main reasons for the decrease in the DC breakdown strength of mineral oil containing particles.

Author Contributions: Designed the experiments, did the motion characteristics experiments and breakdown voltage measurement, wrote the paper, M.D.; improved the simulation method, analyzed the experiment data and revised the paper, J.H.; contributed discussion, R.L., L.C., J.Z., and F.L.

References

1. Wang, X.; Wang, Z. Particle effect on breakdown voltage of mineral and ester based transformer oils. In Proceedings of the IEEE Conference on Electrical Insulation and Dielectric Phenomena, Quebec City, QC, Canada, 26–29 October 2008; p. 598.
2. CIGRE. *Effect of Particles on Transformer Dielectric Strength*; WG 17/SC12; CIGRE: Paris, France, 2000.
3. Lu, W.; Liu, Q. Effect of cellulose particles on impulse breakdown in ester transformer liquids in uniform electric fields. *IEEE Trans. Dielectr. Electr. Insul.* **2015**, *22*, 2554–2564. [CrossRef]

4.   Zhao, T. Research of the Effect of Bubbles and Cellulose Particles on Impulse Breakdown in Transformer Oil. Ph.D. Thesis, North China Electric Power University, Beijing, China, 2017.

5.   Li, Z.; Ma, Q. Monitoring the particle pollution degree of 500 kV transformer's insulation oil. *Transformer* **1999**, *36*, 31–34.

6.   Wu, R.; Liu, J.; Chen, J.; Zhang, Z. Analysis of insulating oil particulate pollution in 500 kV transformer and reactor. *Inn. Mongo. Electr. Power* **2013**, *31*, 35–37.

7.   Xing, L.; Zhang, G. Test and analysis of graininess in 500kV transformer oil. *Transformer* **2009**, *46*, 40–43.

8.   Wei, L.; Su, Z.; Qi, J. Study on particulate contamination detection of insulating oil for 1000 kV transformer and its influence factor. *Insul. Mater.* **2015**, *6*, 50–53.

9.   Zhang, G.; Li, X.; Liu, H.; Lu, L. Study on particle size in 500 kV transformer insulation oil. In Proceedings of the Annual Academic Conference of Tianjin Electric Power Society, Tianjin, China, 20–22 October 2008.

10.   Dan, M.; Hao, J.; Li, Y.; Liao, R.; Yang, L.; Wang, Q.; Zhang, S. Analysis of bridging phenomenon in mineral oil and natural ester contaminated with cellulose particles under different DC electrical field. In Proceedings of the 20th International Symposium on High Voltage Engineering, Buenos Aires, Argentina, 28 August 2017.

11.   Dan, M.; Hao, J.; Qin, W.; Liao, R.; Zou, R.; Zhu, M.; Liang, S. Effect of different impurities on motion characteristics and breakdown properties of insulation oil under DC electrical field. In Proceedings of the 2018 IEEE International Conference on High Voltage Engineering and Application (ICHVE 2018), Athens, Greece, 10–13 September 2018; pp. 1–4.

12.   Zhou, Y.; Hao, M.; Chen, G.; Wilson, G.; Jarman, P. Study of the charge dynamics in mineral oil under a non-homogeneous field. *IEEE Trans. Dielectr. Electr. Insul.* **2015**, *22*, 2473–2482. [CrossRef]

13.   Hao, J.; Liao, R.; Dan, M.; Li, Y.; Li, J.; Liao, Q. Comparative study on the dynamic migration of cellulose particles and its effect on the conductivity in natural ester and mineral oil under DC electrical field. *IET Gener. Transm. Distrib.* **2017**, *11*, 2375–2383. [CrossRef]

14.   Mahmud, S.; Chen, G.; Golosnoy, I.O.; Wilson, G. Bridging in contaminated transformer oil under AC, DC and DC biased AC electric field. In Proceedings of the 2013 IEEE Electrical Insulation and Dielectric Phenomena, Shenzhen, China, 20–23 October 2013; pp. 943–946.

15.   Mahmud, S.; Chen, G.; Golosnoy, I.O.; Wilson, G.; Jarman, P. Experimental studies of influence of DC and AC electric fields on bridging in contaminated transformer oil. *IEEE Trans. Dielectr. Electr. Insul.* **2015**, *22*, 152–160. [CrossRef]

16.   Mahmud, S.; Chen, G.; Golosnoy, I.O.; Wilson, G.; Jarman, P. Bridging phenomenon in contaminated transformer oil. In Proceedings of the International Conference on Condition Monitoring and Diagnosis, Piscataway, NJ, USA, 23–27 September 2012; pp. 180–183.

17.   Li, J.; Zhang, Q.; Li, Y. Generation process of impurity bridges in oil-paper insulation under DC voltage. *High Volt. Eng.* **2016**, *12*, 211–218.

18.   Li, Y.; Zhang, Q.; Li, J.; Wang, T.; Dong, W.; Ni, H. Study on micro bridge impurities in oil-paper insulation at DC voltage: Their generation, growth and interaction with partial discharge. *IEEE Trans. Dielectr. Electr. Insul.* **2016**, *23*, 2213–2222. [CrossRef]

19.   Wang, S.; Shi, J.; Li, J. The effect of a macro-particle on the partial property of transformer oil. *High Volt. Eng.* **1994**, *20*, 26–29.

20.   Fu, S. The acquired charge of macro-particle and its effect on the partial discharge of transformer oil. *High Volt. Eng.* **2000**, *26*, 49–50.

21.   Tang, J.; Zhu, L.M.; Ma, S.X. Characteristics of suspended and mobile micro bubble partial discharge in insulation oil. *High Volt. Eng.* **2010**, *36*, 1341–1346.

22.   Ma, S.X.; Tang, J.; Zhang, M.J. Simulation study on distribution and influence factors of metal particles in transaction transformer. *High Volt. Eng.* **2015**, *41*, 3628–3634.

23.   Wang, Y.Y.; Li, Y.L.; Wei, C.; Zhang, J.; Li, X. Copper particle effect on the breakdown strength of insulating oil at combined AC and DC voltage. *J. Electr. Eng. Technol.* **2017**, *12*, 865–873. [CrossRef]

24.   Wang, Y.; Li, X. Motion characteristic of copper particle in insulating oil under AC and DC voltages. In Proceedings of the 19th IEEE International Conference on Dielectric Liquid, Manchester, UK, 25–29 June 2017; pp. 25–29.

25.   Dan, M.; Hao, J.; Liao, R.; Li, Y.; Yang, L. Different motion and bridging characteristics of fiber particles in mineral oil and natural ester under DC voltage. *Power Syst. Technol.* **2018**, *42*, 665–672.

26. Naciri, N. Finite Element Analysis for Power System Component: Dust Accumulation in Transformer Oil. Ph.D. Thesis, University of Southampton, Southampton, UK, 2011.
27. Butcher, M.; Neuber, A.; Krompholz, H.; Dickens, J. Effect of temperature and pressure on DC pre-breakdown current in transformer oil. In Proceedings of the 31st IEEE International Conference on Plasma Science, Baltimore, MD, USA, 28 June–1 July 2004; pp. 1–4.

7

Mechanism of Saline Deposition and Surface Flashover on High-Voltage Insulators near Shoreline: Mathematical Models and Experimental Validations

Muhammad Majid Hussain [1,*], Muhammad Akmal Chaudhary [2] and Abdul Razaq [3]

[1] Faculty of Computing, Engineering and Science, University of South Wales, Treforest, Cardiff CF37 1DL, UK

[2] Department of Electrical and Computer Engineering, Ajman University, Ajman P.O. Box 346, UAE; m.akmal@ajman.ac.ae

[3] School of Design and Informatics, Abertay University, Dundee DD1 1HG, UK; a.razaq@abertay.ac.uk

[*] Correspondence: muhammad.hussain@southwales.ac.uk

Abstract: This paper deals with sea salt transportation and deposition mechanisms and discusses the serious issue of degradation of outdoor insulators resulting from various environmental stresses and severe saline contaminant accumulation near the shoreline. The deterioration rate of outdoor insulators near the shoreline depends on the concentration of saline in the atmosphere, the influence of wind speed on the production of saline water droplets, moisture diffusion and saline penetration on the insulator surface. This paper consists of three parts: first a model of saline transportation and deposition, as well as saline penetration and moisture diffusion on outdoor insulators, is presented; second, dry-band initiation and formation modelling and characterization under various types of contamination distribution are proposed; finally, modelling of dry-band arcing validated by experimental investigation was carried out. The tests were performed on a rectangular surface of silicone rubber specimens (12 cm × 4 cm × 8 cm). The visualization of the dry-band formation and arcing was performed by an infrared camera. The experimental results show that the surface strength and arc length mainly depend upon the leakage distance and contamination distribution. Therefore, the model can be used to investigate insulator flashover near coastal areas and for mitigating saline flashover incidents.

Keywords: saline mechanism; shoreline; wind speed; outdoor insulators; dry band arcing; flashover

1. Introduction

The performance of outdoor high-voltage insulators near the shoreline is a key factor in the determination of power network systems' stability and reliability. It is well known that contamination is considered a major critical factor responsible for surface flashovers [1]. The process of saline deposition on an insulator surface, associated with flashover and consequent power outages has been a major problem for power network systems since the early 1900s [2]. The time to surface flashover initiation depends on (i) deposition of saline contamination, and (ii) how salt particles penetrate and are diffused on the insulator surface through various wetting agents such as rain, fog, snow, dew or drizzle.

It is recognized that saline deposition and diffusion are affected by various natural processes such as temperature exposure, relatively humidity or moisture level, counter-diffusion of hydroxide ions and environmental load of salts and other adverse weather conditions. Types of contamination deposition on the insulator surface influence surface flashover, which has been extensively studied by several researchers [3–8]. The use of non-ceramic insulators increased significantly in last five decades. Silicone rubber insulators both in service [9] and high-voltage laboratory tests [10] demonstrated better performance than ceramic insulators in contaminated environmental conditions. Initially, non-ceramic insulators prevent water filming on the surface due to their hydrophobic properties, but this resistance gradually decreases due to physical and chemical changes in the silicone materials which can lead to dry-band arcing and surface discharges [11]. The combination of high-voltage stress and a contaminated water film produces dry-band arcing, and the resulting heat can lead to erosion of the insulator surface. The surface is damaged by physico-chemical changes caused by dry-band arcing [12,13]. Most of the previous work on surface flashover of contaminated insulators mainly focused on laboratory or onsite experiments based on alternating current (AC) and direct current (DC) voltage [14–16].

Kim et al. [17] studied chemical changes on the silicone rubber insulators during dry-band arcing but did not investigate the physical changes, for example the behavior of arc lengths with uniform and non-uniform pollution levels, as well as arc resistivity, power and energy. Therefore, there is a need to investigate the effects of dry-band arcing for a better understanding of the physical changes on the surface of silicone rubber insulators.

This paper presents a model, which is based on mechanism of sea salt transportation, deposition and diffusion on outdoor insulators near a shoreline, taking into account the saline concentration and the distance from the shoreline. It also introduces a new mathematical model to investigate the development of dry bands for different types of pollution layers on silicone rubber. A series of simulations and experiments were performed on the model to verify the theoretical results.

2. Mechanism of Salt Transportation and Deposition

The following three sub-models simulating three different processes were combined into one theoretical model of sea salt production, transportation and deposition:

2.1. Production of Saline

There are two major regimes where saline ions and particles are generated and scattered from shoreline to inland. Sea salt particles originate from breaking sea waves, a phenomenon that is followed by a high rate of wave motion and turbulence, air entrainment and surf formation. At high-level oceans, this breaking phenomenon is encountered under higher wind action with the formation of whitecap bubbles. As these bubbles rise, they are forced into the air where they scatter, thus producing saline particles. These particles can be routed to shoreline areas by oceanic wind speeds that exceed 4 to 12 ms^{-1} [18,19], where they tend to settle on outdoor insulators after a certain time and after having covered a certain distance. This mechanism is important in the generation of saline particles at intermediate to high wind speeds. In fact, wind speed is not the only factor to be taken into consideration. However, any factor favouring wave breaking and turbulence in the sea near the coast line must contribute to the formation of saline particles.

2.2. Saline Transportation and Deposition

Pollution near coastal regions is a major source of degradation of power network system equipment. In particular, saline attack is an important and major factor in the deterioration process of high voltage outdoor insulators near the coast. Feliu et al. proposed a complex theoretical model to represent transfer and deposition of saline on testing equipment near coastal areas [20]. However, this was based on constant spray of artificial aerosol, such as haze, dust and smoke, which do not represent natural climate conditions. In the present paper, a sea salt mechanism model representing saline transportation and deposition on outdoor insulators near coastal areas is proposed. The model represents the relationship between sea salt deposition on outdoor insulators and distance from the shoreline. Saline ions and particle changes are also taken into account driven on outdoor insulators by the wind from the sea. The study and experimental implementation of this new model is particularly useful for the investigation of surface degradation and surface flashover of outdoor insulators and substation components near shoreline based on salt concentration, wind speed and direction, and distance from coastal areas. The exposure of this model and experimental work is very similar to that of high-voltage transmission lines running along the seashore of Scotland, where outdoor insulators are exposed to wind, fog and rain, but not to direct saline spray. From various studies [21,22], it was found that near the shoreline salt particles mobilization was based on gravitational settlement and wind speed, and that these salt particles can travel longer distances before deposition. Based on that, a model and mechanism of sea salt transportation and deposition is presented in Figure 1.

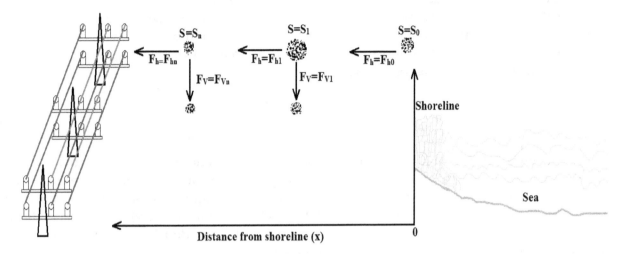

Figure 1. Schematic diagram of sea salt transport and deposition phenomenon.

Figure 1 represents a schematic mechanism of sea salt transportation and deposition, which shows the saline concentration (S) and its variation (S_0, S_1,... and S_n). In this model, oceanic winds, distance from the sea, diffusion and penetration of saline and gravitational settlement of saline on outdoor insulators are taken into account. The resultant vertical settlement flux (F_v) of saline due to gravitational effect and the saline transport near the shoreline on the outdoor insulators' surface is represented by a horizontal flux (F_h), as a significance of saline concentration, saline variation and wind regime. The relationship between saline concentration (S) and its variation (S_0, S_1, ... and S_n) from shoreline to the surface of insulator, vertical resultant deposition flux (F_v) and deposition rate (V_{dep}) is represented in Equation (1) by means of a mathematical simplification of flow velocity fluid mechanics equations:

$$F_v = V_{dep} \, S \tag{1}$$

From Equation (1), it is possible to determine the saline concentration (S) with variation (S_0, S_1 ... , and S_n) from shoreline to the surface of insulator, and deposition variation with time as a function of deposition rate (V_{dep}). It follows that the mass of saline deposited per unit of time is a negative function of the resultant vertical deposition flux. This is represented by Equation (2), where dt is the time variation, h is the thickness of the saline contamination layer and where the negative sign represents the reduction of saline concentration (S) deposition on the insulator surface with time and distance from the sea.

$$\frac{dS}{dt} = -\frac{S V_{dep}}{h} \tag{2}$$

Equation (3) is the solution of Equation (2) on the basis of an environmental natural phenomenon by which saline characteristics change when transported and deposited on outdoor insulators and substations equipment from sea to near shoreline. The saline decreases exponentially as shown in Equation (3), where S_0 is the saline concentration at shoreline, x is the distance from the sea and α is a constant represented by "$\alpha = V_{dep}/v_h$", where v as the wind speed. However, to solve Equation (3) it should be assumed that the deposition rate is constant with time and for any distance from the shoreline and its decay function may be estimated as:

$$S = S_0 e^{-\alpha x} \tag{3}$$

The unit of saline concentration is (mg/cm^2), taking into account the exponential decrease of saline deposition rate with time (saline concentration and deposition velocity has a proportional relation, thus Equation (3) can be rewritten as $V_{dep} = V_{dep0}e^{-\alpha t}$), and that, during a time period (t) as some saline particles are deposited on the surface of insulators installed at certain distance from the sea, while the remaining particles travel from the shoreline to a subsequent distance (x) driven by wind speed (v). As a result, integration of Equation (3) can be expressed by Equation (4), where S is the saline concentration at a distance x from seashore, S_0 is the saline concentration at seashore, V_{dep0} is the initial deposition rate of a saline at shoreline, h is the thickness of the contamination layer on the insulator surface and α is a coefficient of the deposition rate reduction that characterizes saline contamination distribution and its influence on deposition rate.

$$S = S_0 e^{\left(\frac{v_{dep0}}{\alpha h}\right)\left[e^{\left(\frac{-\alpha x}{v}\right)}-1\right]} \tag{4}$$

where $t = \frac{x}{v}$.

The coefficient of deposition rate reduction is directly influenced by saline particles that are distributed near the shoreline. Considering that saline deposition on an insulator surface corresponds to saline particles that influence and remain on its surface during migration of saline from shoreline to inland and assuming that there is a proportional relation between saline flux (Φ), which depends upon insulator weather sheds and environmental characteristics and saline deposition (D) on the insulator surface. This can be expressed by a simplified relationship:

$$D = \Phi(VS) \tag{5}$$

where Φ is capture efficiency constant for the saline, which depends upon climate conditions and outdoor insulators, V is the wind speed and S is the saline concentration. On the basis of these assumptions and as a consequence, Equation (4) can be integrated, simplified and rewritten as:

$$D = D_0 e^{\left(\frac{v_{dep0}}{\alpha h}\right)\left[e^{\left(-\frac{\alpha x}{v}\right)}-1\right]} \tag{6}$$

where D_0 is the saline deposition on the insulator surface at a certain distance from the shoreline and D is the saline deposition on the insulator surface from shoreline to inland. Equations (4) and (6) combine the proposed model of saline transportation and deposition from shoreline to inland on the insulator surface. Thus, Equation (4) represents the saline concentration variation across the seashore regime and Equation (6) represents the saline deposition rate on the insulator surface across the same regime. Contamination deposition mechanisms rather than gravitational settlement, like high speed wind and the scavenging effects due to heavy rainfall or dense fog precipitation were also considered at this stage in the model and related experimental investigations are presented.

The proposed model, which is based on a physical phenomenon, can be used to allow for the action of saline particles generated at sea and to account for the saline concentrations and deposition on the insulator surface near the shoreline. The aim of the proposed model is to observe the action of different input variables individually. Thus, simulations were done taking as a reference Equation (6) to find the saline deposition on insulator surface at different wind speeds and distances from shoreline. Saline deposition on the insulator surface represents the severity of saline transportation from the shoreline. Figure 2 indicates the saline deposition behaviour, when average wind speed varies in a range between 4.0 to 12.0 ms^{-1}. Figure 2 clearly shows that increase in wind speed contributes to migrate more saline ions further from shoreline, however, when wind speed was higher than 12 ms^{-1}, the migration of saline ions tended to be constant. Consequently, saline deposition on the insulator surface increases at fixed distance from shoreline. Moreover, insulator surface deposition significantly decreases with increasing the distance from shoreline to inland. For the simulations displayed in Figure 2, the conditions used were with variable wind speed and initial deposition rate, assuming the constant of coefficient of deposition rate reduction α (0.01 s^{-1}), which depends on insulator surface and environmental characteristics, derived from $\alpha = e^{A.T} \times \frac{V_{dep}}{vh}$, where A is total area of the insulator sheds and T is ambient temperature.

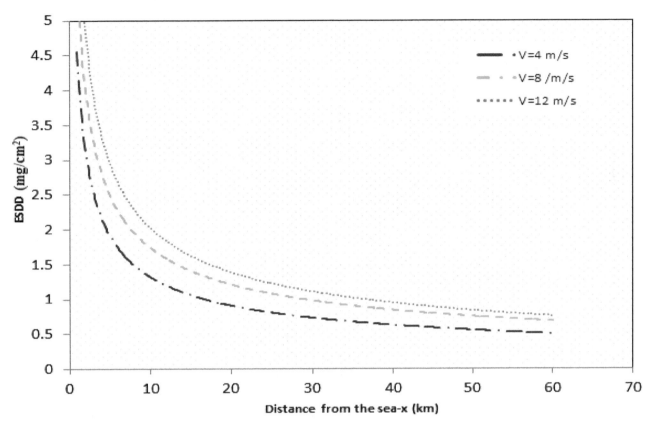

Figure 2. Simulated analysis of the model.

2.3. Saline Penetration and Moisture Diffusion

Saline and moisture migration are two important processes for studying the long-term durability of high-voltage insulators. The interaction between saline penetration and moisture diffusion become more important for high voltage insulator materials subjected to repeated wetting and drying cycles near the shoreline. Various other environmental stresses can also influence these. They will vary depending upon airborne saline particles and moisture levels in a marine environment. The saline penetration and moisture absorption on the insulator surface near the shoreline is usually caused by saline water spray carried by the prevailing winds coming from the coast [23]. It is a complex process that involves heat and mass transfer. Due to constant wetting rate, heat and mass transfer processes are much less important than diffusion and penetration processes. In this case, diffusion and penetration can be considered as one-dimensional, and numerical methods are required to solve it. The simulation was performed using COMSOL Multiphysics software. A 2-dimentional model was developed in COMSOL Multiphysics and two boundary conditions were considered for the simulation, as shown in Figure 3. The technical specifications of the 2-D model are summarized in Table 1. The first condition related to a fiber reinforced plastic (FRP) rod, was impermeability, so that saline concentration and moisture content were considered to be equal to zero. The second boundary was related to the insulator surface, for direct saline penetration and moisture diffusion. During the simulation work, saline ions and moisture struck and penetrated the silicone rubber (SiR), a process which may be described by Fick's Second Law of Diffusion, assuming constant diffusivity and direct saline binding. This law is expressed by Equation (7):

$$\frac{\partial S}{\partial t} = D\frac{\partial^2 S}{\partial^2 X} \tag{7}$$

where S is the saline penetration (mg/cm^2) as a function of time (t) at a distance X from the shoreline and D is the diffusion coefficient of moisture. The solution for saline penetration and moisture diffusion on insulator surface is given by Equation (8). The error function (*erf*) may be determined from the standard table of Fick's law.

$$S(x,t) = S_C(1 - erf\frac{X}{2\sqrt{D}})$$
(8)

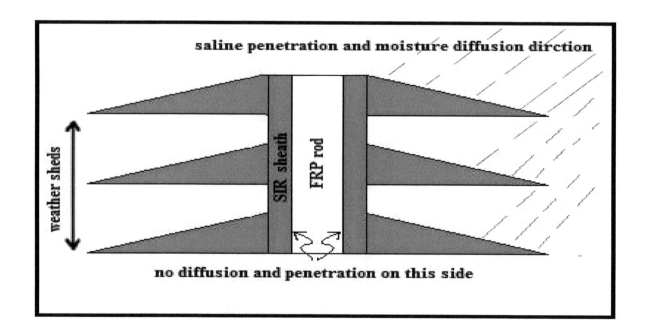

Figure 3. Two-dimensional (2-D) geometry used in simulation.

Table 1. Parameters of 2-D insulator model.

Insulator Type	Leakage Distance (cm)	Shed Spacing (cm)	Shed Diameter
Silicone rubber	40	3	12

The model provides the estimation of the local saline penetration and moisture concentration along the insulator pollution thickness. Figure 4 shows a simulation from the beginning of the penetration and diffusion process to the end of it. The curves of penetration and diffusion into the insulator sample were a two-stage growth process such as that for wetting and drying. The saline penetration in the water film increases non-linearly with the development of moisture concentration. As this process unfolds, the saline dissolved in water steadily gathers at the top of the water film. It can be clearly seen that, at the beginning of penetration and diffusion, the initial rapid growth is followed by a slowdown and ultimately drifts towards saturation. Thus, it can be concluded that the transfer of moisture significantly accelerates the penetration of saline ions on SiR materials, this method can also be used on other types of insulator models such as porcelain and glass. The negative sign with error function (*erf*) in Equation (8) states that increasing the accumulate moisture in wetting condition has the effect of increasing diffusion and penetration of saline on the insulator surface.

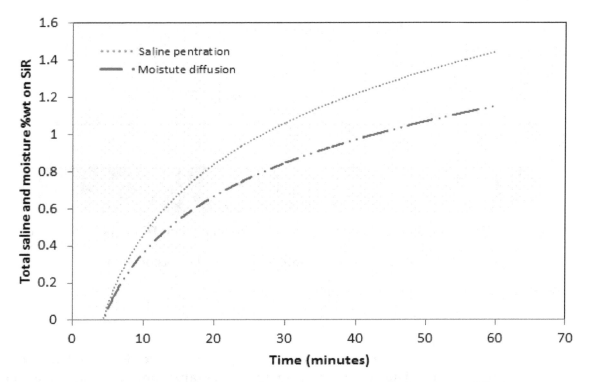

Figure 4. Characteristic curves of saline concentration and moisture diffusion.

3. Modelling of Dry Band Initiation and Formation

The development of a dry band formation on the surface of a silicone rubber model in normal cold fog can be mathematically formulated by considering the energy balance equation, which can be expressed as:

Energy in = Energy out + Energy related to change in and on insulator surface

In our case, the energy balance equation can be expressed as:

$$W_{LC} = W_{\Delta T} + W_{evap.} + W_{cond./conv.} \tag{9}$$

W_{LC} = Energy generated by leakage current
$W_{\Delta T}$ = Energy generated by change in temperature
$W_{evap.}$ = Energy loss by evaporation due to ambient temperature
$W_{cond./conv.}$ = Energy loss by convection and conduction

The model of dry band formation and initiation is shown in Figure 5. Evaporation is an important factor in cold and normal fog conditions in winter and early spring. Before the formation of the dry band, the voltage gradient along the insulator surface is uniform and its relationship to surface resistivity is:

$$E_s = \rho_s \cdot J_s \tag{10}$$

where subscript s represents the surface of insulator, J_s is the surface current density, E_s is the electric field intensity per unit area and ρ_s mass of air per unit volume on insulator surface. Equation (11) shows the relationship between current density and variable dry band length,

$$J_s = \frac{I}{\Delta L} \tag{11}$$

where ΔL is the variable length of dry band with surface current density.

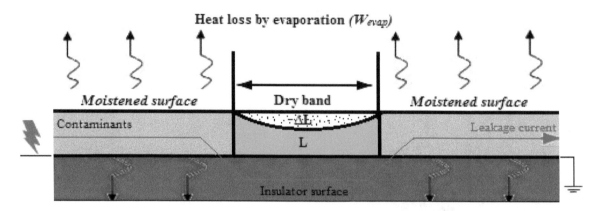

Figure 5. Modeling of energy balance on moistened insulator surface.

The power dissipated per unit area of the insulator surface is:

$$p = \rho_s \cdot J_s^2 \tag{12}$$

Due to the dissipated power the temperature of the pollution layer is rising because of heat transfer with the surroundings. Due to heat transfer, the temperature and dry-band area increased by ΔT and $\Delta s \cdot d(\Delta s)$, respectively, in a short period of time dt. Cold fog is made of condensed water droplets which are the result of a humid air mass being cooled to the dew point where it can no longer hold all of the water vapor.

Therefore, the corresponding volume $d(\Delta L) \cdot l$, and mass of cold fog is $\rho_s \cdot d \cdot (\Delta L) \cdot l$. Heat consumed by the evaporation process is given by:

$$W_{evp.} = -L_e \cdot \rho_s \cdot d(\Delta L) \cdot l \tag{13}$$

where L_e is the latent heat of fog.

The resistance of the dry band region is $l/\Delta s$. $1/\sigma$ due to a small volume of length L. Hence the heat generated by the current in AC system is:

$$W_{LC} = \frac{1}{\sigma \Delta L} \cdot l \cdot (J_S\, \Delta L)^2 \tag{14}$$

The conduction of heat is determined by the temperature difference between cold fog and insulator surface $H_{cond}(T_f - T_s)$ per unit area. The convection of heat between cold fog and the air interface on the insulator surface is $H_{conv}(T_f - T_a)$ per unit area.

The change in temperature of cold fog is in relationship with specific heat as $\Delta T \cdot mCh$. The latent heat of cold fog to moisture due to evaporation is $\Delta m \cdot Lm$. Surface resistivity changes with change of temperature along insulator surface length l in a short period of time dt so that:

$$W_{\Delta T} = C_h.\rho_s \cdot \Delta L \cdot l \cdot d \tag{15}$$

where ρ_s is the medium density.

If the volume of moisture is very small and does not interact with the air or insulator surface, then there is no convection or conduction. If it interacts with the insulator surface, then the area of interaction is $\Delta Am \cdot \rho \cdot l$. In a short period of time dt, the dissipated heat is given by:

$$W_{\underset{conve}{cond}} = H_c\left(T_m - T_p\right) \cdot \Delta A_m \cdot \rho \cdot l \cdot dt \tag{16}$$

By combining Equations (13)–(16), we obtain:

$$\frac{l}{\sigma\Delta L}\cdot l\cdot(J_S\ \Delta L)^2 \cdot dt = -L_e\cdot\rho_s\cdot d(\Delta L)\cdot l + C_h\cdot\rho_s\cdot\Delta L\cdot l\cdot dt + H_c\left(T_m - T_p\right)\cdot\Delta A_m\cdot\rho\cdot l\cdot dt \tag{17}$$

If the distribution of current is uniform along the insulator surface, then Δ can be neglected. Then by dividing both side of Equation (17) by $l\cdot\frac{dt}{L}$ Equation (18) is obtained:

$$\frac{l}{\sigma}\ (J_S)^2 = -L_e\cdot\rho_s\cdot L\cdot\frac{dL}{dt}\cdot l + C_h\cdot\rho_s\cdot L\cdot l\cdot dt + L\cdot H_c\left(T_m - T_p\right)\cdot A_{m-p} \tag{18}$$

From Equation (18), it can be noted that there are five (σ, L, T, ρ_s and t) essential and critical physical parameters involved in the dry-band formation and arcing on the insulator surface. As L is the length of the dry band region, the drying rate can be calculated by using small steps dL/dt until L becomes zero. This is the point where time to dry band arcing on insulator surface will be started. The dry band is completely formed when the area L on the insulator surface becomes zero where the leakage current is intermittent.

4. Partial Arc Electric Model of Dry-Band Flashover

Several researchers have worked to make useful contributions to this subject [1,24]. However, there are several key shortcomings in the models presently available. The present model capable of handling both uniform and non-uniform distributions pollution on the surface of insulators is more relevant. There are several aspects on which arc resistance and dry-band formation depend. The models assume that if a dry band can be formed and if the arc is able to bridge the dry band will continue propagation with different contamination degree. The dry band arcing is modeled mainly in two stages: First the formation of dry band and initial arc and second the arc propagation and arc bridging.

A thorough understanding of all aspects of flashover mechanism on an insulator surface is required to explore the subject further. Such a task would necessarily include an investigation of dry band arcing under different contamination levels along the leakage distance. Propagation of the alternating current (AC) surface flashover on polluted insulators is a complex phenomenon. The length and intensity of arcs may change in milliseconds. The arc is only highly ignited in the period of peak voltage, while during reaming periods the arc ignites and reignites following the voltage. Despite the complexity of the mechanism involved in dry-band arcing, many simplifying assumptions can be made in order to obtain an acceptable mathematical modelling.

The growth of the dry band and the dry-band arcing characteristics method have been well reviewed by Jolly and Poole [25] but their model will be extended here based on fundamental mathematical equations. The model was used for the analysis of the growth of the discharge with different contamination degrees along the leakage distance of specimen. The test procedures were as follows:

CASE 1:

In this case the sample leakage distance is divided into two sections L_1 and L_2 while the corresponding surface conductivities and surface resistances are σ_1, σ_2 and r_{i1}, r_{i2}, respectively, as shown in Figure 6. L_1 is the high voltage side and L_2 the grounded side. Section L_1 was lightly polluted while L_2 was polluted with heavy pollution levels. The hydrophobicity of the high-voltage side L_1, is higher than that of the grounded side L_2. Due to this difference, the arc cannot develop along the entire surface and its length is less than L_1 due to surface strength, as shown in Equation (23). As long as the arc length is smaller than section L_1, very small parts of L_2 can hardly influence arc length X. In this case, it may assume that the probability of subsistence of surface flashover at energized condition is very low.

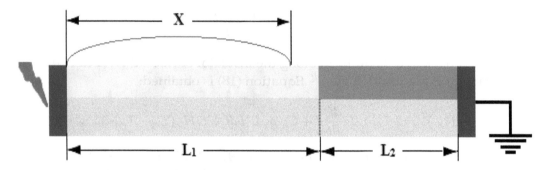

Figure 6. Polluted insulator model for dry band arcing with $X < L_1$.

If the heavily contaminated surface is located at the grounded side, the influence of the totally hydrophilic section (L_2) on arc length X can be described as follows:

$$L = L_1 + L_2 \tag{19}$$

Then the resistance of heavily polluted section L_2 that could not have connected with arc X can be written as:

$$R_i = r_{i1}(L_1 - L) + r_{i2}(L_2) \tag{20}$$

It is a well-known fact that current is necessary to sustain and ignite the arcing process on the insulator surface so that it can be derived from the Obenaus model [26] and rewritten as:

$$I = \left[\frac{nNX}{r_i(L - X)} \right]^{\frac{1}{1+n}} \tag{21}$$

It is assumed that constants n and N for arcs and insulator are comparable. From Equation (21), we can find the critical arc length Xc of the composite insulator that relates to section L_1 as follows:

$$Lim_{(X \to Xc)} = \left[\frac{Xn_1N_1}{r_{i1}(L_1 - L) + r_{i2}L_2} \right]^{\frac{1}{1+n_1}} \tag{22}$$

Equation (23) can be written in terms of sectional lengths and the ratio of surface conductivities and resistivities, and the root of an arc is close to section of L_2 of resistance per unit length. The relationship of Equation (24) shows that the L_1 leakage distance section, bridged with the arc, does not influence arc length X.

$$X_c < \left[\frac{N_1}{r_{i1}} \right]^{\frac{1}{1+n_1}} \tag{23}$$

$$X_c = \left[\frac{L_1 + kL_2}{1 + n_1} \right] \tag{24}$$

The simplified relationship of the surface conductivities and resistivities of sections L_1 and L_2 can be expressed as: $k = \frac{r_{i2}}{r_{i1}} = \frac{\sigma_1}{\sigma_2}$

where σ_1 and σ_2 are surface conductivities and r_{i1} and r_{i2} surface resistance of sections L_1 and L_2.

CASE 2:

In this case the section L_1 was heavily polluted while L_2 lightly polluted. It is assumed that the specimen is totally hydrophilic only near a high-voltage side, as presented in Figure 7. This consideration is due to long-term operating conditions. If section L_1 is hydrophilic as compared to L_2 then what influence of hydrophilic section of the creepage distance L_1 has on the arc length X.

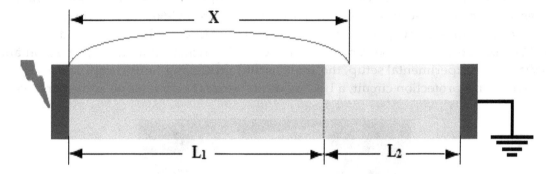

Figure 7. Polluted insulator model for dry band arcing with $X > L_1$.

When arc length X connected with section L_2 with reduced k-times is on the lightly polluted and more hydrophobic section, then the leakage distance of the sample corresponding to L_2 can be written as Equation (25).

$$L_2 = L_2 + kL_1 \tag{25}$$

$$Lim_{(X \to Xc)} = \left[\frac{Xn_2N_2}{r_{i2}(L_2 + kL_1 - L)} \right]^{\frac{1}{1+n_2}} \tag{26}$$

$$X_c > \left[\frac{N_2}{r_{i2}} \right]^{\frac{1}{1+n_2}} \tag{27}$$

$$X_c = \left[\frac{L_2 + kL_1}{1 + n_2} \right] \tag{28}$$

The relationship indicates that the L_1 section of creepage distance connected with the arc but practically L_2 does not have any influence on a critical arc length Xc. To figure out the reason, it can be assumed that discharge current and arc length X is increased resulting in a heavy pollution level and more hydrophilic surface near high voltage end, and at a certain level of contamination, as described in the division of sample creepage distance, uneven potential distribution on the specimen occurs that ignites the arc along section L_1 that causes an immediate extension for section L_2 resulting in a sudden surface flashover occurring.

5. Experimental Setup and Results

Characteristics of Dry-Band Formation

The most significant factors on the surface flashover voltage are the number, location and length of dry bands. The initiation and formation of dry band is due to many factors, such as voltage, contamination type, amount and distribution, and various environmental stresses. It is a well-known fact that, when wet and polluted insulators are energized, discharge current causes Joule heating to form dry bands. Joule heating drives the low resistance layer and evaporates wetting. Drying is more intense where the current density is high. As a result of more evaporation, more small dry bands are formed on the surface. The discharge current is associated with dry band length and if the length is sufficiently long the discharge current will decrease and the arc will extinguish, resulting in a surface flashover. Thus, a comparison is made between the dry band initiation and formation under uniform, non-uniform and discontinuous non-uniform contamination distribution on specimens. The investigation is based on contamination layer parameters, such as conductivity, layer thickness and length, and insulator surface dielectric strength. The reference insulator with no sheds consisted of two main parts. One part was a silicone rubber plate sample of rectangular shape (12 cm × 4 cm × 8 cm), mechanically connected with electrodes, one of them connected to a high voltage alternating current (HVAC) power source with AC voltage of 0–100 kV at 50 Hz, and the other one grounded as shown in Figure 8. The electrodes were made of 0.9 mm thickness copper. The contamination along the leakage distance was achieved by solid layer method has been given in IEC-60507, 1991, by brush applications of different conductivities. For experimental work the 69 kV AC test voltage was produced by a 10 kVA, 100 kV, and 50 Hz transformer. The supplied voltage can be increased manually or automatically at a rate of 1 kV/s. In the experimental setup, the analysis and processing units comprised a high frequency current transformer, a protection circuit, a LeCroy digital storage oscilloscope and a personal computer.

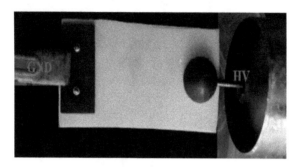

Figure 8. Experimental set up.

For sample contamination, the ratio of equivalent salt deposit density (ESDD) to non-soluble deposit density (NSDD) was 1:4. The values of ESDD and NSDD with contamination values are shown in Table 2.

Table 2. Contamination values.

Contamination Type	ESDD (mg/cm^2)	NSDD (mg/cm^2)
Uniform	0.200	0.850
Non-uniform	0.080	0.350

Initially, the distribution of voltage and layer resistivity was uniform and the sample surface infrared pictures were uniformly heated, as shown in Figure 9. This confirmed that contamination layer conductance was also uniform. As the surface becomes wet, resistivity decreases and discharge current increases. This condition is not resilient due to slightly higher resistance in some segments of the sample surface with exceeding voltage gradient in these sections. The dissipation of heat is higher at these locations such that they become dry more rapidly than the remaining surface, forming dry bands. It is clearly shown that only small dry bands are formed near the high voltage electrode as well as near areas with higher electric field strength. However, as opposed to non-uniform contamination distribution, the dry bands do not tend to elongate towards the ground electrode.

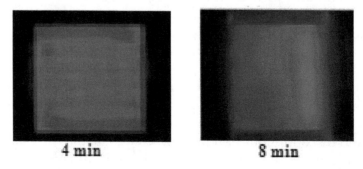

4 min   8 min

Figure 9. Dry-band development with uniform contamination record by infrared camera.

Under non-uniform contamination distribution, however, the leakage current is increased, soon leading to reduced surface conductance, as shown in Figure 10a. This strongly indicates the drying part of the non-uniform contamination distribution by the heated area of the conducting layer, with the expected formation of a complete dry band. Initially the dry band formed at the high voltage electrode is followed by a series of small discharges that gradually scatter on other part of specimen. These discharges move towards near the ground electrode region and take the form of an arc-like discharge.

On the other hand, Figure 10b shows the infrared images of dry band discharges with discontinuous non-uniform contamination distribution. It is clearly shown that the extinction of the discharges is frequently followed by reignition, temporarily bridging the other dry bands. At this state, leakage current is at its maximum value just before the start of the discharge phenomenon. It is also observed that under discontinuous non-uniform contamination the insulating surface achieves its highest dielectric strength when the specimen conductive surface carries a lot of wider dry bands located at different locations of the specimen. Multiple dry bands created at same time on an insulating surface have the ability to weaken the field strength of each other. Therefore, with the existence of multiple dry bands the field strength of each one is normally less than when it is individually created. If a number of dry bands are formed, then after a short period only one will remain and due to its higher resistance almost all the voltage will be dropped across this dry band. The dry bands formed under a discontinuous non-uniform contamination layer may be considered as a potential barrier which may efficiently weaken the dielectric strength and develop a large voltage drop on outer dry bands.

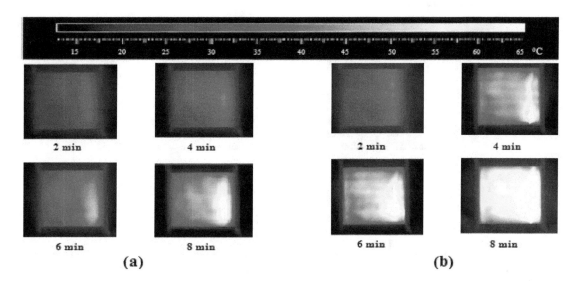

Figure 10. Dry-band development recorded by infrared camera: (**a**) under non-uniform contamination; (**b**) under non-uniform discontinuous contamination.

Results obtained indicate that with discontinuous nonuniform contamination distribution, dry-band elongation on the full leakage distance took 6 minutes, and that after 8 minutes there was a sharp rise in dry band elongation with multiple dry bands. However, with a uniform contamination layer, dry-band elongation took place only up to the 40% of the total length in 6 minutes. After that there was no more elongation of dry-band length.

6. Comparison of Models with Experimental Results

6.1. Inspection of Dry-Band Formation

This section describes the difference between the mathematical model and the test results. It is noted that in the model there is a smooth increase of resistance at the initiation and development of dry bands, while in the test the dry bands outset suddenly as shown in Figures 9 and 10. In the model (Section 3), it is anticipated that the dry band is initiated when the moist layer becomes wet. When the ΔL is very close to zero, it is shown that the dry band width is increased. However, this period is difficult to measure in the experiment. Therefore, the dry bands developed before the moisture is totally vaporized. The critical phenomena of moist film before the dry band forms are illustrate here. The equilibrium of forces between moist contact interfaces on insulator surface can be derived by the Young–Dupre equation:

$$\lambda_m Cos\ \theta = R_s\ (\lambda_s - \lambda_{ms}) \tag{29}$$

In Equation (29), the R_s is the surface roughness coefficient which depend upon the material condition, θ is the contact angle in degrees between moisture and material surface, λ_m is the surface tension of moisture, λ_s surface tension of material surface and λ_{ms} is the interfacial between the moisture and material surface.

This section explains the phenomena that observed during the experiment. For a moisture layer on the material surface, we anticipated the equilibrium state is stretched first. When the surface was hydrophilic, the length of dry bands varied from ΔL to L. However, as the moisture layer is being evaporated, the moisture layer becomes thinner and thinner, and therefore the interfacial force λ_{ms} decreases. At this point the discharge current is cut off, which caused an increase of the local electric field. Therefore, the dry band's initiation and development are started. Once the dry-band arcing is started, the arc temperature would dry the remaining part of moisture layer. At this stage, the surface is not fully dry, the dry-band arcing is weak and causes a number of multiple dry bands, which is observed in tests as shown in Figure 11.

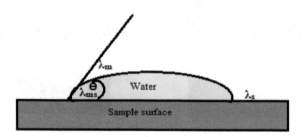

Figure 11. Interfacial equilibrium between water and sample surface.

6.2. Onset of Dry-Band Arcing

To validate the model, the experimental results of dry band arcing, arc extension and surface flashover were compared with those of the proposed mathematical model. For the experimental analysis two scenarios were configured as those on the mathematical model: (1) high-voltage side with heavy pollution, whilst the grounded side was lightly polluted, and (2) high-voltage side with light pollution and grounded side with heavy pollution. These pollution scenarios are commonly seen in the field under the operating conditions to which are exposed insulators located near the shoreline and at sites with dominant winds speed [27–30].

In the first scenario (Figure 12), when voltage is applied to a sample, local arcs are first initiated on the heavily contaminated side/part, which is essentially due to higher discharge current. The discharge current causes ohmic heating to form multiple dry bands. The voltage across the dry bands which were usually the low conductive surface parts caused air break down. This caused the dry bands to be moved towards the lightly contaminated part which became electrical in series with the heavily contaminated part of the surface. Multiple dry bands spread out onto the sample surface and nearby the electrodes, and then some of the dry bands were bridged by local arcs. Local arcs ignited and reignited many times, and then gradually developed over the surface to connect to the other arcs, thus increasing the total length of the arcs.

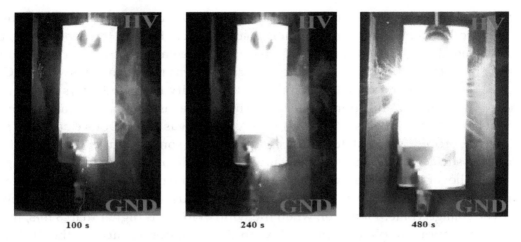

Figure 12. Image of dry-band arcing activity progression.

On the other hand, in the case of the second scenario (Figure 13), the dry-band arcing did not initiate suddenly, as the section with light contamination levels, which is more hydrophobic than the other section, can strongly limit the development of the discharge current. It was found that if the length of the arc is equal to the length of the considered leakage section (such that the considered section is completely bridged) then the arc will propagate over to the next section and that the contamination level of the remaining part of the sample will have little effect on dry-band arcing. The experimental results obtained show that the surface strength and arc length mainly depend on the leakage distance and contamination distribution.

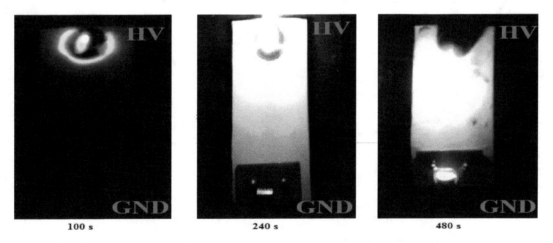

Figure 13. Image of discharge activity progression.

From the above results, it can be concluded that due to voltage drop along an arc channel the arc root transfers the potential of the electrodes next to the lightly contaminated section of the leakage distance. If the potential is sufficiently high, then after attainting the edge of the lightly contaminated section of the leakage distance, arc extension surface flashover will occur on the specimen. Surface flashover instantly occurs and is not connected with the partial arc's development. It was also observed that on the lightly contaminated section, the drop of tangential component of electric field does not exit. However, it can be observed mainly along the heavily contaminated section of the specimen.

6.3. Impact of Wind Velocity on Surface Flashover Characteristics

To obtain the effects of the wind velocity on the surface flashover characteristics of composite insulator. The AC surface flashover process of polluted insulator was observed at 0 m/s, 2 m/s, 4 m/s, 8 m/s 10 m/s and 12 m/s wind velocities, respectively. The tests were carried out under two conditions. One was wind but no contamination deposition condition and other was wind with contamination deposition condition. In order to wet the surface of insulator and contamination distribution uniform, $\theta = 90°$ was selected between wind direction and insulator axis. During the experiment, the relative humidity was sustained in between (80%–90%) in the environmental chamber, which helped to bond the saline mixtures on the insulator surface, and the temperature was maintained between 0 °C and 2 °C.

During tests it was observed that, when the wind velocity bellowed 8 m/s or less as shown in Figure 14, the fog drifts slowly, and approached the surface of the insulator all around it. Simultaneously, the influence of contamination on surface flashover voltage was continuous due to the result of the wetting and drying process on insulator surface. Although at some moment the effect of moistening was larger and the surface flashover voltage was higher; while other, the effect of drying was larger and the surface flashover voltage was lower. Therefore, the value of surface flashover voltage changed by a big margin at 8 m/s and for lower wind speed values. Thus, the lowest value was selected as the surface flashover voltage value. However, when the wind velocity is higher than 8 m/s, the mist drifts very fast. The fog cannot completely approach with the insulator surface, so the drying effect was higher than the wetting effect. At this situation, the flashover voltage was relatively low and stable as compared to lower wind velocities, thus the average value was preferred as the surface flashover voltage value. From Figure 14, it can be clearly seen that when the wind velocity bellows up to 8 m/s, the surface flashover voltage diversifies by a big margin, the lowest higher than the normal down 32.9%, and when the wind velocity higher than 8 m/s, the surface flashover voltage gradually decreases steadily and at 12 m/s surface flashover voltage is lower than the normal flashover voltage 6.2%.

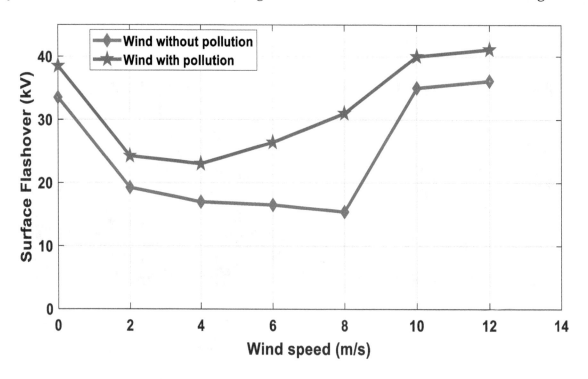

Figure 14. Influence of wind on surface flashover characteristic.

6.4. Effect of Conductivity and Pollution Layer Length of Surface Flashover

The dimensioning of insulators with respect to contamination is always done based on the performance characteristic of real insulators under uniform artificial contamination. However, due to the shape and in-service position, climate conditions and following the action of electrical stresses, the outdoor high-voltage insulators are actually contaminated in a non-uniform manner. The surface flashover of a uniformly contaminated insulator can be defined by a one dry band arc connected in series with a resistance of the contamination layer. In contrast, for insulators with a non-uniform contamination layer, several dry-band arcs can ignite simultaneously over their contaminated surface and may develop to a full surface flashover. This section is made to investigate the effect of uniform, non-uniform and discontinuous non-uniform contamination layer parameters, such as the conductivity and the length of the contamination layer on the material surface. Surface flashover tests are carried out following the same procedure as presented in [30].

The results of the present work and the other investigations [31] are approximately close. However, the surface flashover voltage of the present work and the work carried out by [32] are slightly lower than the surface flashover voltage with natural contamination. The main reason for this is that ions solubility of marine specification salts in artificial contamination almost slightly varies in fog, but it is not in the case of natural contamination deposition such as desert and ash, which contains weaker electrolytes. Also, the other most significant factor on the flashover strength is the length of the pollution layer. These results agree with that mentioned in [31] and would explain the observations that artificially contaminated insulators do allow surface flashover voltage lower than those naturally contaminated for the same conduction and climate conditions. It is clearly shown in Figures 15 and 16 that the length of contamination layer significantly affects the surface flashover voltage. The dielectric strength decreases with increasing of the conductivity layer. Finally, there is another strong effect observed on surface flashover due to the contamination class. It perceived that surface flashover strength is higher with uniform as compared to non-uniform and non-uniform discontinuous contamination distribution. Figure 16 shows the variation of surface flashover voltage with different pollution layer length under different types of contamination deposition. We can observe that surface flashover voltage decreases when the length of the pollution layer increases. It elapses from 36.2 kV with a uniform distribution to 33.1 kV for a surface entirely contaminated with a non-uniform discontinuous.

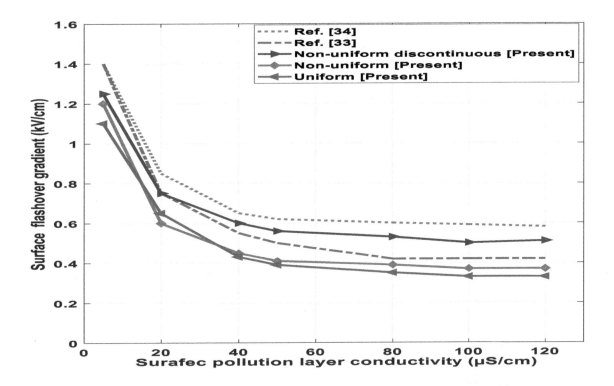

Figure 15.   Surface flashover voltage versus the pollution layer conductivity with various contamination class.

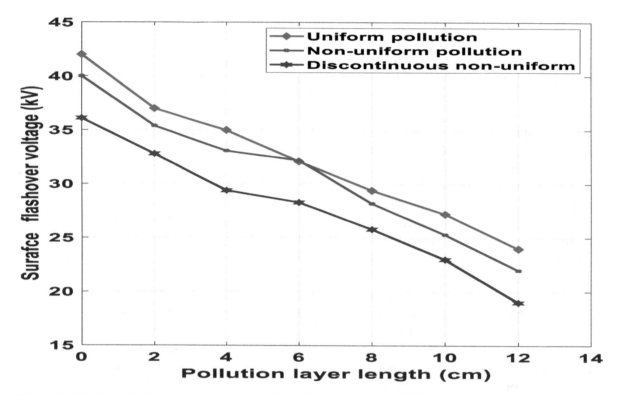

Figure 16. Surface flashover voltage versus the pollution layer length for different contamination class.

7. Conclusions

In this paper, a novel mathematical model is presented to understand the sea salt migration and deposition on high-voltage insulators near shoreline. The model introduced a new phenomenon based on saline transportation and deposition. Based on this, two more mathematical models were presented to understand the phenomenon of dry-band initiation and formation, as well as the behaviour of dry-band arcing with two different conduction states of the contamination layer on the insulator surface. Based on these, the following conclusions could be drawn:

1. From simulation and experimental work it was found that the pollution accumulation rate on the insulator surface increases with the increase of wind velocity and decreases with the increase of distance from shoreline to inland. However, when wind speed was higher than 12 m/s, the contamination density tended to be constant.

2. It was also observed that the transfer of moisture significantly accelerates the saline ions' penetration on the insulator surface.

3. The dry-band initiation and formation model presented is based on the energy balance equation. The integration of this equation results in a variation of dry-band length under moisture evaporation on the insulator surface. With the reduction of dry-band length, the surface resistance increases, which increases the discharge current on the insulator surface. The conduction and convection processes increase the surface resistance. It was observed that under this process single and multiple dry bands appear on the insulator surface.

4. Discontinuous non-uniform contamination distribution of the insulator surface leads to multiple dry bands and lower surface flashover voltage as compared to uniform and non-uniform contamination distribution.

5. A dry band arcing model was developed. It showed that the insulator surface strength and arc length mainly depend on the distribution of the pollution layer and leakage distance. The results obtained show that the configured scenarios are trustworthy to detect the dry band arcing on the insulator surface with different contamination distribution. The obtained dry-band arcing initiation and elongation rate confirm the efficiency of the proposed model.

Author Contributions: Conceptualization, M.M.H.; methodology, M.M.H., M.A.C. and A.R.; analysis, M.M.H., M.A.C. and A.R.; investigation, M.M.H.; writing—review and editing, M.M.H., M.A.C. and A.R.; visualization, M.M.H. and M.A.C.; supervision, M.M.H.; project administration, M.M.H., M.A.C. and A.R.

References

1.　Venkataraman, S.; Gorur, R.S. Extending the applicability of insulator flashover models by regression analysis. *IEEE Trans. Dielectr. Electr. Insul.* **2007**, *14*, 368–374. [CrossRef]

2.　Baker, A.C.; Farzaneh, M.; Gorur, R.S.; Gubanski, S.M.; Hill, R.J.; Schneider, H.M. Insulator selection for overhead AC lines with respect to contamination. *IEEE Trans. Power Deliv.* **2009**, *24*, 1633–1641. [CrossRef]

3.　El-Amine Slama, M.; Hadi, H.; Flazi, S. Investigation on influence of salts mixture on the determination of flashover discharge constant Part I: A Preliminary Study. In Proceedings of the IEEE Conference on Electrical Insulation and Dielectric Phenomena, Quebec, QC, Canada, 26–29 October 2008; pp. 674–677.

4.　Sima, W.; Yuan, T.; Yang, Q.; Xu, K.; Sun, C. Effect of nonuniform pollution on the withstand characteristics of extra high voltage suspension ceramic insulator string. *IET Gen. Trans. Distrib.* **2009**, *4*, 445–455. [CrossRef]

5.　Majid Hussain, M.; Farokhi, S.; McMeekin, S.G.; Farzaneh, M. Dry band formation on HV insulators polluted with different salt mixtures. In Proceedings of the 2015 IEEE Conference on Electrical Insulation and Dielectric Phenomena (CEIDP), Ann Arbor, MI, USA, 18–21 October 2015; pp. 201–204.

6.　Naito, K.; Morita, K.; Hasegawa, Y.; Imakoma, T. Improvement of the dc voltage insulation efficiency of suspension insulators under contaminated conditions. *IEEE Trans. Dielectr. Electr. Insul.* **1988**, *23*, 1025–1032. [CrossRef]

7.　Seta, T.; Arai, N.; Udo, T. Natural pollution test of insulators energized with HVDC. *IEEE Trans. Power Appar. Syst.* **1974**, *PAS-93*, 878–883. [CrossRef]

8.　Kimoto, I.; Fujimura, T.; Naito, K. Performance of insulators for direct current transmission line under polluted condition. *IEEE Trans. Power Appar. Syst.* **1973**, *PAS-92*, 943–949. [CrossRef]

9.　Houlgate, R.G.; Swift, D.A.; Cimador, A.; Pourbaix, F.; Marrone, G.; Nicolini, P. Field experience and laboratory research on composite insulators for overhead lines. *CIGRE Pap.* **1986**, *15*, 12.

10.　Schneider, H.M.; Guidi, W.W.; Burnham, J.T.; Gorur, R.S.; Hall, J.F. Accelerated aging and flashover tests on 138 kV nonceramic line post insulators. *IEEE Trans. Power Deliv.* **1993**, *8*, 325–336. [CrossRef]

11.　Kim, S.H.; Cherney, E.A.; Hackam, R. The loss and recovery of hydrophobicity of RTV silicone rubber insulator coatings. *IEEE Trans. Power Deliv.* **1990**, *5*, 1491–1500. [CrossRef]

12.　Starr, W.T. Polymeric outdoor insulation. *IEEE Trans. Electr. Insul.* **1990**, *25*, 125–136. [CrossRef]

13.　Simmons, S.; Shah, M.; Mackevich, J.; Chang, R.J. Polymer outdoor insulating materials Part 3-silicone elastomer considerations. *IEEE Electr. Insul. Mag.* **1997**, *13*, 25–32. [CrossRef]

14.　Baker, A.C.; Zaffanella, L.E.; Anaivino, L.D.; Schneider, H.M.; Moran, J.H. Contamination performance of HVDC station post insulators. *IEEE Trans. Power Deliv.* **1988**, *3*, 1968–1975. [CrossRef]

15.　Wilkins, R. Flashover voltage of high-voltage insulators with uniform surface pollution films. *Proc. IEE* **1969**, *116*, 457–465. [CrossRef]

16.　Rizk, F.A.M. Mathematical models for pollution flashover. *IEEE Trans. Dielectr. Electr. Insul.* **1981**, *78*, 71–103.

17.　Kim, S.; Cherney, E.A.; Hackam, R.; Rutherford, K.G. Chemical changes at the surface of RTV silicone rubber coating on insulators during dry-band arcing. *IEEE Trans. Dielectr. Electr. Insul.* **1994**, *1*, 106–123.

18.　Morcillo, M.; Chico, B.; Mariaca, L.; Otero, E. Salinity in marine atmospheric corrosion: Its dependence on the wind regime existing in the site. *Corros. Sci.* **2000**, *42*, 91–104. [CrossRef]

19.　Spiel, D.E.; Leeuw, G.D. Formation and production of sea spray aerosols. *J. Aerosol Sci.* **1996**, *27*, S65–S66. [CrossRef]

20.　Feliu, S.; Morcillo, M.; Chico, B. Effect of distance from sea on atmospheric corrosion rate. *Corrosion* **1999**, *55*, 883–889. [CrossRef]

21.　Gustafsson, M.E.R.; Franzen, L.G. Dry deposition and concentration of marine aerosols in a coastal area. *Atmos. Environ.* **1996**, *30*, 977–989. [CrossRef]

22.　Meira, R.; Andrade, C.; Alonso, C.; Padaradtz, I.J.; Borba, J.C. Salinity of marine aerosols in a Brazilian coastal area–influence of wind regime. *Atmos. Environ.* **2007**, *41*, 8431–8441. [CrossRef]

23. Hamada, S.; Hino, S.; Kanyuki, K. Salt measurement in the coastal region. *Fac. Eng. Yamaguchi Univ.* **1986**, *01*, 255.

24. Karady, G.; Amrah, F. Dynamic modeling of AC insulator flashover characteristics. *IEE Symp. High Volt. Eng.* **1999**, *467*, 107–110.

25. Jolly, D.C.; Poole, C.D. Flashover of contaminated insulators with cylindrical symmetry under DC conditions. *IEEE Trans. Dielectr. Electr. Insul.* **1979**, *EI-14*, 77–84. [CrossRef]

26. Obenaus, F. Contamination flashover and creepage path length. *Dtsch. Elektrotechnik* **1958**, *12*, 135–136.

27. Majid Hussain, M.; Farokhi, S.; McMeekin, S.G.; Farzaneh, M. Risk assessment of failure of outdoor high voltage polluted insulators under combined stresses near shoreline. *Energies* **2017**, *10*, 1661. [CrossRef]

28. Mizuno, Y.; Kusada, H.; Naito, K. Effect of climatic conditions on contamination flashover voltage of insulators. *IEEE Trans. Dielectr. Electr. Insul.* **1997**, *4*, 286–289. [CrossRef]

29. Majid Hussain, M.; Farokhi, S.; McMeekin, S.G.; Farzaneh, M. Contamination performance of high voltage outdoor insulators in harsh marine pollution environment. In Proceedings of the IEEE 21st International Conference on Pulsed Power, Brighton, UK, 18–22 June 2017.

30. Majid Hussain, M.; Farokhi, S.; McMeekin, S.G.; Farzaneh, M. Mechanism of saline deposition and surface flashover on outdoor insulators near coastal areas Part II: Impact of various environmental stresses. *IEEE Trans. Dielectr. Electr. Insul.* **2017**, *24*, 1068–1076. [CrossRef]

31. Terrab, H.; Bayad, A. Experimental study using design of experiment of pollution layer effect on insulator performance taking into account the presence of dry bands. *IEEE Trans. Dielectr. Electr. Insul.* **2014**, *21*, 2486–2495. [CrossRef]

32. Zohng, H.P.; Dong, X.C. *Tests and Investigations on Naturally Polluted Insulators and Their Application to Insulation Design for Polluted Areas*; CIGRE Report 33–07; CIGRE: Paris, France, 1982.

8

Assessment of Surface Degradation of Silicone Rubber Caused by Partial Discharge

Kazuki Komatsu [1],*, Hao Liu [1], Mitsuki Shimada [2] and Yukio Mizuno [1]

[1] Department of Electrical and Mechanical Engineering, Nagoya Institute of Technology, Nagoya 466-8555, Japan

[2] Information and Analysis Technologies Division, Nagoya Institute of Technology, Nagoya 466-8555, Japan

* Correspondence: 30413080@stn.nitech.ac.jp

Abstract: This paper reports experimental and analytical results of partial discharge degradation of silicone rubber sheets in accordance with proposed procedures. Considering the actual usage condition of silicone rubber as an insulating material of polymer insulators, an experimental procedure is established to evaluate long-term surface erosion caused only by partial discharge. Silicone rubber is subjected to partial discharge for 8 h using an electrode system with air gap. Voltage application is stopped for subsequent 16 h for recovery of hydrophobicity. The 24 h cycle is repeated 50 or 100 times. Deterioration of sample surface is evaluated in terms of contact angle and surface roughness. It is confirmed the proposed experimental procedure has advantage of no arc discharge occurrence, good repeatability of results, and possible acceleration of erosion. Surface erosion of silicone rubber progresses gradually and finally breakdown of silicone rubber occurs. Alumina trihydrate (ATH), an additive to avoid tracking and erosion by discharge, is not necessarily effective to prevent breakdown caused by partial discharge when localized electric field in air is enhanced by adding ATH. In such a situation, lower permittivity and higher resistance of silicone rubber seem dominant factors to prevent partial discharge breakdown and a careful insulation design should be required.

Keywords: silicone rubber; partial discharge; degradation; breakdown; contact angle; surface roughness; FTIR; ATH

---

1. Introduction

Silicone rubber has been widely used as an electrical insulating material of polymer insulators of power transmission and distribution lines, because of advantages like light weight, easy handling, and water repellency. Electrical and mechanical degradation of silicone rubber is caused by discharges on its surface, exposure to ultraviolet ray, surface contamination, and heavy rain.

With respect to degradation of silicone rubber caused by discharge, arc discharge gives serious and irreversible damage to silicone rubber when compared with partial discharge. To improve resistance to tracking and erosion of silicone rubber caused by arc discharge, alumina trihydrate (ATH) is usually added to silicone rubber [1]. Water of crystallization contained in ATH evaporates by absorbing energy of arc discharge, resulting in good performance. Arc discharge usually occurs on silicone rubber surface under sever condition such as high electric field, heavy contamination and wetting [2]. However, occurrence probability of such conditions may be low.

On the contrary, partial discharge occurs more easily under less sever condition. Partial discharge is a localized breakdown observed in a small portion of an insulation system subjected to high electric field. It does not cause short-circuit of the insulation system. In the case of polymer insulators, partial discharge may be observed at the triple junction point-like interface of silicone rubber/electrode/air and silicone rubber/water droplet/air [3,4]. Partial discharge generated in air attacks the surface of silicone rubber. It is difficult for partial discharge to breakdown silicone rubber by one attack because its energy is not sufficient enough. However, silicone rubber subjected to partial discharge over a long period of time will be eroded gradually from the surface to the bulk and finally breakdown. Knowledge of partial discharge degradation of silicone rubber is considered essential to electrical insulation design of composite insulator. However, researches focused on partial discharge degradation of silicone rubber have not been performed much.

Analysis of contact angle and SEM images of surface after 48 and 96 h partial discharge treatment show that the resistance to partial discharge is improved by adding nano-sized silica to silicone rubber [5]. Effect of partial discharge on rheological and chemical properties of silicone rubber has been discussed in terms of oxidization and localized temperature rise, moisture, and depolymerization [6]. Another paper reports that deposited charge on silicone rubber surface by partial discharge may have an impact on its flashover voltage, but the material degradation caused by partial discharge is limited [7]. To understand partial discharge degradation phenomena of silicone rubber, further fundamental long-term experiments under various conditions in accordance with appropriate procedures seem necessary.

The authors have carried out experiments in rather actual conditions, where silicone rubber was subjected to partial discharge in salt or clean fog [8,9]. To simplify the experimental condition by avoiding the effect of water droplet on silicone rubber, we started experiments in the atmosphere without fog [10]. The present paper reports a series of results obtained in accordance with proposed experimental procedures, which are suitable to study degradation of hydrophobic silicone rubber caused only by partial discharge. The repeatability of results is good and the acceleration rate can be changed in a certain range by adjusting applied voltage and spacing of the gap where partial discharge occurs. The effect of alumina trihydrate (ATH)—a typical additive for mitigating tracking and erosion of silicone rubber—on partial discharge degradation and breakdown is studied by using five kinds of silicone rubber samples including different amount of additives. Results of chemical analysis show that ATH is decomposed by partial discharge as reported in the case of arc discharge. However, it is suggested addition of ATH lowers permittivity and resistivity of silicone rubber, which results in active partial discharge in air gap and shorter lifetime to breakdown.

2. Experimental Procedures

2.1. Sample and Electrode

Five kinds of silicone rubber samples of the same size (45 × 45 × 2 mm) were used as samples. The matrix of each sample was identical. Respective samples had different amount and surface treatment of additives: ATH and silica. The physical properties of samples [11] are summarized in Table 1.

Table 1. Physical properties of samples.

Sample	A	B	C	D	E
ATH [1,2]	0	50 Treated	100 Treated	100 Untreated	100 Untreated
Silica [2]	Treated	Treated	Treated	Untreated	Treated
Density (g/cm^3)	1.14	1.38	1.53	1.53	1.52
Resistivity (Ω-cm)	6.73×10^{17}	5.29×10^{15}	3.38×10^{14}	8.68×10^{12}	1.75×10^{14}
Relative permittivity	2.9	3.7	4.2	4.4	4.4
Breakdown strength (kV/mm)	14.8	15.4	15.1	14.0	14.2
Hardness (Hs)	62	72	83	77	80

[1] Part by weight. [2] Treated stands for rounded surface of particle.

A bundle of four steel round bars of 6 mm diameter was used as high voltage electrode; each bar had a hemispherical tip of 3 mm radius of curvature. The ground electrode was a stainless steel disc of 30 mm in diameter. A silicone rubber sample was placed on the ground electrode. An air gap of 1 or 2 mm spacing was formed between the tip of the round bars (high voltage electrode) and a sample surface. The electrode system is shown in Figure 1. Five electrode systems in total, each sample was placed in one electrode system, were set in an acrylic chamber of 1 m^3.

The level of applied ac voltage was determined so that no arc discharge occurred throughout the experiment and surface degradation progressed as fast as possible only by partial discharge. In the present study, 6.6 and 8.5 kVrms (60 Hz) were selected for air gap spacing of 1 and 2 mm, respectively. Voltage was applied simultaneously to five electrode systems for 8 h and then interrupted for 16 h for recovery of hydrophobicity of sample surface. The 24-h cycle was repeated 50 or 100 times, during these cycles degradation performance of five samples is categorized into 3 groups. Applied voltage and current of each sample were monitored and recorded. A schematic diagram of the experimental setup is shown in Figure 2.

In the present study, samples were not tested under the same partial discharge conditions (for example, amount of charge, and number of pulses). The magnitude of applied voltage was fixed instead, considering the actual usage condition of polymer insulators at the site. It result in different partial discharge activities among samples because localized electric field in the air gap depends on permittivity and conductivity of the sample.

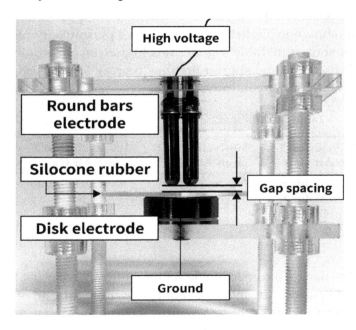

Figure 1. Electrode system.

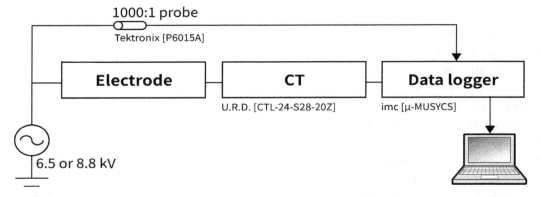

Figure 2. Schematic diagram of experimental setup.

2.2. Analyses of Surface Erosion

Contact angle was measured with a Drop Master DM500 (Kyowa Interface Science Co., LTD.) just before voltage application and just after voltage interruption of every cycle. A water droplet of 1 μl (conductivity: ~50 μS/cm) was dropped on a horizontally placed sample. Measurement was performed at five spots on the eroded area of a sample by partial discharge and the average contact angle was calculated.

Surface erosion of a sample was evaluated after 8-h voltage application with a surface roughness meter (TOKYO SEIMITSU CO., LTD., Surfcom 1400D) every three cycles. Measurement was carried out by 3 mm-long linear motion of a sensor on a sample surface.

Confocal microscope (Lasertec Corporation, OPTELICS HYBRID C3) enabled close observation of a sample surface of the area 1.5 mm × 1.5 mm. Images of samples were obtained every ten cycles. The arithmetic average roughness Ra of the surface in the observed area was also available. Ra gives the average of the absolute values of the roughness profile ordinates [12].

Infrared absorption spectrum was obtained every three cycles with a Fourier transformation infrared spectrophotometer (JASCO Corporation, FT/IR 6300).

3. Results

3.1. Sample Breakdown

No arc discharge was observed during the experiments. Samples C, D, and E were broken-down under some conditions as shown in Table 2. Higher relative permittivity of these samples compared with samples A and B as shown in Table 1 generates higher localized electric field in the air gap, resulting in active partial discharge and shorter lifetime to breakdown. Also, lower resistivity of these samples may have some relation to breakdown.

Table 2. Number of cycles of sample breakdown.

Sample	50-Cycle Test 1 mm Air Gap	50-Cycle Test 2 mm Air Gap	100-Cycle Test 1 mm Air Gap	100-Cycle Test 2 mm Air Gap
C	No breakdown	No breakdown	No breakdown	65
D	No breakdown	30	82	47
E	47	33	80	39

3.2. Contact Angle

Figure 3a,b shows the change in contact angle with number of cycle just before voltage application and just after voltage interruption in the 50-cycle test with 2 mm air gap spacing, respectively. Characteristics obtained in the 100-cycle test with 2 mm gap spacing are shown in Figure 4a,b. The initial value of contact angle is about 100 degrees for any sample. After 8-h voltage application, it decreased at most 20 degrees due to reduction of hydrophobicity of sample surface by partial discharge. Reduction in contact angle of Sample A is much lower than those of other samples. This corresponds to lower surface erosion of Sample A as described below. Recovery of contact angle to almost the initial value is confirmed after 16-h voltage interruption. It is considered that 16 h in the present experimental procedures are appropriate for low-molecular weight silicone oil to migrate from the bulk to the surface of a sample and to recover hydrophobicity. No remarkable reduction is observed with the number of cycle. Samples A and B are superior to the other samples when repeating cycles.

Large fluctuation of contact angle may be attributed to microscopic uneven surface caused by partial discharge, not only erosion like pitting but also accumulation of discharge by-product on the surface as suggested by the results of surface roughness measurement shown later in Figure 7. Contact angle does not necessarily reflect the surface profile in a small area, because the size of a water droplet for evaluation of contact angle is large enough to mask microscopic uneven surface and consequently an average information on surface in a larger area will be given.

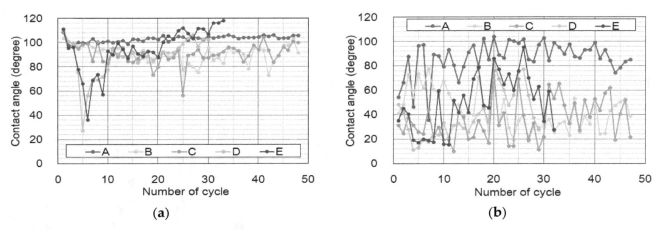

Figure 3. Change in contact angle with number of cycles in 50-cycle test with gap spacing 2 mm. (**a**) Just before commencement of voltage application. (**b**) Just after voltage interruption.

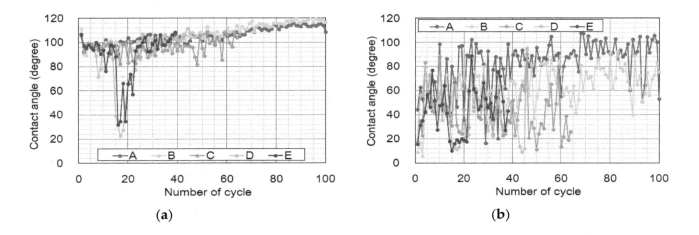

Figure 4. Change in contact angle with number of cycles in 100-cycle test with gap spacing 2 mm. (**a**) Just before commencement of voltage application. (**b**) Just after voltage interruption.

3.3. Observation of Sample Surface

Optical and confocal microscope images of sample surface are shown in Figure 5, which indicate change in surface erosion in the area of 1.5 mm × 1.5 mm of samples A and E in the 100-cycle test with 1 mm air gap. It is clearly shown especially by confocal microscope images that surface erosion progresses gradually with the number of cycle and degree of erosion is much different between samples A and E; surface erosion of sample E is serious and sample A is eroded slightly even after completing 80th cycle.

Figure 6 shows confocal microscopic images of all samples taken after completing 80 cycles in the 100-cycle test with 1 mm air gap. Appearance of surface erosion of Sample B is similar to that of Sample C. Sample D has eroded surface similar to Sample E. Samples D and E are severely damaged, which relates to the shorter lifetime to breakdown as shown in Table 2.

A quantitative discussion of surface erosion will be given in the next section.

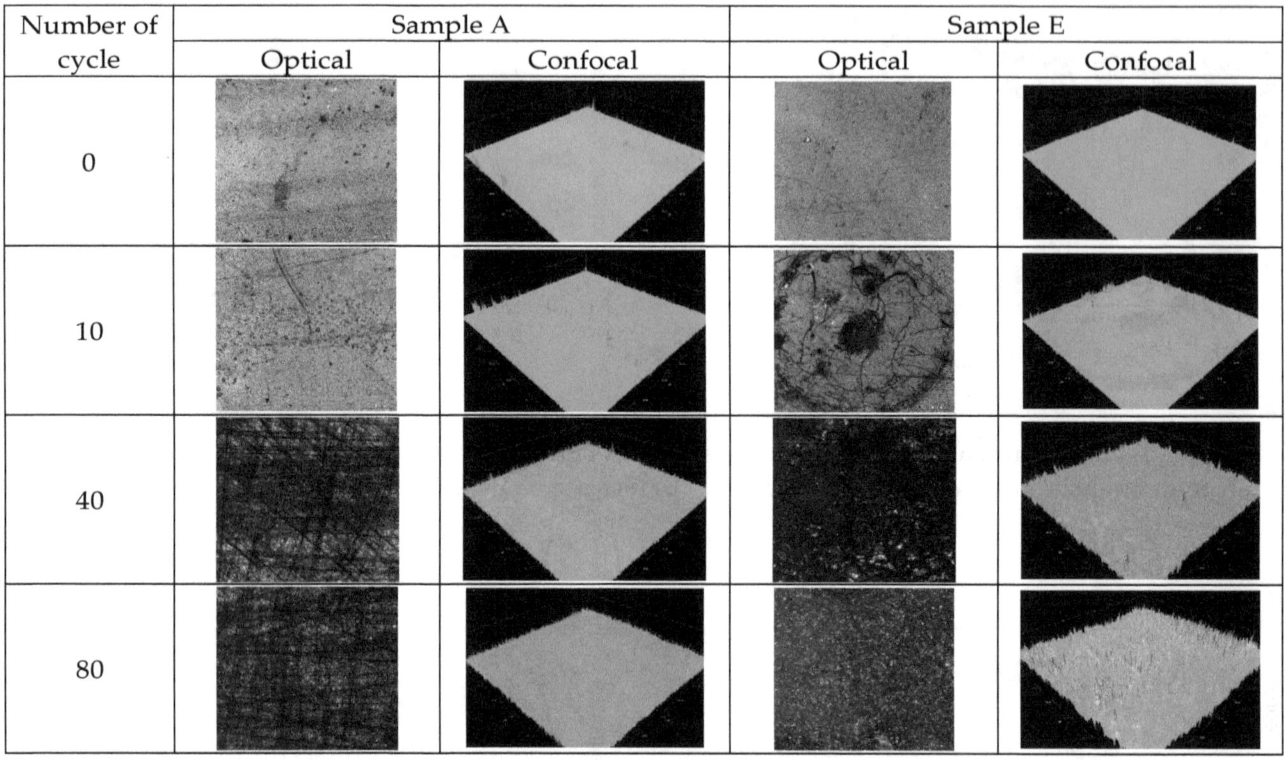

Number of cycle	Sample A		Sample E	
	Optical	Confocal	Optical	Confocal
0				
10				
40				
80				

Figure 5. Optical and confocal microscope images of surface of samples A and E obtained in the 100-cycle test with 1 mm air gap.

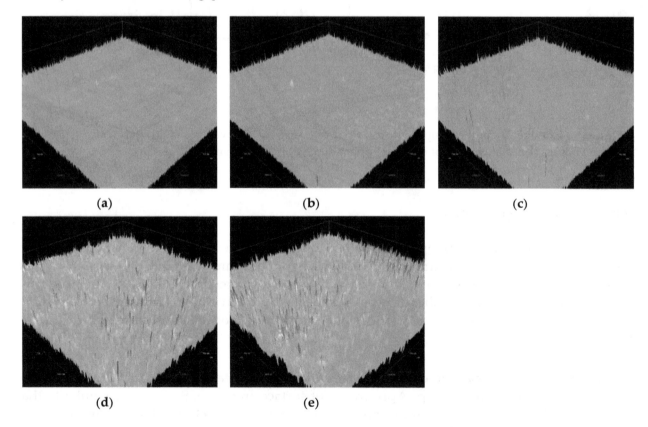

(a)　　　　　　　　　　　(b)　　　　　　　　　　　(c)

(d)　　　　　　　　　　　(e)

Figure 6. Confocal microscope images of sample surface obtained after completing 80 cycles in the 100-cycle test with 1 mm air gap. (**a**) Sample A; (**b**) Sample B; (**c**) Sample C; (**d**) Sample D; (**e**) Sample E.

3.4. Surface Erosion

3.4.1. Analysis Based on Surface Roughness Meter

Figure 7 shows surface roughness profiles of samples A and E, which are obtained after completing 6 and 60 cycles in the 100-cycle test with 1 mm air gap. Negative surface roughness means pitting on a sample surface generated by partial discharge. Discharge by-product accumulated on the sample surface may be a possible reason for the positive surface roughness. No remarkable change in surface roughness of sample A is observed between after 6 and 60 cycles. On the contrary, the surface of sample E is more eroded after completing 60 cycles; surface roughness is ~10 times larger compared with that after completing six cycles. These results correspond well to lower contact angle and shorter lifetime to breakdown of sample E compared with those of sample A, which is described above.

To evaluate quantitatively surface roughness shown in Figure 7, the arithmetic mean roughness Ra is used. Change in Ra in the 50-cycle tests with 1 and 2 mm air gap are shown in Figure 8a,b, respectively. When applied voltage is low, it is clear from Figure 8a that erosion of samples B to E progresses gradually with the number of cycle. Surface erosion is progressed in the order of samples D and E, samples B and C, and A, but difference among samples is not significant. In the case of 2 mm air gap, higher applied voltage intensifies partial discharge activity in the air gap and samples are clearly divided into three groups from standpoint of surface erosion. Samples D and E are eroded seriously by partial discharge and finally break down before completing 100 cycles. Erosion of samples B and C progresses steadily with the number of cycle. Sample A is eroded little. It is understood from Figure 8a,b that acceleration of partial discharge erosion is achieved by increasing applied voltage and gap spacing, especially for samples D and E.

Ra obtained in the 100-cycle test is shown in Figure 9 for 1 mm air gap. Ra increases gradually with the number of cycle except sample A. The characteristics of samples B and C are similar to each other. The same can be said of samples D and E. The result for 2 mm air gap are not shown because only limited data were available due to failure of the measuring instrument. Comparing Figure 9 with Figure 8a, obtained under the same applied voltage and the same gap spacing, surface roughness after completing 40 cycles, for example, is almost the same for any sample. This shows a good repeatability of results, suggesting the proposed experimental procedures are suitable to investigate partial discharge degradation of silicone rubber.

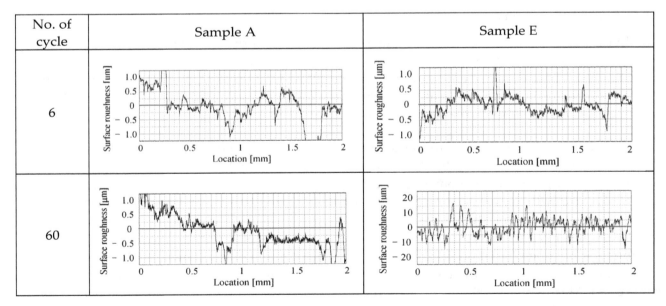

Figure 7. Examples of surface roughness profiles obtained with surface roughness meter in the 100-cycle test with 1 mm air gap.

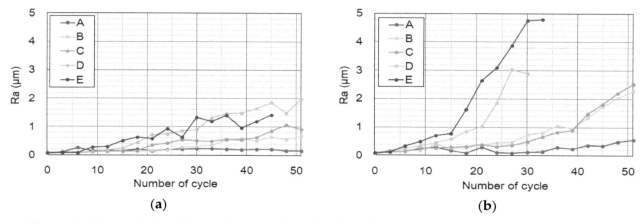

Figure 8. Change in Ra with number of cycle in the 50-cycle tests. (**a**) 1 mm air ga; (**b**) 2 mm air gap.

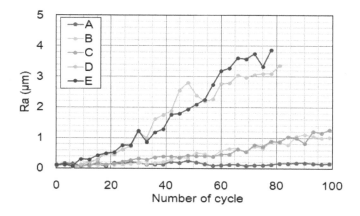

Figure 9. Change in Ra with number of cycle in the 100-cycle test with 1 mm air gap.

3.4.2. Analysis Based on Confocal Microscopic Images

In the previous section, surface roughness along the 3 mm straight line is discussed. Analysis of surface roughness in the area (1.5 mm × 1.5 mm) is available with confocal microscopic images shown in Section 3.2. Three-dimensional coordinate can be obtained at the time of surface scanning, where X- and Y-coordinates indicate the location on the sample surface and Z-coordinate shows height with reference to the sample surface without erosion.

Figure 10a,b shows changes in the arithmetic average roughness Ra over the 1.5 mm × 1.5 mm area with the number of cycle in the 100-cycle tests with 1 mm and 2 mm air gap, respectively. Samples D and E give larger Ra when compared with the other samples, which is the same tendency with results obtained with the surface roughness meter. Ra in Figure 10 is much larger than that in Figures 8 and 9. A possible reason is the difference in resolution of measuring instruments used. As the surface roughness meter is a stylus type apparatus, a fine probe needle is moved on a sample surface. Surface roughness is obtained based on up-and-down movement of the probe. Thus, it is considered difficult to measure precisely the depth of a narrow pitting smaller than the probe diameter, resulting in lower value of Ra. Meanwhile, LASER beam is used to evaluate surface roughness in the confocal microscope. Since the resolution is higher and a long and narrow pitting can be evaluated precisely, larger Ra is obtained.

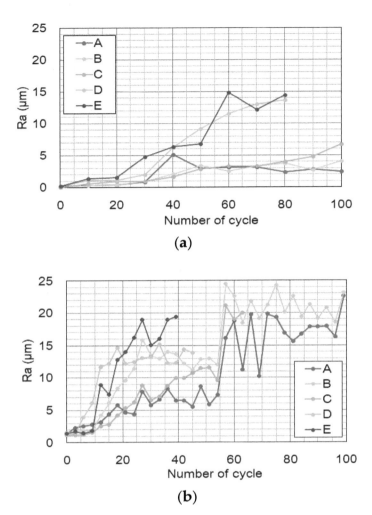

Figure 10. Change in Ra with number of cycle in the 100-cycle tests. (**a**) 1 mm air gap; (**b**) 2 mm air gap.

3.5. FTIR Spectrum

Change in FTIR spectrum with the number of cycle is shown in Figure 11 for all samples, which are obtained in the 100-cycle test with 2 mm air gap. Characteristics of Samples B and C are similar to each other. The same can be said of Samples D and E. These results correspond to progress of surface roughness described above.

In the case of sample A without ATH, absorbance around 786 and 1257 cm^{-1} attributed to hydrophobic Si-CH$_3$ [13] decreases with the number of cycle. Almost no change is observed in absorbance around 1008 cm^{-1} attributed to Si-O-Si. Increase of silica-related absorbance [13] is observed around 1060 cm^{-1} with increase in the number of cycle. A decrease in absorbance of hydrophobic Si-CH$_3$ and almost no change in absorbance of Si-O-Si, shown in Figure 12a, suggest that a side chain of silicone rubber is cleaved by partial discharge but main chain is damaged little, leading to limited erosion.

In sample E, eroded more by partial discharge compared with sample A, absorbance around 3620–3370 cm^{-1} vanishes when the number of cycle is large, which is related to ATH [13]. At the same time, absorbance around 940 cm^{-1} increases, which is attributed to Si-H [13]. It is suggested that ATH changes into aluminum hydroxide by losing hydroxyl groups. Not only absorbance of hydrophobic group Si-CH$_3$ but also that of Si-O-Si of main chain of silicone rubber decrease rather rapidly with the number of cycle as shown in Figure 12b, suggesting progress of silicone rubber decomposition and short lifetime to breakdown.

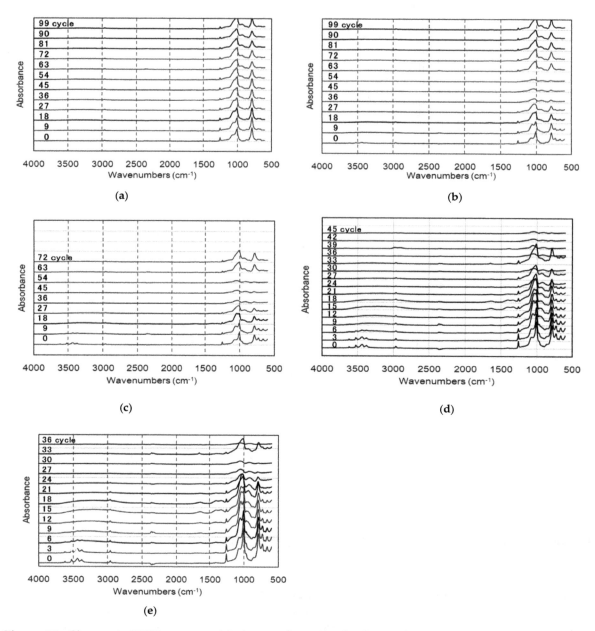

Figure 11. Change in FTIR spectra with the number of cycle obtained in the 100-cycle test with 2 mm air gap. (**a**) Sample A; (**b**) Sample B; (**c**) Sample C; (**d**) Sample D; (**e**) Sample E.

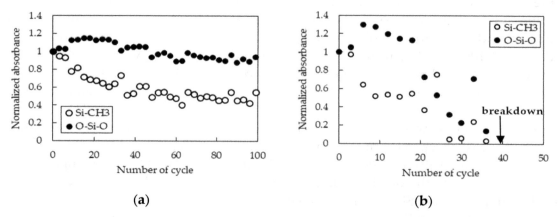

Figure 12. Change in normalized absorption of Si-CH$_3$ and O-Si-O with the number of cycle obtained in the 100-cycle test with 2 mm air gap. (**a**) Sample A; (**b**) Sample E.

4. Discussion

The proposed experimental procedures containing a cycle of 8-h voltage application and 16-h interruption seem suitable to evaluate partial discharge degradation of hydrophobic material like silicone rubber because hydrophobicity recovers almost to the initial value during 16 h. It reflects a practical usage condition of polymer insulators that hydrophobicity recovers if factors affecting hydrophobicity are removed.

Consistent results are obtained in 50- and 100-cycle tests. In both tests, surface erosion is larger in the order of samples D and E, samples B and C, and sample A. Progress of partial discharge degradation of silicone rubber, especially samples D and E, is accelerated in the case of 2 mm gap spacing compared with that of 1 mm in both tests.

Figure 13 shows change in Ra of five kinds of samples up to 50 cycles measured with the surface roughness meter, which are obtained by carrying out experiments 3 times under the condition of 1 mm air gap. Each sample shows almost the same performance in any experiment. Also Ra is larger in the order of samples D and E, samples B and C, and sample A in any experiment. It is considered the proposed experimental procedures give acceptable repeatability of results.

Surface erosion of a sample increases gradually by partial discharge with the number of cycle. In samples D and E, decrease in hydrophobic $Si\text{-}CH_3$ and $Si\text{-}O\text{-}Si$ of main chain is confirmed by FTIR analysis. This is a possible reason for low contact angle, sever surface erosion, and consequently short cycles to breakdown. Samples D and E contain ATH, which is considered effective to enhance resistance to tracking and erosion by discharge because heat is absorbed by releasing the water of hydration from ATH molecule when the temperature of an ATH filled polymer reaches ~200 degrees [1,14]. This role of ATH seems effective for partial discharge because decrease in ATH is suggested through FTIR analysis in the present study. Nevertheless, surface erosion of samples D and E containing ATH is serious compared with sample A without ATH. This is attributed to intensified partial discharge activities in the air gap for samples D and E because their higher permittivity generates higher localized electric field in the air gap of the electrode system. Lower resistivity of samples D and E may also relate to shorter lifetime to breakdown.

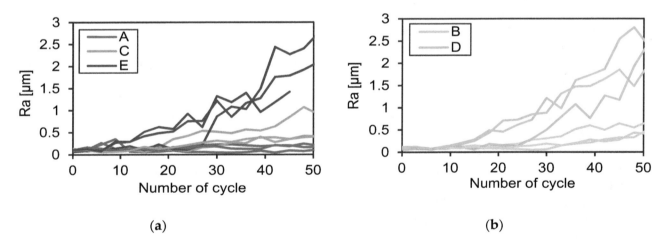

(a) (b)

Figure 13. Repeatability of change in Ra with number of cycle obtained for 1 mm gap spacing. (a) samples A, C, and E; (b) samples B and D.

5. Conclusions

The proposed experimental procedures of partial discharge degradation of silicone rubber are evaluated by experimental results using five kinds of silicone rubber samples. Also the difference in resistance to partial discharge among samples are discussed. The results are summarized as follows.

1. The procedures proposed in this study is considered suitable for investigation of partial discharge degradation of hydrophobic material like silicone rubber. It has advantages of no arc discharge

occurrence, recovery of hydrophobicity during the resting time, acceptable repeatability of results, and possible acceleration of degradation by adjusting gap spacing and applied voltage.

2. It is confirmed that surface erosion of silicone rubber progresses gradually and finally breakdown occurs only by partial discharge, though instantaneous damage is smaller compared with arc discharge. Partial discharge is one of the important factors in the long-term electrical insulation design of polymer insulators.

3. It is suggested ATH is not necessarily effective to prevent breakdown of silicone rubber by partial discharge when localized electric field in air gap is enhanced by adding ATH. In such a situation, lower permittivity and higher resistance of silicone rubber seem more dominant factors to prevent damage by partial discharge and a careful insulation design should be required.

4. Considering the actual usage condition of polymer insulators, applied voltage is fixed in the present study. Quantification of partial discharge activities and discussion of erosion from this standpoint are also important. Furthermore, degradation phenomena when discharge is kept constant across different sample is another important topic. The authors would like to investigate them in the near future.

Author Contributions: Conceptualization, K.K. and Y.M.; Methodology, K.K., H.L. and M.S.; Validation, K.K.; Formal Analysis, K.K., H.L. and M.S.; Investigation, K.K. and H.L.; Resources, Y.M.; Data Curation, K.K.; Writing-Original Draft Preparation, K.K.; Writing-Review & Editing, Y.M.; Visualization, K.K. and M.S.; Supervision, Y.M.; Project Administration, K.K. and Y.M.; Funding Acquisition, Y.M.

Acknowledgments: The silicone rubber samples were supplied in connection with activities of an Investigating R & D Committee of the Institute of Electrical Engineers of Japan. The authors would like to thank the members of the committee.

References

1. Kumagai, S.; Yoshimura, N. Tracking and Erosion of HTV Silicone Rubber and Suppression Mechanism of ATH. *IEEE Trans. Dilectr. Electr. Insul.* **2001**, *8*, 203–211. [CrossRef]

2. Aeshad; Nekahi, A.; McMeekin, S.G.; Farzaneh, M. Flashover Characteristics of Silicone Rubber Sheets under Various Environmental Conditions. *Energies* **2016**, *9*, 683. [CrossRef]

3. Nazemi, M.H.; Hinrichsen, V. Experimental Investigations on Water Droplet Oscillation and Partial Discharge Inception Voltage on Polymeric Insulating Surface under the Influence of AC Electric Field Stress. *IEEE Trans. Dielectr. Electr. Insul.* **2013**, *20*, 443–453. [CrossRef]

4. Sarathi, R.; Mishra, P.; Gautam, R.; Vinu, R. Understanding the Influence of Water Droplet Initiated Discharges in Damage Caused to Corona-Aged Silicone Rubber. *IEEE Trans. Dielectr. Electr. Insul.* **2017**, *24*, 2421–2431. [CrossRef]

5. Nazir, M.T.; Phung, B.T.; Hoffman, M. Performance of Silicone Rubber Composites with SiO_2 Micro/Nano-filler under AC Corona Discharge. *IEEE Trans. Dielectr. Electr. Insul.* **2016**, *23*, 2084–2815. [CrossRef]

6. Habas, J.P.; Arrouy, J.M.; Perrot, F. Effect of Electric Partial Discharges on the Rheological and Chemical Properties of Polymers Used in HV Composite Insulators after Railway Service. *IEEE Trans. Dielectr. Electr. Insul.* **2009**, *16*, 1444–1454. [CrossRef]

7. Gubanski, S.M. Outdoor Polymeric Insulators: Role of Corona in Performance of Silicone Rubber Housings. In Proceedings of the IEEE International Conference on Electrical Insulation and Dielectric Phenomena (CEIDP), Ann Arbor, MI, USA, 18–21 October 2015; pp. 1–9.

8. Ueda, F.; Oue, K.; Mizuno, Y. Degradation of Silicone Rubber Caused by Partial Discharge in Clean Fog. In Proceedings of the 11th International Conference on the Properties and Applications of Dielectric Materials, Sydney, Australia, 11–22 July 2015. Paper No. AL0-5.

9. Ueda, F.; Tamamura, M.; Mizuno, Y. Degradation of Silicone Rubber Caused by Partial Discharge in Clean Fog and in Atmosphere. In Proceedings of the Conference on Electrical Insulation and Dielectric Phenomena, Toronto, ON, Canada, 16–19 October 2016. Paper No. 8A-12.

10. Komatsu, K.; Shimada, M.; Mizuno, Y. Interaction of Partial Discharge in Air with Silicone Rubber. In Proceedings of the International Conference on High Voltage Engineering, Athens, Greece, 10–13 September 2018. Paper No. P-AM-4.

11. Hikita, M.; Hishikawa, S.; Kondo, T.; Yaji, K. Surface Modification and Performance of Polymer Composite Materials for Outdoor Insulation. In Proceedings of the Annual Meeting of IEEJ, Nagoya, Japan, 20–22 March 2013. Paper No. 2-S1-4.

12. Surface Roughness Terminology and Parameters. Available online: https://www.predev.com/pdffiles/surface_roughness_terminology_and_parameters.pdf (accessed on 18 June 2019).

13. Gao, Y.; Wang, J.; Liang, X.; Yan, Z.; Liu, Y.; Cai, Y. Investigation on Permeation Properties of Liquids into HTV Silicone Rubber Materials. *IEEE Trans. Dielectr. Electr. Insul.* **2014**, *21*, 2428–2437. [CrossRef]

14. Meyer, L.; Grishko, V.; Jayaram, D.; Cherney, E.; Duley, W.W. Thermal Characteristics of Silicone Rubber Filled with ATH and Silica under Laser Heating. In Proceedings of the Conference on Electrical Insulation and Dielectric Phenomena, Cancun, Mexico, 20–24 October 2002; pp. 848–852.

9

Thermal Effect of Different Laying Modes on Cross-Linked Polyethylene (XLPE) Insulation and a New Estimation on Cable Ampacity

WenWei Zhu [1,2,†], YiFeng Zhao [1,†], ZhuoZhan Han [1], XiangBing Wang [2], YanFeng Wang [2], Gang Liu [1,*], Yue Xie [1,*] and NingXi Zhu [1]

[1]   School of Electric Power, South China University of Technology, Guangzhou 510640, China
[2]   Grid Planning Research Center, Guangdong Power Grid Co., Ltd., Guangzhou 510000, China
*   Correspondence: liugang@scut.edu.cn (G.L.); epxieyue@mail.scut.edu.cn (Y.X.)
†   These authors contributed equally to this work.

Abstract: This paper verifies the fluctuation on thermal parameters and ampacity of the high-voltage cross-linked polyethylene (XLPE) cables with different insulation conditions and describes the results of a thermal aging experiment on the XLPE insulation with different operating years in different laying modes guided by Comsol Multiphysics modeling software. The thermal parameters of the cables applied on the models are detected by thermal parameter detection control platform and differential scanning calorimetry (DSC) measurement to assure the effectivity of the simulation. Several diagnostic measurements including Fourier infrared spectroscopy (FTIR), DSC, X-ray diffraction (XRD), and breakdown field strength were conducted on the treated and untreated specimens in order to reveal the changes of properties and the relationship between the thermal effect and the cable ampacity. Moreover, a new estimation on cable ampacity from the perspective on XLPE insulation itself has been proposed in this paper, which is also a possible way to judge the insulation condition of the cable with specific aging degree in specific laying mode for a period of time.

Keywords: thermal effect; cable; XLPE; laying modes; Comsol Multiphysics; thermal parameters; cable ampacity

1. Introduction

Thermoplastics are widely used in the insulation of power cables, such as polyvinyl chloride (PVC), low density polyethylene (LDPE), and cross-linked polyethylene (XLPE), which possess higher physical, electrical, and heat-resistant properties compared with oil-paper insulation [1]. The PVC insulation is commonly applied in low-voltage and medium-voltage cables, of which the permissible long-term load temperature is 70 °C [2]; the LDPE power cables can be operated at higher voltage levels by adding special additives [3]. In addition, the permissible long-term load temperature is 75 °C and the permissible maximum short-circuit temperature is 130 °C. Due to the crosslinking treatment of PE, the permissible long-term load temperature and the permissible maximum short-circuit temperature have been elevated to 90 and 250 °C, respectively [4]. The XLPE cables have been widely applied in high-voltage (HV) and extra-high voltage (EHV) transmission systems due to the its high performance on electricity and thermal properties. Therefore, it is meaningful to expand the research frontiers about the XLPE cables in the actual operation to elevate the permissible long-term load capacity and extend their service life.

XLPE is a high-molecular polymer with crystal and amorphous phase, which is subjected mostly to the thermal effect because the crystal structure of XLPE is prone to be affected by the melting and cooling process [5]. It is well evidenced that long-term thermal effect can change the morphology of the insulation distinctly and lead to degradation on the insulation [6,7]. In fact, the influence of thermal effect on polymers is a complicated process and the insulation properties can be improved in some certain situations [8]. Much research has been carried out on this subject. The inverse temperature effect is presented to indicate that the degradation of semi-crystalline polymers is obvious at the low temperatures and a significant recovery at elevated temperatures [9]. In the study about the non-isothermal melt-crystallization kinetics of polymers [10,11], it is well known that the ability of XLPE to crystallize is highly dominated by the cooling rate and the current melting condition of XLPE, because different melting and cooling rates can change the form of spherulites, the size and distribution level of which determine some specific conductive properties of XLPE [12,13]. Therefore, it is meaningful to investigate the changes of properties on XLPE with different melting and cooling rates.

During its practical operation, cable ampacity is influenced by many factors, such as the kind of laying modes and cable specifications [14,15]. Although the factor of the laying mode has been taken into account in International Electrotechnical Commission (IEC) and Institute of Electrical and Electronics Engineers (IEEE) standards in the calculation of cable ampacity, the influence of cable aging on the cable ampacity was rarely concerned. It is universal to calculate the XLPE cable ampacity by the thermal-circuit method and simulation software, which follows the basic rule of calculating the maximum current through the conductor corresponding to the steady-state temperature of 90 °C in different environments [16]. The precondition of the cable ampacity calculation is setting the volumetric thermal capacity and thermal resistance of XLPE as fixed values [17–19]. To a large extent, these methods neglect the changes on XLPE insulation itself caused by the different aging factors and the influence of the heating and cooling process to the material during the cable operating condition, which would totally change the morphology of XLPE [20–22]. It can be inferred that the thermal resistance and thermal capacity of XLPE in the thermal-circuit model are variable for the reason that the microstructure of XLPE is changing with many factors, such as heat and electricity. Moreover, papers [23,24] have found that the measured values of thermal resistance and thermal capacity of XLPE are different distinctly from the IEC standard. Further, it can be considered that the cable ampacity is a fluctuating value rather than a constant as to the same specification XLPE cable in the same environment with the passage of time. Therefore, it is meaningful to pursue a new way to assess the cable ampacity.

This paper has focused on two aspects of the XLPE insulation in order to reveal the changes on XLPE properties and the relationship between thermal effect and cable ampacity. The first aspect concerns the verification of fluctuation on thermal parameters and ampacity of the cable with different insulation conditions. The second aspect provides a new estimation on cable ampacity from the perspective of the changes of XLPE insulation condition, considering the factors of the different operating years and the different laying modes of the cables. Such correlations between the cable insulation and the cable ampacity should be researched further for a better understanding of the cable ampacity and the aging mechanism of XLPE.

2. Experimental

2.1. Preparation of XLPE Specimens

Two retired high-voltage AC XLPE cables with service years of 15 and 30, and a spare high-voltage AC XLPE cable were selected in this paper. For convenience, they are named by their service year: XLPE-0, XLPE-15, and XLPE-30. Some critical parameters of these cables are listed in Table 1.

Overheated operation has not been reported for these two retired cables, which means that the temperature in the insulation layer remained below 90 °C during the cable operation. Each cable insulation was peeled parallel to the conductor surface, and the tape-like XLPE peels were obtained. Peels near the inner semi-conductive layer were taken as the specimens, because these positions of the insulation endured the most severe electrical and thermal stresses. These obtained specimens were all cleaned by alcohol to remove the surface impurities.

Table 1. Critical parameters of the cables.

Cable	V_L	M_I	M_C	d_I	O_P
XLPE-0	110/63.5	XLPE	Cu	18.5	1998–
XLPE-15	110/63.5	XLPE	Cu	18.5	1999–2015
XLPE-30	110/63.5	XLPE	Cu	18.5	1985–2015

V_L—voltage level in kV, M_I—insulation material, M_C—conductor material, d_I—insulation thickness in mm, O_P—operation period. Other parameters such as cross-linking method are unknown.

2.2. Cable Thermal Parameters and Ampacity Measurement

By means of the high-voltage cable thermal parameter detection control platform, as illustrated in Figure 1, the three cables with the length of 1.5 m were taken to test its ampacity under the air-laying mode. The critical thermal parameters and ampacity of these three cables are listed in Table 2, where volumetric thermal capacity δ of XLPE was measured by differential scanning calorimetry (DSC) at the temperature of 30 °C; thermal resistance R of XLPE and thermal conductivity λ of XLPE were deduced by thermal circuit model and its Matlab program, and results are shown in the Appendix A; ambient temperature θ_O and conductor steady-state temperature θ_C were measured by thermocouples, and the ampacity under the air-laying mode I_A was tracked by the Matlab program.

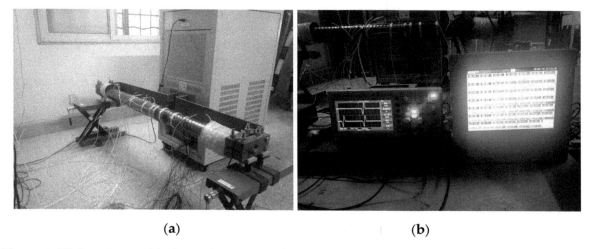

(a)   (b)

Figure 1. High-voltage cable thermal parameter detection control platform. (**a**) Ampacity experimental cable and AC constant current source, (**b**) thermal compensation system and paperless recorder.

Table 2. Critical thermal parameters and ampacity of the cables.

Cable	δ (J/K·m³)	R (K·m/W)	λ (W/K·m)	θ_O (°C)	θ_C (°C)	I_A (A)
XLPE-0	1.94×10^6	3.23	0.31	17.7	89.6	1244.07
XLPE-15	1.96×10^6	3.45	0.29	17.7	89.7	1222.99
XLPE-30	2.07×10^6	5.26	0.19	17.7	90.2	1096.47

δ—volumetric thermal capacity of cross-linked polyethylene (XLPE), R—thermal resistance of XLPE, λ—thermal conductivity of XLPE, θ_O—ambient temperature, θ_C—conductor steady-state temperature and I_A—ampacity under the air-laying mode.

From the results in Table 2, it can be verified that the thermal parameters and ampacity fluctuated with different conditions (thermal history, operating years or material characteristic) of the XLPE insulation. Therefore, it is meaningful to provide new estimations on cable ampacity to improve the current research on cable ampacity.

The structural model and thermal parameters in Table 2 of these cables with the same specification is the foundation of the subsequent analysis on simulation of different laying modes based on Comsol Multiphysics. The configuration of the experimental cables is shown in Figure 2.

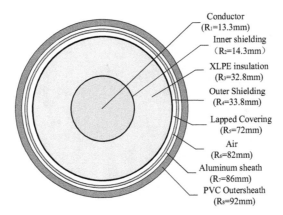

Figure 2. Configuration of the ampacity experimental cable and structural model used for Comsol Multiphysics. R_x is the corresponding radius of each layer.

2.3. Simulation of Different Laying Modes of Cables

Comsol Multiphysics was adopted to simulate different laying modes of the XLPE cables. The thermal parameters of the three cables listed in Table 2 were extracted to construct the simulation model, respectively, and we found that the simulation results had the same changing tendency. The permanents of XLPE-0 were used to construct the following simulation model. The establishment of thermal field model and determination of the boundary conditions are not the highlight in this paper, the detail can be referred to literature [25–27].

Firstly, we constructed the structural model (Figure 2) under the air-laying mode on the Comsol Multiphysics and we found that the simulative ampacity approached to the practical experiment, which was 1152 A. Therefore, it is assured that the following simulative results are effective.

Secondly, we defined cable ampacity under air-laying mode as reference cable ampacity I_R ($I_R = 1200$ A) and constructed the structural models into three different laying modes, including pipeline, ground, and tunnel-laying modes based on Comsol Multiphysics, as shown in Figure 3.

Thirdly, we applied the 1.2 I_R and the 0 I_R into the simulated cables, which were laid in pipeline, underground, and in the tunnel, respectively, and obtained the simulative melting and cooling curves of three different laying modes that are shown in Figure 4a,b.

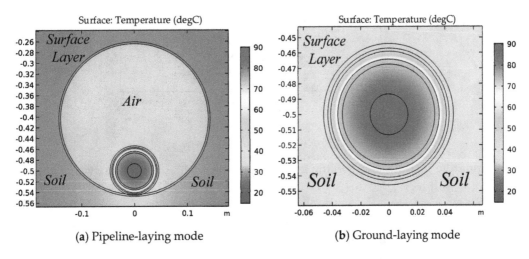

(a) Pipeline-laying mode   (b) Ground-laying mode

Figure 3. *Cont.*

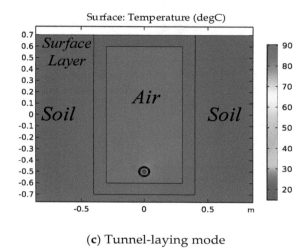

(c) Tunnel-laying mode

Figure 3. Schematic diagrams of the structural models of three different laying modes based on Comsol Multiphysics.

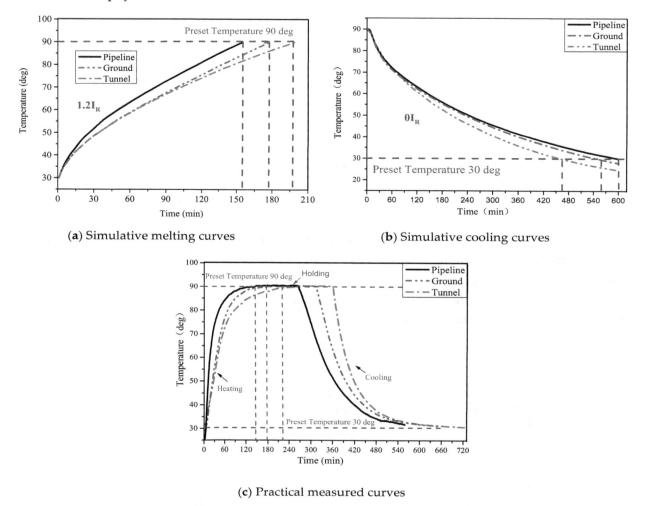

(a) Simulative melting curves

(b) Simulative cooling curves

(c) Practical measured curves

Figure 4. Simulative melting and cooling curves and practical measured curves of three different laying modes.

2.4. Thermal Aging

The cable specimens with different operation years were formed into a group, and there were four groups in total. The first group was the untreated group which played a role of reference. The second group simulated the cable laying in the pipeline with the lowest heat dissipation efficiency. The third group simulated the cable laying under the ground with the moderate heat dissipation efficiency. The last group simulated the cable laying in the tunnel with the highest heat dissipation efficiency.

Thermal aging experiments were performed on the last three groups of the specimens. Three modified aging ovens were used to control the heating and cooling process, which fitted the corresponding simulative curves, and a 2 h constant temperature phase of 90 °C between the heating and cooling phases were added to the thermal aging. The three phases, including heating, holding, and cooling, were repeated 80 times in order to enlarge the differences among the specimens. The practical temperature curves of the three different laying modes are shown in Figure 4c, measured by the thermo-couple sensors.

2.5. Diagnostic Measurements

The Fourier transform infrared spectroscopy (FTIR), the differential scanning calorimetry (DSC) measurement, X-ray diffraction (XRD), and the breakdown field strength measurement were adopted to analyze the properties changes on the specimens after the thermal effects.

Micro-structure changes on the specimens were analyzed by the VERTEX 70 infrared spectrometer manufactured by German. Each specimen was tested at 32 scans in the range of 600~3600 cm^{-1} with a resolution of 0.16 cm^{-1} and the signal-to-noise ratio of 55,000:1. The obtained spectra were analyzed by OPUS software.

Thermal properties changes on the specimens were analyzed by the DSC NETZSCH-DSC 214 instrument manufactured by German. Five milligram specimens were prepared to test, with the program of two heating phases and a cooling phase under nitrogen atmosphere to avoid thermal degradation. The temperature was increased from 25 to 140 °C at a rate of 10 °C/min and maintained at 140 °C for 5 min, and then cooled to 25 °C. This scanning was repeated twice per measurement, and the first cooling and second heating phases were analyzed in this paper.

Crystal structure changes on the specimens were analyzed by Bruker D8 ADVANCE X-ray diffractometer manufactured by Germany. The experimental interval of Bragg angle was $2\theta = 5°-90°$ by step size of 0.02° with 0.1 s/step scan rate. The obtained data were analyzed by software of DIFFIAC plus XRD Commander.

Electric properties changes of the specimens were analyzed by the ZJC-100 kV voltage breakdown tester manufactured by China. Each specimen was cut into $50 \times 50 \times 0.5$ mm to test, with the voltage rising rate of 1 kV/s under the transformer oil. The valid AC breakdown field strength of each specimen was measured 5 times to obtain the average breakdown field strength.

3. Results and Discussions

In the process of the thermal effects, the features of each laying mode from the heating and cooling curves were observed in the results of Figure 4. Among the three laying modes, the pipeline mode possesses the quick heating and slow cooling features, the ground mode possesses the moderate heating and slow cooling features, and the tunnel mode possesses the slow heating and quick cooling features. These differences led to the different statuses among the specimens with different operating years. In the following parts, we will focus on the connection among the changes of micro-structure, crystal structure, and external electrical property of each specimen in order to reveal the relationship between the thermal effects and the cable ampacity.

3.1. Result of FTIR Spectroscopy Measurement

The oxidation process of XLPE under the thermal effects can generate major by-products of thermal-oxidation, such as carbonyl groups and unsaturated groups, which can signify the aging status of the specimens. Figure 5 is the FTIR spectra of each specimen whose wavenumber and absorbance represent the kind and content of the corresponding group. The wavelengths of 720, 1471, 2856, and 2937 cm^{-1} are all caused by the vibration of the methylene band (-CH$_2$-). Absorption peaks ranging from 1700 cm^{-1} to 1800 cm^{-1} can be considered as the thermo-oxidative products [3]. Among of them, carboxylic acid absorption appears at 1701 cm^{-1}, ketone absorption locates at 1718 cm^{-1}, and aldehyde absorption situates at 1741 cm^{-1}. The peak at 1635 cm^{-1} is assigned to the unsaturated groups absorption, which can indicate the decomposition process.

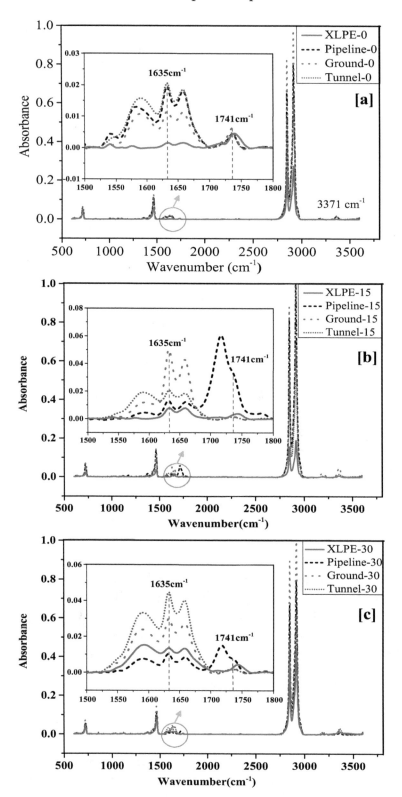

Figure 5. Fourier transform infrared spectroscopy (FTIR) spectra of the untreated specimens and the treated specimens under thermal aging of three different laying modes. (**a**) the spare cable; (**b**) the cable with service years of 15; (**c**) the cable with service years of 30.

From Figure 5a, it is clear that some peaks in the range of 1500 to 1700 cm^{-1} are present after the thermal effects. These peaks reflect a slight decomposition process on the backbone of the XLPE macromolecules. This phenomenon can be associated with the thermal activation to the XLPE molecular chain, which, as a consequence of the adequate motion of the molecular chain, leads to the generation of a certain quantity of small chain segments, free polar groups, etc. On the other aspect, there are no excessive displacement in the position of the peaks in the range of 1700 to 1800 cm^{-1} among the specimens after thermal aging, which signifies the process of thermo-oxidation is moderate. Therefore, it can be considered that the different thermal aging modes mainly have activated the motion of macromolecular chains, but hardly cause the oxidative degradation in regard to the spare cable.

From Figure 5b, we can see the similar phenomenon happening on the specimens except for Pipeline-15. Significant peaks in the range of 1700 to 1800 cm^{-1} are presented in Pipeline-15, which is responsible for the aggravation of oxidative degradation. Moreover, there was a pronounced increase in peaks in the range of 1600 to 1650 cm^{-1} in Ground-15, which indicates a dominant process of chains scission occurs under the thermal effect of ground mode.

From Figure 5c, we can notice that quantities of small molecular chains have emerged on the untreated one (XLPE-30) due to the long-term operation in actual condition. After the thermal aging with different laying modes, it can be deduced that the thermal effects have sped up the decomposition process on the molecular backbone of Ground-0 and Tunnel-0, and a certain quantity of broken molecular chains have already transformed into oxidative groups in Pipeline-30. This indicates severe oxidative degradation occurs in Pipeline-30. It may be admitted that the potential to resist oxidation of XLPE-30 was weakened compared with XLPE-0.

In order to quantify the situation of oxidation and decomposition under the thermal effects, carbonyl index and unsaturated band index were chosen for research. The definition of these two indexes are as follows [3]:

$$CI = I_{1741}/I_{1471}, \tag{1}$$

$$UBI = I_{1635}/I_{1471}, \tag{2}$$

where carbonyl index (CI) is the relative intensities of the carbonyl band at 1741 cm^{-1} (aldehyde absorption) to the methylene band at 1471 cm^{-1}; unsaturated band index (UBI) is the relative intensities of the unsaturated group at 1635 cm^{-1} to the methylene band at 1471 cm^{-1}.

The two indexes are shown in Figure 6. It can be stated that the carbonyl index is decreased in all the specimens compared with the untreated ones, except for Pipeline-15 and Pipeline-30, and the unsaturated band index is increased except for Tunnel-15, Pipeline-15, and Pipeline-30. That means the thermal effects aggravate basically the decomposition process of the specimens to generate a certain quantity of broken molecular chains. The oxidation process was followed by chains scission and formation of smaller chain segments, which would easily react with oxygen (O_2) to transform into oxidative groups. For Pipeline-15 and Pipeline-30, the broken molecular chain segments are prone to transform into oxidative groups, probably due to the heating and cooling features of pipeline mode. With regard to Tunnel-15, the two indexes are decreased, which symbolizes the optimization of the micro-structure. As we demonstrated above, the status of the degradation and the stability of the specimens after thermal effects are listed in Table 3.

Table 3. Status of the degradation and stability of the specimens after thermal effects.

Specimen	Pipeline	Ground	Tunnel
XLPE-0	Improved	Improved	Improved
XLPE-15	Degraded	Maintained	Improved
XLPE-30	Degraded	Improved	Maintained

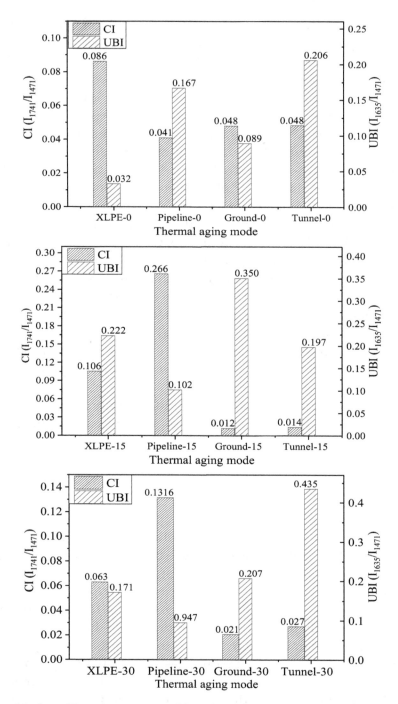

Figure 6. Carbonyl index, *CI*, and unsaturated band index, *UBI*, of the untreated specimens and the treated specimens under thermal aging of three different laying modes.

3.2. Result of DSC Measurement

With the program running, two consecutive scans were conducted on each specimen. The first scan began by heating from 30 to 140 °C and then followed by cooling from 140 to 30 °C. The second scan was excerpted in the same way as the first scan. The thermos-gram of the first heating phase is commonly used for observing the thermal history and the current crystalline condition of the specimens, which can also be analyzed by XRD measurement in Section 3.3. In order to avoid long arguments, two phases were analyzed, including the first cooling and the second heating phase, which can be analyzed on the ability to crystallize and the quality of the crystal structure of the specimens after the thermal effects.

Figures 7 and 8 show the thermos-grams of the two phases corresponding to the operating years of each cable. The obtained parameters are listed in Table 4, where T_c is the crystallizing peak temperature, ΔH_c is the enthalpy of crystallization, T_m is the melting peak temperature, and $\Delta T = T_m - T_c$ is the degree of supercooling, which is proportional inversely to the crystallization rate [22].

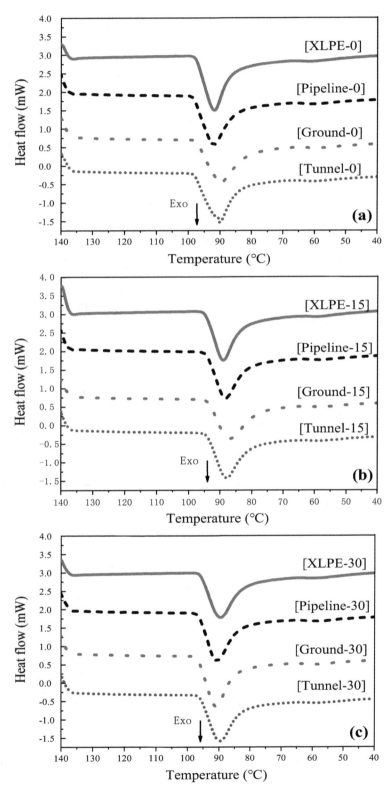

Figure 7. Heat flow as a function of measurement temperature in the first cooling of differential scanning calorimetry (DSC). (**a**) the spare cable; (**b**) the cable with service years of 15; (**c**) the cable with service years of 30.

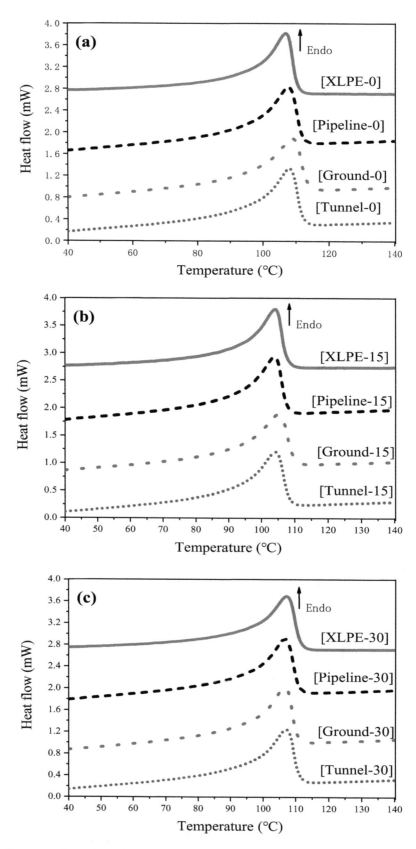

Figure 8. Heat flow as a function of measurement temperature in the second heating of differential scanning calorimetry (DSC). (**a**) the spare cable; (**b**) the cable with service years of 15; (**c**) the cable with service years of 30.

Table 4. Parameters obtained from first cooling and second heating phases.

Specimen	T_c (°C)	ΔH_c (J/g)	T_m (°C)	ΔH_f (J/g)	ΔT (°C)
XLPE-0	91.8	−99.0	106.9	100.8	15.1
Pipeline-0	92.0	−108.8	107.6	116.6	15.6
Ground-0	89.7	−102.8	109.0	111.1	19.3
Tunnel-0	90.0	−110.8	108.0	117.2	18.0
XLPE-15	89.0	−93.4	104.0	98.7	15.0
Pipeline-15	88.2	−99.6	103.8	108.1	15.6
Ground-15	86.9	−93.4	105.0	101.9	18.1
Tunnel-15	87.8	−99.1	104.1	101.1	16.3
XLPE-30	89.5	−103.9	107.2	103.4	17.7
Pipeline-30	90.6	−103.0	106.6	110.5	16.0
Ground-30	90.9	−105.1	106.6	111.3	15.7
Tunnel-30	89.8	−104.5	107.1	106.4	17.0

T_c—crystallizing peak temperature, ΔH_c—enthalpy of crystallization, T_m—melting peak temperature, ΔH_f—enthalpy of fusion and ΔT—super-cooling degree.

In Figure 7a, it is clearly observed that the exothermic peaks of Ground-0 and Tunnel-0 appear at a slight lower temperature and the shapes become broader, compared to XLPE-0 and Pipeline-0, in the first cooling phase. In Figure 8a, it can be found that melting peaks of all the treated specimens move towards higher temperatures, but the endothermic peaks are expanded. That means that although the thermal effects disperse the crystalline region, especially for Ground-0 and Tunnel-0, the main crystal structure of each treated specimen is improved.

In Figure 7b, exothermic peaks of all the treated specimens displace slightly toward lower temperatures. In Figure 8b, the melting peaks locate at higher temperatures for Ground-15 and Tunnel-15 but at lower temperature for Pipeline-15. The endothermic peaks become broader especially for Pipeline-15. It can be stated that the crystal structure becomes more deteriorated for a long time in the mode of the pipeline.

In Figure 7c or Figure 8c, the exothermic peaks of the treated specimens move towards a higher temperature and the endothermic peaks move towards a lower temperature, which means the thermal effects disrupt the distribution of crystal structure [28]. Therefore, the ability to crystallize is increased, but the quality is declined relatively. Among the specimens, Ground-30 presents the highest crystallization performance.

In addition, the DSC endotherms indicate a range of melting processes that can be related to the crystallinity and lamellar thickness variations. The crystallinity and average lamellar thickness [3,8] can be calculated by Formulas (1) and (2). The calculating formulas are as follows:

$$\chi(\%) = \Delta H_f / \Delta H_f^0 \times 100, \tag{3}$$

$$T_m = T_{m0}(1 - 2\sigma_e / \Delta H_m L), \tag{4}$$

where $\chi(\%)$ is crystallinity; ΔH_f^0 is the enthalpy of fusion of an ideal polyethylene crystal per unit volume; T_m is the observed melting temperature (K) of lamellar of thickness L; T_{m0} is the equilibrium melting temperature of an infinitely thick crystal; σ_e is the surface-free energy per unit area of basal face; ΔH_m is the enthalpy of fusion of an ideal polyethylene crystal per unit volume; and L is the lamellar thickness. The used values for calculation were as follows: $T_{m0} = 414.6$ K, $\Delta H_m = 2.88 \times 10^8$ J/m^3, and $\sigma_e = 93 \times 10^{-3}$ J/m^2. Changes on crystallinity and lamellar thickness with different thermal aging modes are depicted in Figure 9.

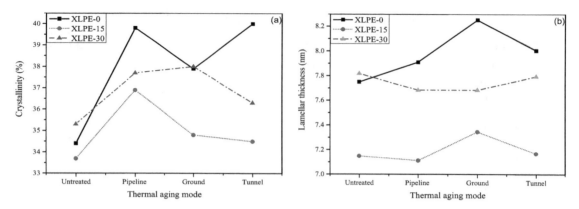

Figure 9. Changes on crystallinity and lamellar thickness of the main endothermic peak with different thermal aging modes: (**a**) Crystallinity; (**b**) lamellar of thickness.

We can observe that the crystallinity of all the specimens are increased in various degrees after the thermal aging in Figure 9a. This phenomenon is ascribed to adequate movement of the molecular chains and the generation of short chain segments, which are inclined to form the secondary crystal [3]. It is hard to judge the status of crystal structure just from the crystallinity. Therefore, the analysis should be combined with the change of lamellar of thickness obtained at the main endothermic peak shown in Figure 9b. We can find that both Pipeline-15 and Pipeline-30 have a high crystallinity but the lamellar thickness of main endothermic peaks are low, which indicates the crystal structures are mainly made up of secondary crystal. On the contrary, the Ground-0 and the Ground-15 have a relatively low crystallinity with solid lamellar of thickness, which means the crystal structures are firm and compact. Analogously, we can deduce the status of crystal structure of each specimen as shown in Table 5.

Table 5. Status of the crystal structure of the specimens after thermal effects.

Specimen	Pipeline	Ground	Tunnel
XLPE-0	Improved	Improved	Improved
XLPE-15	Degraded	Improved	Improved
XLPE-30	Degraded	Improved	Maintained

3.3. Result of XRD Measurement

The changes on the crystal structure of each specimen before and after the thermal effect were analyzed by X-ray diffractometer, to observe the influence on the crystalline phase of each layer under the different aging conditions.

Figure 10 displays the X-ray spectrum of each position in cable insulation before and after the accelerated aging test. It can be observed that two main crystalline peaks of each specimen appear at $2\theta = 21.22°$ and $2\theta = 23.63°$, which correspond to the (110) and (200) lattice planes. There is a one small peak observed at $2\theta = 36.5°$ which corresponds to the (020) lattice plane as Miller demonstrates [29]. There is no excessive displacement in the position of the peaks or in their splitting among the specimens, but the intensity and the shape of the peaks are different. It is indicated that accelerated aging hardly produces any new crystalline phase in the crystal structure, but results in the changes on the crystallinity and the grain size of the specimens.

From Figure 10, it can be noticed remarkably that a turnover phenomenon occurs in the thermal aging of tunnel mode, which is reflected by the fact that the crystal of the (110) lattice plane modifies into the (200) lattice plane. It is commonly assumed that the (200) lattice plane corresponds to the deformed spherulites, whose optical axes are oriented parallel to the radial direction [29]. This transformation indicates the shape of spherulite is associated with the heating and cooling process under different thermal effects. Therefore, we may admit the crystal structure changes on XLPE should be in the light of the features of different laying modes. For more precision on the objective to analyze the crystal structure changes among the specimens, the crystallinity percentage and the grain size are introduced by Formulas (5) and (6).

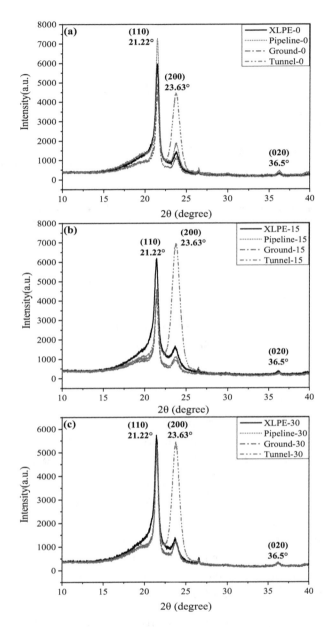

Figure 10. X-ray diffraction (XRD) spectrum of the untreated specimens and the treated specimens under thermal aging of three different laying modes. (**a**) the spare cable; (**b**) the cable with service years of 15; (**c**) the cable with service years of 30.

The crystallinity percentage can be calculated by the Hinrichsen method [30]. The X-ray spectrum of each specimen is fitted by Gaussian functions. Figure 11 displays the corresponding three Gauss fit peaks obtained by using the original 9.1. The calculating process is given as follows:

$$\chi(\%) = (S_2 + S_3)/(S_1 + S_2 + S_3) \times 100, \tag{5}$$

where χ (%) is crystallinity percentage, S_1 is the area of the amorphous halo, S_2 is the area of the main crystallization peak at $2\theta = 21.22°$, and S_3 is the area of the secondary crystallization peak at $2\theta = 23.63°$.

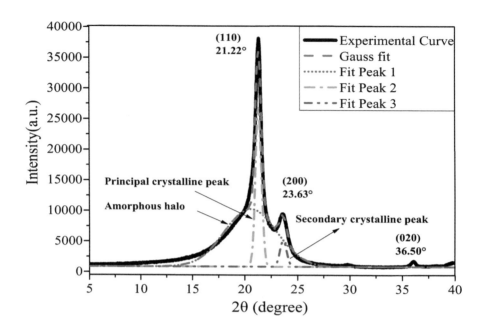

Figure 11. Gaussian fitting of the crystalline peaks and the amorphous halo by the Hinrichsen method.

In addition, the grain size of the different diffraction peaks corresponding to different crystal lamellar can be calculated by Scherrer equation [31]. The calculating formula is as follows:

$$D_{hkl} = (K \cdot \lambda_X)/(\beta \cos \theta), \tag{6}$$

where D_{hkl} is the grain size (A) perpendicular to the (hkl) crystal face; λ_X is experimental X-ray wavelength (nm), which is 0.15418 nm; β is the broadening of the diffraction peak (khl) (Rad) producing by grain refinement; and K is 0.89 when β is the full width at half maximum of the diffraction peak.

Basing on Formulas (5) and (6), we present the crystallinity percentage and the grain size perpendicular to the (110) and (200) crystal faces in Figure 12. With regard to XLPE-0 in Figure 12a, the crystallinity was lifted up in varying degrees by the disruption of the crystalline order after the thermal effect in each laying mode. Especially for Tunnel-0, a significant increase in crystallinity and distinct shrink in grain size perpendicular to the (200) crystal face indicate that the spherulites are tightly distributed in the insulation and the crystal structure is relatively perfect, even though most of the spherulites were deformed by thermal effect.

With regard to XLPE-15 in Figure 12b, it shows that the changes of the crystallinity and the grain sizes of Ground-15 and Tunnel-15 are similar to the corresponding ones of XLPE-0. The decrease in crystallinity and increase in crystallite size of the principal crystalline area for Pipeline-15 reflect that the spaces among the crystal structures are expanded and the integral crystal structure is not closely arranged.

With regard to XLPE-30 in Figure 12c, minor differences in the crystallinity and the grain sizes between Pipeline-30 and Ground-30 demonstrate that the influence of the features in pipeline and ground modes on the current crystalline condition is not pronounced. For Tunnel-30, we can also find high crystallinity with tight, deformed spherulites, as shown in Figure 12a,b. As we demonstrated above, the status of the current crystal structure of the specimens after thermal effects are listed in Table 6.

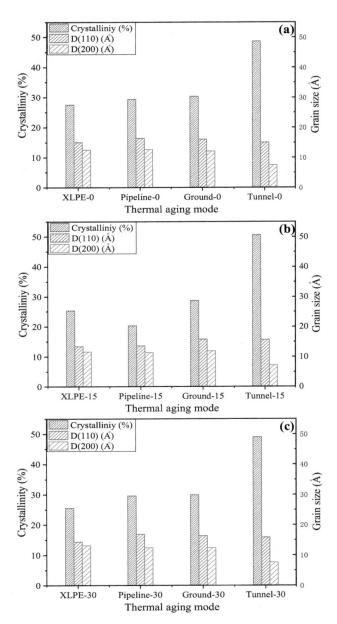

Figure 12. Changes on crystallinity and grain size perpendicular to the (110) and (200) crystal faces. (**a**) the spare cable; (**b**) the cable with service years of 15; (**c**) the cable with service years of 30.

Table 6. Status of the current crystal structure of the specimens after thermal effects.

Specimen	Pipeline	Ground	Tunnel
XLPE-0	Improved	Improved	Improved
XLPE-15	Degraded	Improved	Improved
XLPE-30	Maintained	Maintained	Improved

3.4. Results of the Breakdown Field Strength Measurement

The breakdown field strength test was conducted for each specimen and the valid average values are listed in Table 7. It is obvious that the electric properties were improved in varying degrees among XLPE-0 specimens after the different thermal effects. This phenomenon is probably attributed to the flexible movement of the molecular chain which activates the development of crystalline order and the annealing effect [6].

Table 7. Breakdown field strength (kV/mm).

Specimen	Untreated	Pipeline	Ground	Tunnel
XLPE-0	66.106	69.273	68.860	71.325
XLPE-15	73.489	63.941	75.640	81.650
XLPE-30	68.382	65.463	67.609	69.600

With regard to XLPE-15 and XLPE-30, the electric properties were improved under thermal effects in the ground- and tunnel-laying modes, but were degraded in the pipeline-laying mode. It may be considered that the features of the pipeline-laying mode are not relatively suitable for the aged cables with high loading for a long time.

3.5. Assessment of the Structural and Electrical Properties of XLPE under Thermal Effects

From the results of four diagnostic measurements, the correlations between structural and electrical properties were connected to assess the insulation condition.

For the spare cable under the three different thermal effects, the change of temperature activates the adequate movement of the molecular chain, leading to higher crystallinity and electrical performance of the treated specimens. The increase in crystallinity among the specimens is probably ascribed to the scission of the molecules traversing the amorphous regions, followed by rearrangement of the chains imperfectly crystallized at the manufacturing step to increase crystallinity [32]. Moreover, the improvement of crystal structure has a positive influence on the electrical breakdown and the resistivity of the insulation [33].

For the cable with service years of 15, thermal effect in tunnel mode presented a higher performance in structural and electrical properties, which can be identified by the diminution of oxygen-containing groups, improvement of crystalline morphology, and elevation of breakdown field strength. Properties indicators from the diagnostic measurements show a certain degradation happening on thermal effect in pipeline mode with high carbonyl content, inferior crystal structure, and drop in electrical breakdown.

For the cable with service years of 30, the flexibility of the molecular chain and the potential to recrystallize were weakened during long-term practical operation. Thermal effect in pipeline mode reflects degradation from the aspects of carbonyl content, crystal structure, and breakdown field strength compared with XLPE-30. A slight improvement of properties occurs on thermal effect in ground mode and maintaining of properties occurs on thermal effect in tunnel mode.

In this case, compared with the untreated ones, the status of the specimens under different thermal effects can be assessed comprehensively in Table 8.

Table 8. Comprehensive assessment on the specimens under the thermal effects.

Specimen	Pipeline	Ground	Tunnel
XLPE-0	Improved (greatly)	Improved (greatly)	Improved (greatly)
XLPE-15	Degraded (severely)	Maintained	Improved (greatly)
XLPE-30	Degraded (slightly)	Improved (slightly)	Maintained

3.6. A Proposal of New Estimation on Cable Ampacity

With the combination of the reference cable ampacity I_R and the results of diagnostic measurements, the relationship between the cable ampacity and the insulation properties was revealed according to the connection among the change of micro-structure, crystal structure, and external electrical property of each specimen. A new estimation on cable ampacity can be proposed from the perspective of changes of XLPE insulation properties to prolong the cables' service life. The specific regulation strategy of cable ampacity is as follows:

From the results of Table 8, the improvement and maintenance of overall insulation properties indicate that the margin of reference cable ampacity I_R can be lifted properly to 1.2 I_R. Analogously, the degradation of overall insulation properties indicates that the reference cable ampacity I_R should not be lifted to 1.2 I_R or should be decreased properly based on I_R.

4. Conclusions

The verification of fluctuation on thermal parameters and ampacity of the cable, with different insulation conditions and thermal effects of different laying modes, on XLPE insulation, with different operating years, was analyzed in this paper. Moreover, a new estimation on cable ampacity in regard to the specific aging status of XLPE cable with a specific laying mode was proposed. The following regulation strategy of cable ampacity based on I_R can be made to prolong the cables' service life from the perspective of XLPE insulation itself:

1. For the spare cable, it has a high potential to adapt to thermal effect of the three laying modes, so the margin of ampacity can be elevated to 1.2 I_R.

2. For the cable which had operated for 15 years, the margin of ampacity can be elevated to 1.2 I_R in the tunnel, should not be lifted to 1.2 I_R for a long time in ground mode, and should be decreased properly in the pipeline based on I_R.

3. For the cable which had operated for 30 years, the margin of ampacity can be elevated to 1.2 I_R in ground mode, should be reduced from 1.2 I_R in tunnel mode, but the margin of ampacity should not be elevated in the pipeline based on I_R.

In the future, more work will be devoted to assessing the relationship between the cable ampacity and XLPE insulation properties in the cable actual working condition. Establishing more effective methods on estimating cable ampacity and prolonging the cables service life are of great significance to the electric power industry.

Author Contributions: This paper is a result of the collaboration of all co-authors. G.L. and Y.X. conceived and designed the study. W.Z. and Y.Z. established the model and drafted the manuscript. Z.H. refined the language and provided statistical information. N.Z. helped with the corrections. X.W. and Y.W. designed and performed the experiments. All authors read and approved the final manuscript.

Acknowledgments: The authors gratefully acknowledge Guangdong Power Grid Co., Ltd. Grid Planning Research Center and this research is truly supported by the foundation: Technical Projects of China Southern Power Grid (No. GDKJXM20172797).

Appendix A

The Matlab program of calculating the ampacity under the air-laying mode of the cables is shown as follows (XLPE-0):

```
function y = XLPE-0()
global k dki dko Rk Ck Rk1 Ck1 l p i A B P T1u E To I h2 R12 number number1 T1 number2 C0 Tou r0p R11
load('XLPE-0.mat');                    %introduce the matrix (column 1: conductor temperature;
                                        column 2: current)
I = XLPE(:,2);                          %reading the current in the matrix
T1 = XLPE(:,1);                         %reading the temperature in the matrix
t1 = 0:60:54900;                        %calculating time range (total data volume in the matric)
hold on                                 %keeping the drawing interface
```

```
number = 30;                              %insulation hierarchical Number
number1 = number + 1;                     %lapped covering
number2 = number + 2;                     %outer-sheath
d1 = 26.6*10^-3;                          %conductor diameter
d2 = 65.6*10^-3;                          %insulation diameter
d6 = 92.0*10^-3;                          %outer-sheath diameter
h1 = (d2-d1)/(2*number);                  %insulation hierarchical thickness
h2 = 3.0*10^-3;                           %outer-sheath thickness
a = 0.00393;                              %resistance temperature coefficient of copper conductor
p1 = 3.23;                                %thermal resistance coefficient of insulation layer
p6 = 3.5;                                 %thermal resistance coefficient of outer-sheath
Dc = 344.312*10^4;                        %volumetric thermal capacity of Cu
DPE = 194.12*10^4;                        %volumetric thermal capacity of XLPE
D6 = 242.54*10^4;                         %volumetric thermal capacity of outer-sheath
r0p = 3.482e-5;                           %DC Resistance of Conductor at 20 °C
R11 = 0.1095;                             %thermal resistance of lapped covering
dki = zeros(1,number);                    %inner diameter array of each insulation layer
dko = zeros(1,number);                    %outer diameter array of each insulation layer
Rk = zeros(1,number);                     %thermal resistance array of each insulation layer
Ck = zeros(1,number);                     %thermal capacity array of each insulation layer
Rk1 = zeros(1,number2);                   %thermal resistance array
Ck1 = zeros(1,number2);                   %thermal capacity array
A = zeros(number2,number2);               %A matrix
B = zeros(number2,number2);               %A matrix
P = zeros(number2,1);                     %P matrix
E = zeros(1,number2);                     %temperature initial matrix
T1c = zeros(1,915);                       %calculating point number
R12 = 0.0299;                             %outer-sheath thermal resistance
C0 = Dc*pi*0.25*d1*d1;                        %conductor thermal capacity
C12 = D6*pi*0.25*(d6^2-(d6-2*h2)^2);          % outer-sheath thermal capacity
for k = 1:number
    dki(k) = d1 + 2*h1*(k-1);
    dko(k) = d1 + 2*h1*k;
    Rk(k) = p1/(2*pi)*log(1 +
2*h1/dki(k));                              %thermal resistance of each insulation layer
Ck(k) = DPE*pi*0.25*(dko(k)^2-dki(k)^2);      %thermal capacity of each insulation layer
    end
C11 = 5549.4;                             %lapped covering thermal capacity
for n = 1:number2                         %array of initial cable internal temperature
E(n) = 17.7;
    end
    T1u = 17.7;                           %initial conductor temperature
    for u = 0:60:54900;                   %time horizon
    for l = 1:number2
        if(l==1)
            Rk1(l) = Rk(l);
            Ck1(l) = Ck(l) + C0;
                else if(l==number1)
            Rk1(l) = R11;
            Ck1(l) = C11;
                else if(l==number2)
```

```
        R_k1(l) = R12;
        C_k1(l) = C12;
    else
        R_k1(l) = Rk(l);
        C_k1(l) = Ck(l);
    end
end
for i = 1:number2
for j = 1:number2
if(j==i)
    if(j==1)
        A(i,j) = -(C_k1(i)*R_k1(i))^-1;B(i,j)=C_k1(i)^-1;
    else
            A(i,j) = -C_k1(i)^-1*(R_k1(i-1)^-1+R_k1(i)^-1);B(i,j)=C_k1(i)^-1;
    end
else if(j==I + 1)
        A(i,j) = (C_k1(i)*R_k1(i))^-1;B(i,j)=0;
else if(j==i-1)
        A(i,j) = (C_k1(i)*R_k1(j))^-1;B(i,j)=0;
else
        A(i,j) = 0;B(i,j) = 0;
end
end
end
tt = [u,u+60];
T_ou = T_o(u/60+1);                           %initial surface temperature
i = I(u/60 + 1);                              %initial current
T_1c(u/60 + 1) = E(1);                        %calculating conductor temperature
r_p = r_0p*(1+a*(T_1u-20));                   %DC resistance of conductor
x = pi*400e-7/r_p;
Y = x^2/(192 + 0.8*x^2);                      %skin effect factor of conductor
r = r_p.*(1 + Y);                             %AC resistance of conductor
p = i^2*r;                                    %heating power of conductor
[t,x] = ode23t(@odefun3,tt,E);               %solution of differential equation
E = x(end,:);                    %updating the initial temperature array for the next period
T_1u = E(end,1);                % updating the conductor temperature array for the next period
end
y = T_1c';                                    %calculating conductor temperature
plot(t1,y,'g')                                %plotting curves
hold on
end
function dx = odefun3(t,x)                    %P array solution
global A B P p T_ou R12 number2
dx = zeros(number2,1);
for m = 1:number2
    if(m==1)
    P(m) = p;
        else if(m==number2)
        P(m) = T_ou/R12;
    else
        P(m) = 0;
    end
end
dx = A*x + B*P;
end
```

The calculating result of the XLPE-0 cable ampacity through the Matlab program above is shown in Figure A1.

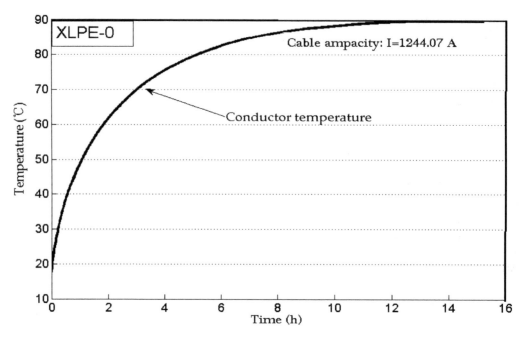

Figure A1. The calculating result of the XLPE-0 cable ampacity.

References

1. Orton, H. Power cable technology review. *High Volt. Eng.* **2015**, *41*, 1057–1067. [CrossRef]
2. Shwehdi, M.H.; Morsy, M.A.; Abugurain, A. Thermal aging tests on XLPE and PVC cable insulation materials of Saudi Arabia. In Proceedings of the IEEE Conference on Electrical Insulation and Dielectric Phenomena, Albuquerque, NM, USA, 19 November 2003; pp. 176–180. [CrossRef]
3. Liu, X.; Yu, Q.; Liu, M.; Li, Y.; Zhong, L.; Fu, M. DC electrical breakdown dependence on the radial position of specimens within HVDC XLPE cable insulation. *IEEE Trans. Dielectr. Electr. Insul.* **2017**, *24*, 1476–1486. [CrossRef]
4. Ouyang, B.; Li, H.; Li, J. The role of micro-structure changes on space charge distribution of XLPE during thermo-oxidative ageing. *IEEE Trans. Dielectr. Electr. Insul.* **2017**, *24*, 3849–3859. [CrossRef]
5. Diego, J.A.; Belana, J.; Orrit, J.; Cañadas, J.C.; Mudarra, M.; Frutos, F.; Acedo, M. Annealing effect on the conductivity of xlpe insulation in power cable. *IEEE Trans. Dielectr. Electr. Insul.* **2011**, *18*, 1554–1561. [CrossRef]
6. Xie, Y.; Zhao, Y.; Liu, G.; Huang, J.; Li, L. Annealing Effects on XLPE Insulation of Retired High-Voltage Cable. *IEEE Access.* **2019**. [CrossRef]
7. Xie, Y.; Liu, G.; Zhao, Y. Rejuvenation of Retired Power Cables by Heat Treatment. *IEEE Trans. Dielectr. Electr. Insul.* **2019**, *26*, 668–670. [CrossRef]
8. Celina, M.; Gillen, K.T.; Clough, R.L. Inverse temperature and annealing phenomena during degradation of crosslinked polyolefins. *Polym. Degrad. Stab.* **1998**, *61*, 231–244. [CrossRef]
9. Kalkar, A.K.; Deshpande, A.A. Kinetics of isothermal and non-isothermal crystallization of poly (butylene terephthalate) liquid crystalline polymer blends. *Polym. Eng. Sci.* **2010**, *41*, 1597–1615. [CrossRef]
10. Wang, Y.; Shen, C.; Li, H.; Qian, L.; Chen, J. Nonisothermal melt crystallization kinetics of poly (ethylene terephthalate)/clay nanocomposites. *J. Appl. Polym. Sci.* **2010**, *91*, 308–314. [CrossRef]
11. Xie, A.S.; Zheng, X.Q.; Li, S.T.; Chen, G. The conduction characteristics of electrical trees in XLPE cable insulation. *J. Appl. Polym. Sci.* **2010**, *114*, 3325–3330. [CrossRef]
12. Xie, A.; Li, S.; Zheng, X.; Chen, G. The characteristics of electrical trees in the inner and outer layers of different voltage rating XLPE cable insulation. *J. Phys. D Appl. Physic* **2009**, *42*, 125106–125115. [CrossRef]
13. Insulated Conductors Committee of the IEEE Power Engineering Society. *IEEE GUIDE for Soil Thermal Resistivity Measurements*; IEEE Std 442, Reaffirmed 2003; IEEE: Piscataway, NJ, USA, 1981.

14. Frank, D.W.; Jos, V.R.; George, A.; Bruno, B.; Rusty, B.; James, P.; Marcio, C.; Georg, H. *A Guide for Rating Calculations of Insulated Cables*; Cigré TB # 640; Cigré: Paris, France, 2015.

15. International Electrotechnical Commission. *Calculation of the Current Rating of Electric Cables*; IEC Press: Geneva, Switzerland, 2006.

16. Wang, P.; Liu, G.; Ma, H. Investigation of the Ampacity of a Prefabricated Straight-Through Joint of High Voltage Cable. *Energies* **2017**, *10*, 2050. [CrossRef]

17. Meng, X.K.; Wang, Z.Q.; Li, G.F. Dynamic analysis of core temperature of low-voltage power cable based on thermal conductivity. *Can. J. Electr. Comput. Eng.* **2006**, *39*, 59–65. [CrossRef]

18. Del Pino Lopez, J.C.; Romero, P.C. Thermal effects on the design of passive loops to mitigate the magnetic field generated by underground power cables. *IEEE Trans. Power Deliv.* **2011**, *26*, 1718–1726. [CrossRef]

19. Sharad, P.A.; Kumar, K.S. Application of surface-modified XLPE nanocomposites for electrical insulation-partial discharge and morphological study. *Nanocomposites* **2017**, *3*, 30–41. [CrossRef]

20. Andjelkovic, D.; Rajakovic, N. Influence of accelerated aging on mechanical and structural properties of cross-linked polyethylene (XLPE) insulation. *Electr. Eng.* **2001**, *83*, 83–87. [CrossRef]

21. Fu, Q.; Liu, J.P.; He, T.B. Non-isothermal Crystallization Behavior and Kinetics of Metallocene Short Chain Branched Polyethylen. *Chem. Res.* **2002**, *6*, 1183–1188. [CrossRef]

22. Olsen, R.; Anders, G.J.; Holboell, J.; Gudmundsdottir, U.S. Modelling of dynamic transmission cable temperature considering soil-specific heat, thermal resistivity, and precipitation. *IEEE Trans. Power Deliv.* **2013**, *28*, 1909–1917. [CrossRef]

23. Han, Y.J.; Lee, H.M.; Shin, Y.J. Thermal aging estimation with load cycle and thermal transients for XLPE-insulated underground cable. In Proceedings of the IEEE Conference on Electrical Insulation and Dielectric Phenomenon (CEIDP), Fort Worth, TX, USA, 1 October 2017; pp. 205–208. [CrossRef]

24. Xiaobin, C.; Zhixing, Y.I.; Kui, C.; Xianyi, Z.; Jia, Y. Laying mode and laying spacing for single-core feeder cable of high speed railway. *China Railw. Sci.* **2015**, 85–90. [CrossRef]

25. Chen, Y.; Duan, P.; Cheng, P.; Yang, F. Numerical calculation of ampacity of cable laying in ventilation tunnel based on coupled fields as well as the analysis on relevant factors. In Proceedings of the IEEE Conference on Intelligent Control and Automation, Shenyang, China, 29 June–4 July 2014; pp. 3534–3538. [CrossRef]

26. Wang, Y.; Chen, R.; Li, J.; Grzybowski, S.; Jiang, T. Analysis of influential factors on the underground cable ampacity. In Proceedings of the IEEE Conference on Electrical Insulation Conference, Annapolis, MD, USA, 5–8 June 2011; pp. 430–433. [CrossRef]

27. Wang, P.; Ma, H.; Liu, G. Dynamic Thermal Analysis of High-Voltage Power Cable Insulation for Cable Dynamic Thermal Rating. *IEEE Access.* **2019**, *7*, 56095–56106. [CrossRef]

28. Xu, Y.; Luo, P.; Xu, M.; Sun, T. Investigation on insulation material morphological structure of 110 and 220 kv xlpe retired cables for reusing. *IEEE Trans. Dielectr. Electr. Insul.* **2014**, *21*, 1687–1696. [CrossRef]

29. Bin, Y.; Adachi, R.; Tong, X. Small-angle HV light scattering from deformed spherulites with orientational fluctuation of optical axes. *Colloid Polym. Sci.* **2004**, *282*, 544–554. [CrossRef]

30. Li, J.; Li, H.; Wang, Q. Accelerated inhomogeneous degradation of XLPE insulation caused by copper-rich impurities at elevated temperature. *IEEE Trans. Dielectr. Electr. Insul.* **2016**, *23*, 1789–1797. [CrossRef]

31. He, K.; Chen, N.; Wang, C.; Wei, L.; Chen, J. Method for determining crystal grain size by X-ray diffraction. *Cryst. Res. Technol.* **2018**, *53*, 1700157. [CrossRef]

32. Rabello, M.S.; White, J.R. The role of physical structure and morphology in the photodegradation behavior of polypropylene. *Polym. Degrad. Stab.* **1997**, *56*, 55–73. [CrossRef]

33. Martini, H.; Zhao, S.; Friberg, A.; Jabri, Z. Influence of electron beam irradiation on electrical properties of engineering thermoplastics. In Proceedings of the IEEE Conference on Electrical Insulation Conference (EIC), Montreal, QC, Canada, 19–22 June 2016; pp. 305–308. [CrossRef]

10

Offshore Wind Farms On-Site Submarine Cable Testing and Diagnosis with Damped AC

Edward Gulski [1], Rogier Jongen [1,*], Aleksandra Rakowska [2] and Krzysztof Siodla [2]

[1] Onsite hv solutions AG, Lucerne, Toepferstrasse 5, 6004 Lucerne, Switzerland; e.gulski@onsitehv.com

[2] Institute of Electric Power Engineering, Poznan University of Technology, Piotrowo 3A, 60-965 Poznan, Poland; aleksandra.rakowska@put.poznan.pl (A.R.); krzysztof.siodla@put.poznan.pl (K.S.)

* Correspondence: r.jongen@onsitehv.com

† This paper is an extended and updated version of our conference paper published in 2018 IEEE International Conference on High Voltage Engineering and Application (ICHVE), 10–13 September 2018, Athens, Greece.

Abstract: The current power cables IEC standards do not provide adequate recommendations for after-laying testing and diagnosis of offshore export and inter-array power cables. However the standards IEEE 400 and IEEE 400.4 recommend partial discharge monitored testing, e.g., by continuous or damped AC voltages (DAC). Based on the international experiences, as collected in more than 20 years at different power grids, this contribution focuses on the use of DAC for after-laying testing and diagnosis of submarine power cables both the export and inter-array cables. Higher risk of failure, long unavailability, higher repair costs, and maintenance costs imply that advanced quality control is becoming more important. The current state of the existing and drafting international standards are based on onshore experiences and not related to the actual serious problems experienced with failures on export up to 230 kV and inter-array cables up to 66 kV. The application of damped AC as a testing solution in this concern is specially discussed. The advantages of this testing technique, in combination with actual testing examples, show the findings on export and inter-array cables at offshore wind farms.

Keywords: offshore; export cables; inter-array cables; damped AC voltage (DAC), after-laying cable testing; on-site diagnosis; condition assessment; partial discharges; and dissipation factor

---

1. Introduction

Considering that reliable energy transport is fundamental for on- and offshore infrastructures, the aspects of maintaining the quality control regulations for newly installed and service aged cable connections are of importance. As a result, important questions about maintaining/updating internal procedures for a reliable network operation are:

1. How to perform, in a sensitive and non-destructive way, the detection of poor workmanship defects of newly installed cable circuits?

2. How to perform non-destructive diagnostics of cable circuits in service to determine the actual condition?

Following present IEC standards for power cables [1–3], the after-installation testing protocols for power cables are limited to manufacturer's minimum recommendations and therefore do not cover the present needs to keep possible failure risk during operation as small as possible. As a result considering responsible operation and asset management of offshore power cables the following aspects have to be considered:

1. After installation, testing of newly installed cable systems to find:

    a.   Manufacturing related defects → due to the high level of quality control less probable;

    b.   Accessories parts delivery problems → due to the diversification in the supply chains more probable;

    c.   Installation related defects → due to the diversification in the installation supply chains highly probable.

2.   Maintenance and diagnostic testing of cable systems in operation to estimate:

    a.   Operational damages and electrical and thermal over-stresses → cannot be neglected e.g., transients and over-voltages;

    b.   Aging processes → depends on many operational and local factors, e.g., presence of installation defects, load constriction or load increase works;

    c.   The remaining life → goal of most asset managers to keep capital expenditure (CAPEX) and operational expenditure (OPEX) on an optimal level.

Unfortunately, regarding testing of offshore power cables, parties involved are not aware of risk management of those cable circuits and in their testing, specifications are simply referring to IEC procedures to ensure their quality testing procedures. About 30 years ago these standards were introduced for onshore application by manufacturers only e.g., the IEC 60502, IEC 60840, and IEC 62067 [1–3]. These cable manufacturers' standards are extensively discussing the factory testing aspects. However, only basic tests for the after-installation test are mentioned. Furthermore, no guidelines are provided regarding the maintenance of cable circuits and maintenance/diagnostic testing. In contrast to the IEC standards, the standards IEEE 400 [4] and IEEE 400.4 [5] recommend partial discharge monitored testing, e.g., by continuous or damped AC voltages (DAC).

The number of onshore and offshore wind farms is growing worldwide due to the increase in demand for renewable energy. An important aspect of an offshore wind farm is the submarine or subsea cables. They play a vital role in bringing the generated power from the wind turbine to the offshore substation and eventually to shore. In case of damage to this critical infrastructure repairing any damages can be challenging and costly [6].

In the past years, an average of at least 10 subsea cable failures is declared to wind farm insurers each year by the wind park owners. The financial severity of such a cable failure continues to grow. It is stated that the cable failures are accounting for 77% of the total global cost of offshore wind farm losses [7]. Almost 70% of the cable faults recorded in the claims database can be attributed to contractor errors during installation. However, those errors do not always become evident until the wind farms start operations or are operational for a certain time.

As the development of (offshore) wind turbines results in the increase of the physical dimensions as well as the generating capacity, the operating voltage of export cables up to 230 kV and the inter-array cable up to 66 kV networks needs to be increased. Both voltage classes have their own technical challenges. Besides the difference in the network components, also the quality assessment of the installed cable systems is different. High voltage cable testing is among the challenges facing this relatively new part of the offshore wind farm industry.

Advanced risk management and quality control are becoming more and more important. The main challenge during the installation phase is to have specialized teams for cable installation and testing activities. To reduce possible risks or to exclude failures during operation, it is important that systematic testing and diagnosis is performed during manufacturing, transportation, installation, and operation. It has been noticed that the after-installation testing protocol for offshore wind farms is still under consideration. The current state of the international standards is not based on long term experiences as obtained in the offshore sector, but more or less a copy of existing onshore standards and recommendations. As a result, the provided solutions do not cover the actual problems experienced on inter-array cables [6,7].

Evaluating the risks and the obtained valuable experiences, it shows that the offshore industry needs to set up their own reliable specifications for submarine cable testing and diagnosis. The first steps for this are done, for instance, in the new published CIGRE Technical Brochure 722 and the upcoming IEC 63026. Unfortunately, a serious gap in those recommendations is present in the practical implementation regarding offshore after installation testing.

2. Integral Quality Fingerprint

As offshore cables are often produced in very long lengths compared to onshore cables and a lot of typical handling is taken place from manufacturing up to the offshore installation, there are the risks of cable damages. Those can be related to manufacturing, transportation, and the installation stages of the cable. Therefore, there is a need to use an integral approach where the condition of the cables will be verified. Having such fingerprints as obtained during these stages, a comparison could be made to verify the cable quality. This can also be used as a good basis for the cable maintenance activity later on, see Figure 1.

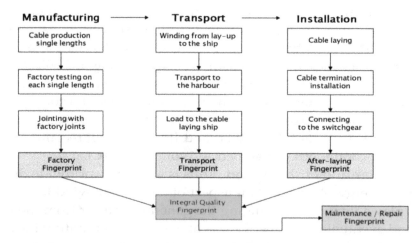

Figure 1. Integral fingerprint for quality control during the manufacturing, transportation, and installation of offshore power cables.

The applied technologies for such fingerprinting for quality control have to be suitable [8,9]:

- For offshore testing, taking into account restrictions regarding e.g., size, weight, weather protection;
- To provide adequate information: voltage testing and fingerprinting (e.g., partial discharges (PD) and dissipation factor (tan δ TD)) during the whole installation and operation process;
- To ensure reliable operation of the submarine power cables;
- To provide contractors the basis for lowering the risks during the warranty period;
- To enable service providers during operation a good basis for condition-based maintenance.

Moreover, the criteria for the risk management for the contractor (e.g., 5-year warranty), system operators and insurers have to be related to the quality control system applied during construction. This means that the various testing techniques provided in the international standards differ significantly in the effectiveness of cable quality assessment after the installation. Therefore the indicated testing options could be classified in different warranty levels:

a. Soak Test ((non-) monitored): Due to lack of information about operational reliability = No warranty.

b. Non-monitored voltage withstand test only: due to showing extreme defects only = Limited warranty.

c. Monitored voltage withstand test (damped AC voltage (DAC) resp. AC Voltage test) with sensitive (PD, TD) fingerprinting: Providing complete information = Full warranty.

Several testing technologies are available to perform on-site testing of power cables. In Table 1 an overview of several test techniques is given including weighting of important aspects. As can be seen from the table DAC monitored testing of offshore cables has several advantages.

Table 1. Overview of test technologies for monitored withstand test of 66 kV subsea inter-array cables.

Test Technology	Technical Acceptance (Meets Offshore Wind Recognized International Standards, e.g., IEC, IEEE)	Capability of Testing High Capacitive Load (Long and Multiple Cables)	Offshore Application (Size, Weight, Weather Protection)	Sensitive PD Detection (System PD level, PD Characteristics)	Costs (Investment /Rental Costs)
ACRT + PD	++	+/− [1]	−−	− [3]	−−
VLF + PD	+	+ [2]	+	− [3]	++
DAC + PD	+	++	+	++	+

ACRT—AC Resonance Test; VLF—Very Low Frequency; DAC—Damped AC; PD—Partial Discharges. [1] when connecting a second resonance reactor; [2] when lowering the VLF test frequency to e.g., 0.01 Hz; [3] Voltage source produces high interference during the PD measurement (>10 pC).

3. Damped AC

Damped AC (DAC) is a technology that is already for 20 years commercially available for testing all types of distribution and transmission cables [9–19]. With DAC it is possible to energize very long lengths of power cables with high capacitance, due to its low input power demand. Moreover this technology can be combined with diagnostic measurements like partial discharges (PD). Damped AC is applicable for factory PD monitored acceptance testing and it is already in use for on-site after-laying/commissioning, maintenance, and diagnostic testing. It is an approved testing methodology, in accordance with relevant testing parameters from international standards and recommendations (IEEE, IEC, and CIGRE).

3.1. DAC Principle

The application of damped AC (DAC) voltages including standardized conventional PD detection and analysis is accepted worldwide for on-site testing and diagnosis of (Extra) High Voltage power cables [5]. The DAC technology has been first introduced on the Jicable conference in 1999 and is now already 20 years worldwide in use to test and diagnose MV and HV power cables [19]. In addition to the equivalence of sinusoidal DAC voltages (in the frequency range of 20–300 Hz) compared to the 50/60 Hz network stresses, the characteristics of the applied technology meet the specification of modern on-site testing system:

- Lightweight modular system;
- Compactness in relation to the output voltage;
- Low effort for system assembling;
- Low power demand incl. long cable lengths;
- Low level of EM noises and the possibility of sensitive PD detection and localization as well as dissipation factor measurements.

DAC testing is used almost always in combination with partial discharge (PD) and dissipation factor (tan δ TD) measurements for new installed and service-aged cables. The use of DAC voltages for testing power cables is in compliance with relevant testing parameters derived from IEC, IEEE, and CIGRE international standards and guidelines. The DAC method is used to energize and to test on-site power cables with sinusoidal AC frequencies [5]. The system consists of a digitally controlled high voltage power supply to energize capacitive load of power cables with large capacitance (e.g., 10 μF), see Figure 2.

The energizing time depends on the maximum available load current of the high voltage power supply, the test voltage, and the capacitance of the test object and has to stay below 100 s [2]. During a number of AC voltage cycles (of several hundred milliseconds), the PD signals are initiated in a way similar to 50/60 Hz inception conditions [5,20–22]. In accordance with [5], no DC stresses are applied to the test object, and the DAC stress can be considered similar to factory partial discharge testing conditions, i.e., a 50 Hz AC test combined with a PD measurement. Due to the continuous voltage increase and immediate transition to the DAC voltage after the maximum test voltage is reached, no steady-state condition occurs, and the low electric field strength in the insulation (typically <20 kV/mm), and short durations (less than a second up to tens of seconds) of bipolar stresses ensures no space charge accumulation.

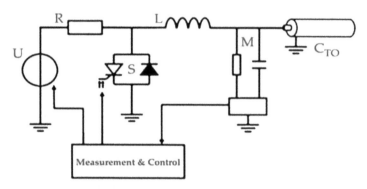

Figure 2. Schematic overview of a damped AC (DAC) system connected to a power cable, with U the high voltage source, R the protective resistor, S the semiconductor switch, L the air core inductor, M the voltage divider, and coupling capacitor and C_{TO} the cable under test.

During the AC resonance phase, the DAC voltage is characterized by a decaying sine wave with a frequency given by the inductance used and the capacitance of the cable under test. Inductor values of DAC systems are chosen such that the DAC voltage frequency is in the near power frequency range of 10-500 Hz, see Figure 3. The maximum cable length is only limited by the energizing time and the maximum current capabilities of the semiconductor switch, therefore DAC systems can be easily used to test cable lengths up to tens of kilometers.

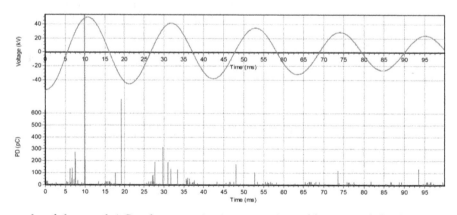

Figure 3. Example of damped AC voltage excitations monitored by partial discharge (PD) detection. The PD activity can be used to localize the breakdown site.

By applying DAC voltage-sensitive PD detection and PD localization of the fault in the power cables is possible. Using time-domain reflectometry (TDR), PD presences in cable terminations, joints, or cable parts can be localized, see Figure 4. According to IEEE 400.4 [5] to execute voltage withstand test this procedure should be repeated for 50 excitations followed after each other to perform a voltage withstand test on the maximum test voltage. Considering the time from the PD initiation until breakdown and the shorter duration of the excitation and decaying characteristics of the voltage, DAC test results obtained may differ from those obtained by continuous AC withstand voltage testing [5]. Assuming that testing should not necessarily be destructive and that the PD inception indicates presence of defects, this difference in the time until breakdown has to be considered as an advantage of DAC testing. In practice, a monitored damped AC hold test is performed to determine whether the cable passes or fails the damped AC test. Due to additional information as provided by PD detection, the monitoring insulation properties during a damped AC withstand test, and the effect of the test voltage during its application can improve the evaluation of the insulation condition.

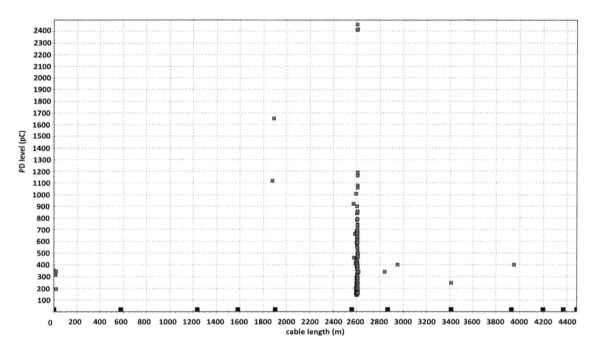

Figure 4. Monitored voltages withstand testing of a 150 kV XLPE cable underground circuit (4.5 km) up to 1.7 *Uo*. Example of PD mapping as obtained during DAC voltage testing up to 1.0 *Uo* shows that joint No. 5 has concentrated PD activity at operating voltage level.

3.2. DAC Diagnostic Testing

3.2.1. Partial Discharge Measurement (PD)

The DAC voltage source does not produce EM interference signals during the PD measurements, thus a sensitive PD detection and localization of the fault in power cables is possible. Therefore the PD monitored voltage withstand testing using damped AC voltage is a very effective method to detect most insulation weak-spots. The PD measurement can be used to pinpoint the exact location of insulation defects at an early stage [19–21], by means of time-domain reflectometry (TDR). With TDR, the PD presences in cable terminations, joints, or cable parts can be made, see Figure 4.

3.2.2. Dissipation Factor Estimation (tan δ TD)

DAC systems are able to estimate the dissipation factor from the damping of the decaying sine wave during the LC resonant phase [22,23]. The degradation of oil-impregnated insulation of HV power cable can be investigated with this parameter.

Applying DAC voltages the dissipation factor tan δ can be estimated using the decay characteristics of the damped AC voltage at different testing voltage levels and the change in the dissipation factor (delta tan delta) in relation to the increasing voltage can be especially valuable for finding insulation ageing development in power cables [5], see Figure 5.

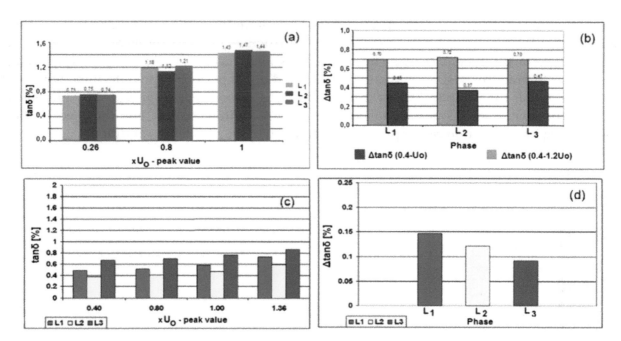

Figure 5. Example of dissipation factor diagnosis data as obtained for two different cable circuits: (**a,b**) A 150 kV power cable with self-contained fluid-filled (SCFF) insulation, length: 850 m, aged 49 years, (**c,d**) A 230 kV power cable with low-pressure fluid-filled (LPFF) insulation, length 13.3 km, aged 33 years.

4. On-Site Testing and PD Detection of Long Lengths Export Cables up to 230 kV

Detection and localization of PD in export cable system with long lengths e.g., 30 km can be improved by performing PD measurements at both sides of the cable circuit. This will, for the worst-case situation (PD at the near end), reduce the traveling distance for PD pulses by a factor of 2. In a single-sided measurement setup, a near end partial discharge has to travel through the whole cable length to the far end and the whole cable length back to the near end. The overall traveling distance is, therefore, two times the cable length. In a dual-sided measurement, the near end PD only has to travel to the far end to be detected there, so it only has to travel the cable length once. For this dual-sided measurement system, a damped AC system for energizing the cable system is used, see Figure 6. This system uses a coupling capacitor with PD detector on the near end side and a second PD detector at the far end side [11].

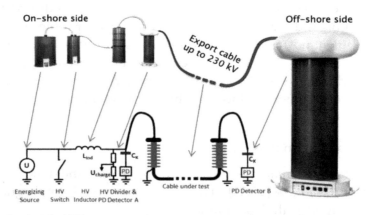

Figure 6. Setup for dual-sided PD monitored on-site testing techniques for offshore export cables of long lengths: DAC excitation circuit and synchronized PD detector units supported by PD measurement with localization feasibility with C_k the coupling capacitor (1–2 nF), L_{ind} is the air core inductance (5–9 H).

The main advantage of the dual-sided measurement is the better sensitivity compared to the single-sided measurement. The PD pulses have to propagate only to the ends of the cable. No reflection at the far end is necessary. As a result there is no additional attenuation due to imperfect reflection or attenuation for traveling through the whole cable once again. The attenuation decreases the amplitude of measurable partial discharge pulses.

Due to the attenuation, the decreasing PD amplitude has about 70% of the original amplitude. In a single-sided measurement setup, the reflected pulse over the far end decreases to 5%. A dual-sided measurement increases the remaining PD amplitude to about 26% [24].

5. On-Site Testing of HVDC Cables

Besides testing HVAC cables with DAC, the technology is also suitable for testing HVDC cables. In particular, the HVDC (submarine) power cable circuits have some typical characteristics:

- Installing a (submarine) HVDC cable is a costly and challenging activity and the technical interventions for its repairing in the case of faults are also costly and difficult;

- High capacitance that requires extremely high-power demand for conventional HVAC test systems;

- A large number of (factory) joints are installed.

Besides the enormous and unrealistic effort required to generate, on-site, the requested power (numbers of required AC resonance test (ACRT) sets) the testing with AC resonant systems only provides a go/no go based on the occurrence of a breakdown and does not provide the desired selection criteria to obtain an overall cable condition assessment where diagnostic parameters like PD and dissipation factor are included. When compared to HVAC cable systems and accessories, the HVDC cables are designed differently, which might result in the damage of the insulation of the HVDC cable and accessories in case of a defect breakdown under the regular AC voltage over-stresses. As the cable field design of HVDC cable insulation is based on the resistive field distribution, which is strongly temperature-dependent, the use of continuous AC stresses as produced by the AC resonant testing is related to a temperature increase (in particular by long lengths), which in its turn could be destructive for the HVDC insulation.

DAC testing makes it possible to energize very long lengths of AC as well as HVDC power cable with high capacitance, due to its low input power demand. Actually, possible defects can be detected in the installed HVDC cable length, as introduced in the factory or after the installation and transportation, and located by means of DAC voltage testing including single- or dual-side partial discharge detection at the cable terminations [19,22]. It is known that about 80% of the insulation defects in power cable circuits are visible through to partial discharges as detected at AC voltage stresses. As a result non-destructive DAC testing avoids the unnecessary temperature load and providing IEC conform sensitive PD detection to detect the insulation defects [11].

6. Field Example Diagnostic Test Offshore Export and Inter-Array Cables

During a maintenance outage of an offshore wind farm, diagnostic testing was performed using DAC testing. Ten complete strings of inter-array cables from the offshore substation (OSS) to the wind turbine generator (WTG) were DAC tested. For this purpose, the switchgear in each wind turbine was switched in such a way that all the inter-array cables between the wind-turbines were connected with each other, see Figure 7. The DAC test system was connected via a special test adapter in the switchgear on the offshore substation (OSS) to the cable string under test. The maximum total lengths of the strings were up to almost 10 km length.

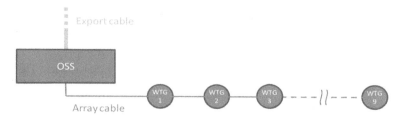

Figure 7. The layout of the array cable strings, each string consist of a total of nine wind turbine generators (WTGs). DAC test was performed from the offshore substation (OSS).

The benefit, in this case, was that the test system did not need to be transported between the tests, as all tests could be performed from the OSS, which required only a single mobilization of the test equipment.

In a string of 8.4 km length of inter-array cables between the OSS and nine wind turbines, partial discharges were detected and localized. Each phase was individually tested using a damped AC offshore system with a DAC frequency of 100 Hz (total cable capacitance 1.85 µF). The test voltage level was ramped up in steps up to the maximum applied test voltage level of 1.4 U_o. PD was monitored during the voltage ramp-up phase and during the withstand testing of 50 DAC excitations at the maximum test voltage. No breakdown was observed during the withstand test, however as the damped AC test voltage was increased up to 1.4 U_o, PD activity was observed in phases L2 and L3 of the string, see Figure 8. PD mapping revealed the PD concentration at WTG 5 in phase L3 and at WTG 6 km in phase L2.

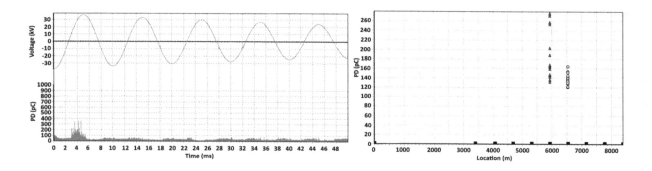

Figure 8. Phase-resolved PD pattern at 1.4 U_o (left), PD mapping showing the complete string length with the localized PD concentrations at the wind turbines (WTG) at WTG 5 and WTG 6. The squares are the locations of the WTGs in the string (right).

A second field example is from an after-laying test of a newly installed 13.3 km long, 220 kV XLPE insulated submarine cable circuit. This cable was tested using a damped ac system at 49 Hz, applying up to 1.3 U_o, (Figures 9–11). Monitored withstand testing was performed. As the damped ac test voltage was increased starting from 0.2 U_o, PD activity was observed in phase L1. An increase in the test voltage resulted in an increase of PD activity, At 0.4 U_o test voltage, a breakdown at the discharging site occurred. PD mapping revealed the PD concentration at 5.3 km indicated the breakdown position in the cable. The defect produced PD before an actual breakdown occurred, and with TDR analysis, the PD defect location could be determined. The other two phases fulfilled the after laying conditions and successfully passed the test. No internal PD activity in the cable insulation and accessories and no breakdown occurred during the tests of the other phases. The breakdown occurred during the first test, however the failure was high ohmic and the PD test could be continued on a lower voltage. Therefore the measurement could be repeated from the other end of the cable (phase L1). The PD activity occurring before the breakdown could be localized at 8 km, which is the same location seen from the original tests (13.3 − 8 = 5.3 km).

Figure 9. On-site testing of a 220 kV 13.3 km long XLPE cable circuit: The damped AC system HV300 is connected to one of the cable section phases (left) and the transition point between land and submarine cable (right).

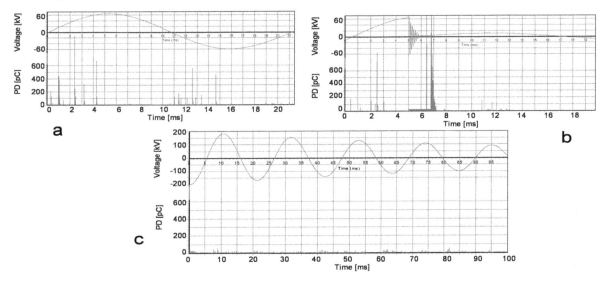

Figure 10. Damped AC voltages and PD patterns as observed during withstand testing of a 220 kV XLPE cable underground circuit (13.3 km): (**a**) example of PD pattern at 0.2 U_o of phase L1, (**b**) example of PD pattern at breakdown voltage of 0.4 U_o of phase L1, (**c**) PD pattern at 1.3 U_o of phases L2 and L3.

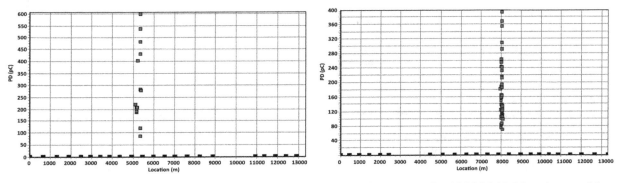

Figure 11. PD Mapping as made up to 1.3 U_o during DAC testing of a 220 kV 13.3 km long cable circuit. The PD concentration at 5.2 km distance indicates the breakdown site of phase L1 (left). Measurement from the other end confirmed this location at 8.1 km (right).

The cable was repaired at the indicated location. After repair, the cable was re-tested and no further PD activity was detected.

7. Failure Statistics and Financial Impact

On an offshore wind farm (OSWF), partial discharge was monitored using voltage withstand tests using DAC and was performed to obtain a full condition assessment. During the test sequence on both wind farms, PD activity was detected and localized in the inter-array cables. It could be verified that on multiple cable terminations, installation issues were the cause for the PD activity. If the number of affected cable terminations in relation to the total number of terminations installed is observed, it can be seen that in this case up to 3.7% of the terminations are affected, see Table 2.

Table 2. Results of DAC condition assessment on an offshore wind farm (OSWF).

No. WTGs	48
Strings	8
No. Terminations in WTGs	268
Terminations with PD	10
Percentage	3.7%

It can be concluded that a considerable number of wind turbines have increased risk of a failure due to the presence of PD at voltages higher than U_o. These PD sites are probably related to poor workmanship during the installation.

The costs for the testing and the costs for a possible failure of a cable termination during operation can be evaluated. The typical costs for testing are in the range of € 10,000 to € 15,000 per turbine dependent on the scope of works of testing; this cost reduces with volume.

Financial losses due to turbines out of operation can very quickly reach more than 100,000 € within three days, i.e., the mean time to repair (MTTR), which depends on fault location in a string (number of wind turbines which are affected), inter-array cable topology, and the weather conditions during the fault, see Table 3. With consideration of failure rates for terminations and other relevant factors, e.g., the MTTR for a 33 kV cable termination, an after-installation test can have a positive financial effect if it prevents the first cable termination fault in a wind farm.

Table 3. Losses of a single inter-array 33 kV or 66 kV termination failures. Approximate failure costs for a 100 WTG (800 MW) wind farm: (German calculation taking average 50% efficiency of the wind farm with a price of EUR 0.16 kWh).

Outage	1 WTG Affected (End of String)	8 WTG Affected (Start of String)
2 weeks stop	EUR 224,000	EUR 1,568,000
3 weeks stop	EUR 336,000	EUR 2,352,000
4 weeks stop	EUR 448,000	EUR 3,136,000
Repair of 1 termination	EUR 152,000–303,000 *	EUR 152,000–303,000 *

* Depending on the complexity of the repair.

However, the size (i.e., the power rating) of the turbine also plays an important role. In the future, larger turbines will be built were a termination fault will lead quickly to higher financial loss due to the greater loss of energy. As a result, a proper after-installation testing program is beneficial to verify the cable quality and can prevent costly failures during operation. The economic impact of an HVAC or HVDC export cable failure can be examined under the following categories:

1.   The cost of repairing the cable;
2.   The cost of lost electric power transmission over a period that can range from several weeks (for a pre-emptive repair) up to 3 to 9 months for an unexpected fault.

The cost of cable repair includes:

(a)   The precise location of the fault;
(b)   Mobilization of a suitable repair vessel, equipment, and personnel;
(c)   de-burial of the failed cable;

(d)   Removal of a length of cable (often several hundred meters long—that includes the fault location);

(e)   Jointing in a length of spare cable to replace the removed section;

(f)   Reburying or otherwise protecting the repaired cable.

HVAC and HVDC export cable links costs will depend on market conditions and weather conditions during the repair, but can be extremely high. Based on several export cable failures (which can be compared to a submarine HVDC cable) in the UK, it is calculated that the average repair cost from those failures is £12.5 million (ranging from £5.3 million to £15.5 million).

It has been estimated that the cost of a pre-emptive repair is substantially less than the cost of an unplanned repair. It is so much lower than the costs for the unplanned repairs that it is safe to state that the cost of a pre-emptive (estimated on £ 3.5 million) repair will be much lower than that of an unplanned repair.

As a result, based on the repair costs, the actual repair times, and the typical price of a loss of supplied energy (£/MWh), a rough estimate of the total cost to the industry due to cable failures can be calculated. This means that the above-stated costs for a failure repair can easily be doubled if the loss of supplied energy is also taken into consideration.

8. Conclusions

In the past 20 years field testing experiences have been obtained with damped AC testing. For several years now, this technology has also been applied to testing submarine cables at offshore wind farms. Besides this, the basic aspects of failure risks of export and inter-array cables for offshore wind farms have been discussed in this contribution. The following conclusions may be drawn after-installation testing:

- Based on the international failure statistics, it can be seen that the cable systems are the most critical parts of complete wind farm installations with a failure impact rate of 30%. However, up until now, the present international standards guidelines are not covering the needs of high-quality after-installation testing of export and inter-array cabling;

- Performing non-destructive PD monitored sensitive testing is a good basis for condition assessment and future life estimation of export and inter-array cables. With this after-installation testing, the faults that can occur as a result of stresses during manufacturing, transportation, and installation can be detected;

- For on-site cable testing, the PD monitored DAC voltage withstand testing is, in many countries, a common practice. PD measurement, including PD-pattern information and time domain reflectometry (PD localization), helps to detect and locate discharging defects in the insulation and accessories of power cables;

- DAC is a very suitable test technology to test long lengths of onshore and offshore export HVAC and HVDC power cable with a low demand of on-site power needed and with the possibility of sensitive PD detection and localization. This testing method makes it possible to obtain an integral cable fingerprinting by damped AC testing, which provides an assessment of the cable circuit integrity for the installation of both newly installed and under operation cables;

- Presented case studies have shown the value of applying the damped AC testing including PD detection and dissipation factor estimation to find upcoming failures prior to service operation. It is shown in the presented example that up to 3.7% of the installed cable terminations had partial discharges in this particular case. Those defects could not be found during traditional un-monitored (i.e., without PD measurements) voltage withstand testing, at which is only tested if the cable system withstands the over-voltage for a certain duration;

- Although testing of cables involves costs, it has shown that the testing with DAC is a cost-effective solution compared to the costs, for example, of a termination failure in an operational offshore wind farm.

Author Contributions: Conceptualization, R.J. and E.G.; validation, R.J. and E.G.; investigation, R.J.; data curation, R.J.; writing—original draft preparation, K.S., R.J., and E.G.; writing—review and editing, A.R. and K.S.; visualization and data submission, R.J.; supervision, E.G.

References

1.   IEC. *IEC 60840: Power Cables with Extruded Insulation and the Accessories for Rated Voltages Above 30 kV up to 150 kV Test Methods and Requirements*; IEC: Geneva, Switzerland, 2011.

2.   IEC. *IEC 62067: Power Cables with Extruded Insulation and the Accessories for Rated Voltages Above 150 kV*; IEC: Geneva, Switzerland, 2011.

3.   IEC. *IEC 60502: Power Cable with Extruded Insulation and the Accessories for Rated Voltages from 1 kV up to 30 kV*; IEC: Geneva, Switzerland, 2011.

4.   IEEE. *IEEE 400-2012: Guide for Field Testing and Evaluation of the Insulation of Shielded Power Cable Systems Rated 5 kV and Above*; IEEE: Piscataway, NJ, USA, 2012.

5.   IEEE. *IEEE 400.4-2015: Guide for Field-Testing of Shielded Power Cable Systems Rated 5 kV and Above with Damped Alternating Current Voltage (DAC)*; IEEE: Piscataway, NJ, USA, 2015.

6.   Hodge, N.; Maurer, R. Power under the Sea, Allianz Global Risk Dialogue, Autumn 2014, pp. 26–29. Available online: https://www.agcs.allianz.com/assets/PDFs/GRD/GRD_02_2014_EN.pdf (accessed on 8 August 2018).

7.   Tisheva, P. Cable Failures Account for Most of Offshore Wind Losses, June 2016. Available online: https://www.renewablesnow.com/news/cable-failures-account-for-most-of-offshore-wind-losses-528959 (accessed on 26 June 2018).

8.   Gulski, E.; Jongen, R.; de Heus, M.; Rakowska, A.; Siodla, K.; Gaal, H. On-Site Acceptance and Diagnostic Testing of Submarine Inter-Array Cables at Offshore Wind Farms using Damped AC. In Proceedings of the 2018 IEEE International Conference on High Voltage Engineering and Application (ICHVE), Athens, Greece, 10–13 September 2018.

9.   Jongen, R.; Gulski, E.; Rakowska, A.; Siodla, K.; Gaal, H. After Installation Testing of Inter-Array Cables at Offshore Wind Farms using Damped AC Voltages. In Proceedings of the 10th International Conference on Insulated Power Cables, Versailles, France, 23–27 June 2019.

10.   Cejka, G.; Gulski, E.; Jongen, R.; Quak, B.; Parciak, J.; Rakowska, A. Integrated Testing and Diagnosis of Distribution Cables using Damped AC and Very Low Frequency Voltages. In Proceedings of the 10th International Conference on Insulated Power Cables, Jicable 2019, Versailles, France, 23–27 June 2019.

11.   Gulski, E.; Jongen, R.; Quak, B.; Parciak, J.; Rakowska, A.; Siodla, K. Fifteen Years Damped AC On-site Testing and Diagnosis of Transmission Power Cables. In Proceedings of the 10th International Conference on Insulated Power Cables, Versailles, France, 23–27 June 2019.

12.   Gulski, E.; Wester, F.J.; Smit, J.J.; Seitz, P.N.; Turner, M. Advanced PD diagnostic of MV power cable system using oscillating wave test system. *IEEE Electr. Insul. Mag.* **2000**, *16*, 17–25. [CrossRef]

13.   Gulski, E.; Smit, J.J.; Petzold, F.; Seitz, P.P.; Quak, B.; de Vries, F. Advanced Solution for On-Site Diagnosis of Distribution Power Cables. In Proceedings of the Jicable 2007, Versailles, France, 24–28 June 2007.

14.   Gulski, E.; Wester, F.J.; Schikarski, P.; Seitz, P.N. PD diagnoses and condition assessment of distribution power cables using damped AC voltages. In Proceedings of the XIII International Symposium on HV, Delft, The Netherlands, 25–29 August 2003; p. 776.

15.   Gulski, E.; Jongen, R.; Patterson, R. *Modern Testing and Diagnosis of Power Cables using Damped AC Voltages*; NETA World: Portage, MI, USA, 2015.

16.   Gulski, E.; Rakowska, A.; Siodla, K.; Jongen, R.; Minassian, R.; Cichecki, P.; Parciak, J.; Smit, J. On-Site Testing and Diagnosis of Transmission Power Cables up to 230 kV Using Damped AC Voltages. *IEEE Electr. Insul. Mag.* **2014**, *3*, 27–38. [CrossRef]

17.   Gulski, E.; Chojnowski, P.; Rakowska, A.; Siodla, K. Importance of sensitive on-site testing and diagnosis of transmission power cables. *Przeglad Elektrotechniczny* **2009**, *2*, 171–176.

18.   Gulski, E.; Rakowska, A.; Siodla, K.; Chojnowski, P. On-site testing and diagnosis of transmission power cables. *Przeglad Elektrotechniczny* **2009**, *4*, 195–200.

19.   Wester, F.J.; Gulski, E.; Smit, J.J. Electrical and acoustical PD on-site diagnostics of service aged medium voltage power cables. In Proceedings of the 5th International Conference on Power Insulated Cables, Jicable 1999, Versailles, France, 20–24 June 1999.

20. Bodega, R.; Morshuis, P.H.; Lazzaroni, M.; Wester, F.J. PD recurrence in cavities at different energizing methods. *IEEE Trans. Instrum. Meas.* **2004**, *53*, 251–258. [CrossRef]

21. Wester, F.J.; Guilski, E.; Smit, J.J. Detection of partial discharges at different AC voltage stresses in power cables. *IEEE Electr. Insul. Mag.* **2007**, *23*, 28–43. [CrossRef]

22. Wester, F.J. *Condition Assessment of Power Cables using PD Diagnosis at Damped AC Voltages*; Optima Grafische Communicatie: Rotterdam, The Netherlands, 2004; ISBN 90-8559-019-1.

23. Houtepen, R.; Chmura, L.; Smit, J.J.; Quak, B.; Seitz, P.P.; Gulski, E. Estimation of dielectric loss using damped AC voltages. *IEEE Electr. Insul. Mag.* **2011**, *27*, 20–25. [CrossRef]

24. Wild, M.; Tenbohlen, S.; Gulski, E.; Jongen, R. Basic aspects of partial discharge on-site testing of long length transmission power cables. *IEEE Trans. Dielectr. Electr. Insul.* **2017**, *24*, 1077–1087. [CrossRef]

11

A Novel Partial Discharge Detection Method based on the Photoelectric Fusion Pattern in GIL

Yiming Zang [1], Yong Qian [1], Wei Liu [2], Yongpeng Xu [1,*], Gehao Sheng [1] and Xiuchen Jiang [1]

[1]   Department of Electrical Engineering, Shanghai Jiao Tong University, 800 Dongchuan Road, Minhang, Shanghai 200240, China; zangyiming@sjtu.edu.cn (Y.Z.); qian_yong@sjtu.edu.cn (Y.Q.); shenghe@sjtu.edu.cn (G.S.); xcjiang@sjtu.edu.cn (X.J.)

[2]   Key Laboratory for Sulfur Hexafluoride Gas Analysis and Purification of SGCC, Anhui Electric Power Research Institute of SGCC, Hefei 230022, China; sgccliu@163.com

*   Correspondence: xyp3525@sjtu.edu.cn.

Abstract: Optical detection and ultrahigh frequency (UHF) detection are two significant methods of partial discharge (PD) detection in the gas-insulated transmission lines (GIL), however, there is a phenomenon of signals loss when using two types of detections to monitor PD signals of different defects, such as needle defect and free particle defect. This makes the optical and UHF signals not correspond strictly to the actual PD signals, and therefore the characteristic information of optical PD patterns and UHF PD patterns is incomplete which reduces the accuracy of the pattern recognition. Therefore, an image fusion algorithm based on improved non-subsampled contourlet transform (NSCT) is proposed in this study. The optical pattern is fused with the UHF pattern to achieve the complementarity of the two detection methods, avoiding the PD signals loss of different defects. By constructing the experimental platform of optical-UHF integrated detection for GIL, phase-resolved partial discharge (PRPD) patterns of three defects were obtained. After that, the image fusion algorithm based on the local entropy and the phase congruency was used to produce the photoelectric fusion PD pattern. Before the pattern recognition, 28 characteristic parameters are extracted from the photoelectric fusion pattern, and then the dimension of the feature space is reduced to eight by the principal component analysis. Finally, three kinds of classifiers, including the linear discriminant analysis (LDA), support vector machine (SVM), and k-nearest neighbor (KNN), are used for the pattern recognition. The results show that the recognition rate of all the photoelectric fusion pattern under different classifiers is higher than that of optical and UHF patterns, up to the maximum of 95%. Moreover, the photoelectric fusion pattern not only greatly improves the recognition rate of the needle defect and the free particle defect, but the recognition accuracy of the floating defect is also slightly improved.

Keywords: partial discharge; optical-UHF integrated detection; photoelectric fusion pattern; GIL; NSCT

---

1. Introduction

In recent years, gas-insulated transmission lines (GIL) are widely used in the power transmission of hydropower stations and nuclear power plants because of their high efficiency, large transmission capacity, high reliability, and small footprint [1–3].

In the operation of GIL, partial discharge (PD) is a precursor in the deterioration of insulation performance, which is the main cause of a breakdown. In addition, the severity of PD is closely related to the type of discharge defect. Therefore, PD detection and pattern recognition are particularly important in the GIL [4,5]. In order to improve the reliability of PD detection and pattern recognition, some scholars have proposed a method of combining the fluorescent fiber detection and the ultrahigh frequency (UHF) detection in the PD detection of gas-insulated equipment, which has the characteristics of high sensitivity, strong anti-interference ability, and wide application range [6].

Although optical-UHF integrated detection can effectively detect the occurrence of PD, optical detection and UHF detection have certain limitations for the pattern recognition of PD. UHF detection is limited by its detection bandwidth and is susceptible to influence from the external electromagnetic interference, resulting in the loss of UHF signals [7]. For optical detection, the propagation of the optical signal is affected by the structure of the GIL, the location of sensors, the distance of sensors, and the position of the PD source, which will weaken or even shield optical signals reaching the sensor. This can also result in the loss of optical signals [8]. Therefore, both UHF detection and optical detection have the phenomena of signals loss, which will lead to incomplete feature information in the PD pattern. Therefore, if the UHF pattern and the optical pattern are separately used for pattern recognition in the optical-UHF integrated detection, both detections will be affected by the pattern aliasing or the false pattern in the recognition process due to the missing information of the PD characteristic. These adverse effects will reduce the accuracy of PD pattern recognition.

In order to improve the accuracy of PD pattern recognition in the GIL, this study proposes an image fusion algorithm based on improved non-subsampled contourlet transform (NSCT) [9], which can gain the photoelectric fusion phase-resolved partial discharge (PRPD) pattern by fusing the optical PRPD pattern with the UHF PRPD pattern. By applying the photoelectric fusion PRPD pattern to the pattern recognition of PD, the two detection methods are complemented, avoiding the problem of missing PD characteristic information in a single type of the detection pattern.

Compared with the traditional NSCT method [10], the improved NSCT image fusion algorithm with the multiscale, multidirectional, and translation-invariant characteristics can retain more image edge and texture information, which can more completely contain the PD characteristic information of optical and UHF signals. In order to better characterize the PD characteristics of the PD pattern, 28 characteristic parameters, such as moment features and texture features, are extracted from the photoelectric fusion PD pattern and, then, the principal component analysis (PCA) method is used to reduce the feature vector space to eight characteristic parameters. After the process of PCA, linear discriminant analysis (LDA), support vector machine (SVM), and k-nearest neighbor (KNN) were used to verify the pattern recognition accuracy by photoelectric fusion patterns. Therefore, through the method of photoelectric image fusion, it provides a novel idea for the optical-UHF integrated detection of PD.

2. PD Experiment of GIL

2.1. Experimental Platform and Defect Model

The GIL experimental platform used in this study is shown in Figure 1. In order to isolate external electromagnetic interference, the whole experimental platform is located in the metal shielded room. In the platform, the GIL test tank is made of aluminum alloy, and the seal is good without light injection. The photoelectric integrated sensor that is composed of a fluorescent fiber intertwined on the cylindrical UHF detection is installed on the tank wall for PD signals acquisition. The detection frequency band of the UHF sensor is 300–500 MHz. The photomultiplier tube (PMT) adopts the HAMAMATSU-H10722-01 model whose corresponding range of spectrum is 230,920 nm, and the voltage/current conversion coefficient is 1V/μA. The function of PMT is to convert the collected optical signals into electrical signals, which is helpful for signal processing. The digital PD detector (Haffley DDX 9121b) collects standard PD signals as a reference to confirm the loss of optical signals and UHF signals. The oscilloscope uses LeCroy-HDO6000A [11].

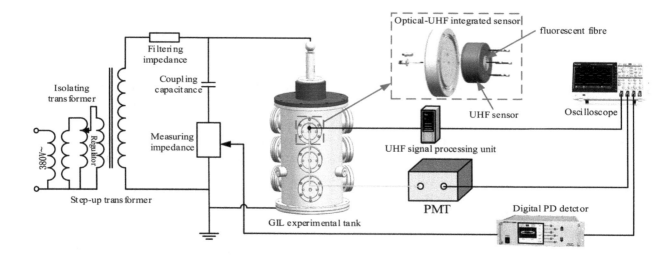

Figure 1. Partial discharge (PD) detection platform of gas-insulated transmission lines (GIL).

In order to simulate the PD defects in GIL, Figure 2 shows three aluminum typical defect models of the needle discharge defect, the floating discharge defect, and the free particle discharge defect designed in this study [12].

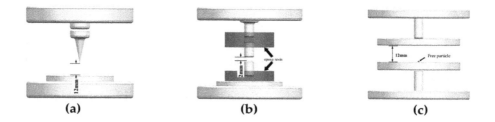

Figure 2. Schematic diagram of three insulation defect models. (**a**) The needle discharge defect, (**b**) the floating discharge defect, and (**c**) the free particle discharge defect.

2.2. Experimental Method

Before the experiment, PD defects and the tank were dedusted. The SF_6 gas was then filled into the GIL experimental tank that was sealed and vacuum until the gas pressure reached 0.2 MPa. According to the above operation, three typical PD defects were placed into the GIL experimental tank for PD experiments. During the experiment, the center of the optical-UHF integrated sensor was facing the defect on the level.

For the data acquisition, PD detection was performed for each defect at several voltage levels. For each defect, 120 detection samples were collected and each sample included 50 PD signals of a power frequency cycle.

2.3. Analysis of the Experimental Results

According to the time domain waveform of Figure 3, comparing the UHF signals and optical signals with the PD detector signals that act as the standard signals confirms that the signal loss is relative to the detection method and discharge defects, rather than external signal interference.

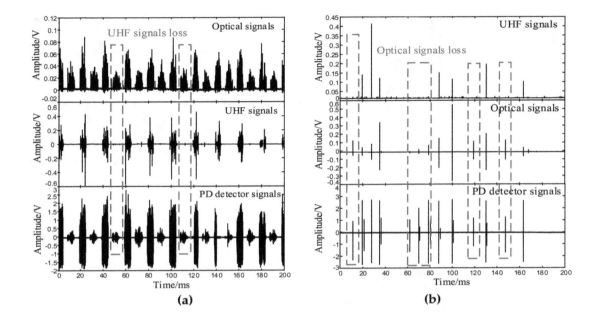

Figure 3. Optical and ultrahigh frequency (UHF) PD signals with different defects. (**a**) Time domain waveform of needle defect and (**b**) time domain waveform of free particle defect.

In the UHF detection, the sensitivity of the UHF detection is low due to the limitation of the detection band and the interference of the environmental noise. Thereby, it is shown that the PD detection of the needle discharge defect has a phenomenon of UHF signal loss in the experiment.

In the optical detection, the uncertainty of the free particle discharge is strong, and the intensity of the free particle discharge is not uniform. In addition, the optical detection is easily affected by the photon propagation path and the reflection condition of GIL inner wall. Therefore, the PD detection of the free particle discharge also has a phenomenon of optical signal loss.

In order to represent the characteristic information of PD more effectively, this study uses the two-dimensional PRPD pattern to describe the phase information of PD. The PRPD pattern is represented by a φ–u two-dimensional pattern, where φ represents the power frequency phase at which PD occurs and u represents the intensity of PD signals [13]. The color of the PRPD pattern represents the discharge density at certain phase and amplitude, which is shown by the color bar. Each PRPD pattern is drawn from a PD signal sample, including 50 PD cycles acquired by the oscilloscope above, which can guarantee that the loss of signal is not an accident.

For the needle defect, the PRPD pattern of the optical-UHF integrated sensor at 20 kV is shown in Figure 4. Optical signals are mainly concentrated near the peaks of positive and negative half cycles, while most UHF signals appear only in the area of the negative half cycle. It can be concluded that the detection sensitivity of the UHF sensor is lower than that of the optical sensor. When the PD intensity of the positive half cycle is small, the UHF sensor cannot detect the PD signals. Therefore, there is a phenomenon of missed UHF signals in the positive half cycle, which can cause the PD characteristic information to be incomplete and interfere with the PD pattern recognition.

For the floating defect, as shown in Figure 5, optical signals and UHF signals are mainly distributed near the peak of the positive and negative half cycles at 20 kV, which has good distribution characteristics of phase concentration. Therefore, the optical and UHF PD signals collected by the optical-UHF integrated sensor has good correspondence.

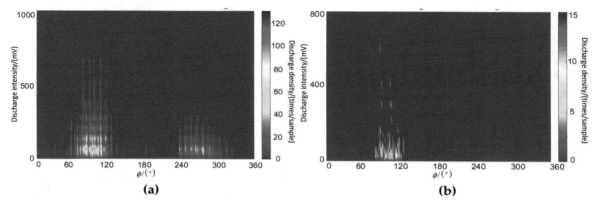

Figure 4. Phase-resolved partial discharge (PRPD) patterns of the needle defect under 20 kV. (**a**) A PRPD pattern of the optical detection and (**b**) a PRPD pattern of the UHF detection.

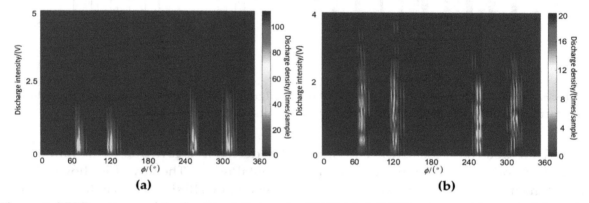

Figure 5. PRPD patterns of the floating defect under 20 kV. (**a**) A PRPD pattern of the optical detection and (**b**) a PRPD pattern of the UHF detection.

For the free particle defect, the PRPD pattern of the optical-UHF integrated sensor at 20 kV is shown in Figure 6. Under the action of the electric field force, the free metal particle undergo random collision movement between the plates and the discharge repetition rate of it is low, which causes the phase distribution of UHF signals to be relatively random in the PRPD pattern. It can be seen from the optical pattern that the randomness of the optical signal distribution is weak. The light spots in the optical pattern are mainly concentrated in the negative half cycle, instead of being distributing randomly. Therefore, the distribution of optical signals has a large difference from the UHF signal pattern. It is indicated that optical signals are attenuated by the occlusion and the reflection phenomenon of the propagation path during the propagation process, which causes optical signals to be missed and incompletely collected [8]. As a result, the loss of PD information leads to a failure to fully represent the PD characteristics of the free particle defect.

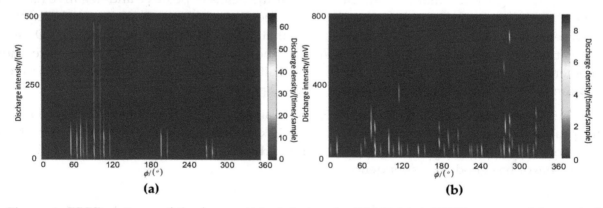

Figure 6. PRPD patterns of the free particle defect under 20 kV. (**a**) A PRPD pattern of the optical detection and (**b**) a PRPD pattern of the UHF detection.

The above experiments prove that the PD characteristic information of UHF and optical PRPD patterns is incomplete. The UHF sensor has the loss of signals on the needle defect, and the optical sensor has the loss of signals on the free particle defect. Therefore, if the two types (UHF patterns and optical patterns) of PRPD patterns are separately used for pattern recognition, the lack of the characteristic information may result in a decrease in recognition accuracy.

3. Image Fusion Algorithm Based on Improved NSCT

3.1. NSCT Structure

For PD patterns, the edge texture plays a key role in identifying the type of PD. Therefore, this study uses an image fusion algorithm based on the improved NSCT method to fuse the optical pattern with the UHF pattern. Its structure can be divided into two parts, non-subsampled pyramid filter banks (NSPFB) and non-subsampled directional filter bank (NSDFB). The structural framework of the NSCT is shown in Figure 7.

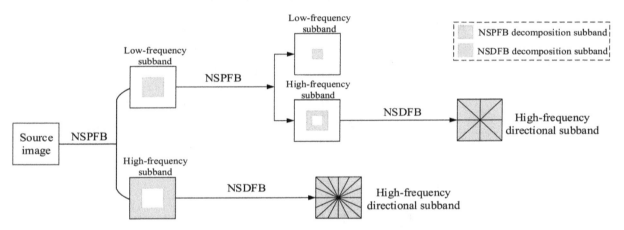

Figure 7. Structure block diagram of non-subsampled contourlet transform (NSCT) decomposition.

NSPFB is used as a filter that is up sampled. In order to achieve multiscale decomposition of the image, NSPFB performs up-sampling processing by the matrix $D = 2I$ iteratively, which can obtain the filter $H(Z^{2I})$. The NSPFB filters the low-frequency subband image of the upper level by the lowpass filter $H_0(Z^{2I})$ and the bandpass filter $H_1(Z^{2I})$, so that each level is decomposed into a low-frequency subband image and a high-frequency subgraph. The definition of the decomposition scale is j. In the filtering process, the ideal frequency domain of the lowpass filter at j scale is $[-\pi/2j, \pi/2j] \times [-\pi/2j, \pi/2j]$. The corresponding ideal frequency domain of the bandpass filtering is $[-\pi/2j + 1, \pi/2j + 1] \times [-\pi/2j-1, \pi/2j-1]$ [14]. Thus, after the image is decomposed by the j-level NSPFB, the image can obtain $j + 1$ subgraph of the same size as the original decomposition image, including one low-frequency subgraph and j high-frequency subgraph. Taking the three-level NSPFB decomposition as an example, its structure is shown in Figure 8 [15].

The NSDFB used in the NSCT is based on a fan-shaped filter bank. The two-channel directional filters, $U_0(Z)$ and $U_1(Z)$, with the fan-shaped frequency domain are up-sampled by the sampling matrix D to obtain filters $U_0(Z^D)$ and $U_1(Z^D)$. Then $U_0(Z^D)$ and $U_1(Z^D)$ are used to filter the subgraph decomposed in the upper level, which can achieve more accurate directional decomposition of the image in the image of corresponding frequency domain. As shown in Figure 9, taking the two-level directional decomposition as an example, the NSDFB decomposes the two-dimensional frequency domain into several wedge-shaped regions representing directionality [16]. Each wedge-shaped region contains detailed direction features of the image. Therefore, by performing k-level directional decomposition on the subgraph at one level of the NSPFB, it is possible to obtain a 2^k directional subgraph with the same size as the source image [17].

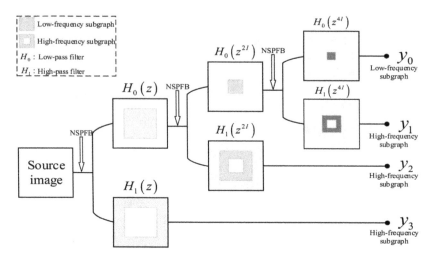

Figure 8. Three-level NSPFB pyramid filter bank.

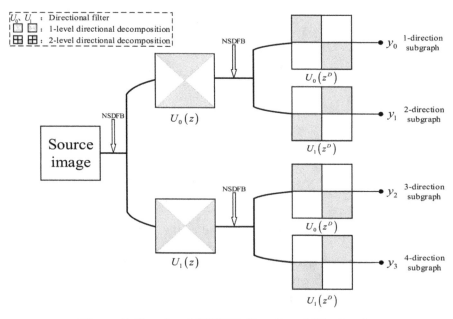

Figure 9. Two-level NSDFB directional filter bank.

3.2. *Improved Image Fusion Rules Based on NSCT*

According to the NSCT decomposition above, when k_j represents the NSDFB directional decomposition levels of the image on the jth-level of the J-level NSPFB decomposition, the number of subgraph generated by the decomposition can be expressed as $1 + \sum_{j=1}^{j} 2^{k_j}$, including one low-frequency subgraph and $\sum_{j=1}^{j} 2^{k_j}$ high-frequency subgraph. In order to ensure the anisotropy in the NSCT image fusion process, we change the k_j on each level of the NSPFB, which can make high-frequency subgraph on each level of the NSPFB have different directional decomposition. The structural flow of image fusion is shown in Figure 10 [18].

The source images A and B are subjected to grayscale processing before the NSCT conversion. By performing NSCT decomposition on the grayscale images of the source images A and B, the high-frequency subband coefficients, $G_{j,r}^A(x, y)$, $G_{j,r}^B(x, y)$, and the low-frequency subband coefficients, $L_J^A(x, y)$, $L_J^B(x, y)$, of each source image can be obtained. Among them, $j = (1, 2, \ldots, J)$ is the number of decomposition levels of the NSPFB, r is the rth-direction of the NSDFB decomposition on the jth-level ($r = 1, 2, \ldots, 2^{k_j}$), and the subband coefficient represents the gray value at location (x, y).

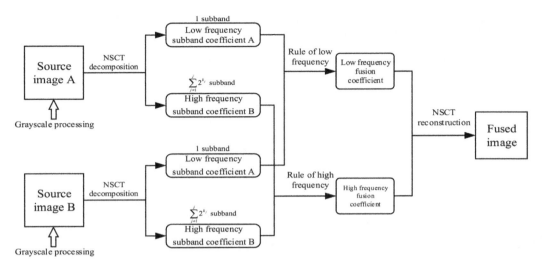

Figure 10. NSCT image fusion structure.

3.2.1. Fusion Rule of Low-Frequency Subgraph

After being decomposed by the NSCT, the contour information of the source image is mainly retained in the low-frequency subgraph. For the PRPD pattern, the contour of PD signals distribution is especially important for pattern recognition. Image fusion of the decomposed low-frequency subgraph is to preserve the contour feature information of the PRPD pattern as completely as possible. Therefore, we proposes a fusion rule of low-frequency subgraph, combining Canny operator and local entropy to better preserve the contour information of the PRPD pattern.

It is known that $L_J^A(x, y)$ and $L_J^B(x, y)$ are two low-frequency subgraphs to be fused, which are the same size. First, the low-frequency subgraph is edge-extracted by the Canny operator to obtain the edge contour binary graphs, $L_{J,Canny}^A(x, y)$ and $L_{J,Canny}^B(x, y)$. Thereby, the contour information of the signal distribution in the low-frequency subgraph is better preserved during the fusion process, which reduces the influence of image texture and sparseness.

In this study, we introduce the concept of image local entropy [19]. The local entropy can reflect the extent of gray dispersion of the image. In the image region with large local entropy, the gray value of this certain region is relatively uniform and contains less feature information. In the image region with small local entropy, the grayscale difference of this certain region is large and contains more feature information. Therefore, the local entropy is larger in the smooth region of the PRPD pattern, while the local entropy is smaller in the boundary contour region of the signal distribution for the PRPD pattern.

$f(x, y)$ is defined as the gray value of location (x, y) in the image, then, an image of size $X \times Y$ whose local entropy $H_{f(x,y)}$ is defined as [20]:

$$H_{f(x,y)} = \sum_{x=1}^{X} \sum_{y=1}^{Y} p_{xy} lg p_{xy} \tag{1}$$

where p_{xy} is the probability of gray distribution at location (x, y), and its expression is as follows:

$$p_{xy} = \frac{f(x, y)}{\sum_{x=1}^{X} \sum_{y=1}^{Y} f(x, y)} \tag{2}$$

Therefore, the fusion method of low-frequency coefficient adopted in this study is summarized as:

1)   The local entropy $H_{f(x,y)}^A$ and $H_{f(x,y)}^B$ at location (x, y) of the edge contour binary patterns $L_{J,Canny}^A(x, y)$ and $L_{J,Canny}^B(x, y)$ are calculated by traversing the sampling window of size 3×3.

2) By comparing the magnitude of the local entropy at each location, it is determined how much the sample window contains image contour information. According to this, the fusion weight coefficients, $c_A(x,y)$ and $c_B(x,y)$, of the images, $L_J^A(x,y)$ and $L_J^B(x,y)$, are calculated.

$$c_A(x,y) = \frac{H_{f(x,y)}^A}{H_{f(x,y)}^A + H_{f(x,y)}^B} \tag{3}$$

$$c_B(x,y) = \frac{H_{f(x,y)}^B}{H_{f(x,y)}^A + H_{f(x,y)}^B} \tag{4}$$

3) According to the local entropy of the image and the fusion weight coefficient, the fused low-frequency subgraph $L_J^{fusion}(x,y)$ is calculated. The fusion rules are as follows:

$$L_J^{fusion}(x,y) = c_A \times L_J^A(x,y) + c_B \times L_J^B(x,y) \tag{5}$$

3.2.2. Fusion Rule of High-Frequency Subgraph

After the NSCT decomposition, the detailed texture information of the source image is mainly retained in the high-frequency subgraph, which represents the density of PD signals. Therefore, the key point of the high-frequency subgraph fusion is to enhance the image features, making the high-frequency subgraph more informative.

In this study, phase congruency (PC) is applied to the fusion rule of high-frequency coefficient. PC analyzes the feature points of the grayscale image from the perspective of the frequency domain. The theoretical basis is that the image is subjected to Fourier transform decomposition, and then the points with the most consistent phase of each harmonic component correspond to the feature point of the image [21]. Thus, PC can measure the importance of subgraph features with a dimensionless measurement.

In the fusion of high-frequency subgraphs, the PC value can represent the sharpness of high-frequency subgraphs. Because the subgraph can be regarded as a 2D signal [22], the PC value of the subgraph at location (x,y) can be calculated by Equation (6).

$$PC(x,y) = \frac{\sum_k E_{\theta_k}(x,y)}{\varepsilon + \sum_n \sum_k A_{n,\theta_k}(x,y)} \tag{6}$$

where A_{n,θ_k} is the amplitude of the n-th Fourier component and angle θ_k, θ_k denotes the orientation angle at k, and ε is a positive constant to offset the DC components of subgraph. In this study, the value of ε is set to 0.001 [23]. $E_{\theta_k}(x,y)$ can be calculated by Equation (7).

$$E_{\theta_k}(x,y) = \sqrt{F_{\theta_k}^2(x,y) + H_{\theta_k}^2(x,y)} \tag{7}$$

where $F_{\theta_k}(x,y)$ and $H_{\theta_k}(x,y)$ can be calculated by Equations (8) and (9) respectively.

$$F_{\theta_k}(x,y) = \sum_n e_{n,\theta_k}(x,y) \tag{8}$$

$$H_{\theta_k}(x,y) = \sum_n o_{n,\theta_k}(x,y) \tag{9}$$

where $e_{n,\theta_k}(x,y)$ and o_{n,θ_k} are convolution results of subgraph at location (x,y), which can be calculated by Equation (10).

$$\left[e_{n,\theta_k}(x,y), o_{n,\theta_k}(x,y)\right] = \left[I(x,y) \times M_n^e, I(x,y) \times M_n^o\right] \tag{10}$$

where $I(x, y)$ denotes the pixel value of subgraph at location (x, y). M_n^e and M_n^o represent the even- and odd-symmetric filters of 2D log-Gabor at scale n [23].

PC is a contrast invariant, which cannot reflect local contrast changes [22]. In order to compensate for the lack of PC, a measure of sharpness change (SCM) is introduced as below:

$$SCM(x, y) = \sum_{(x_0, y_0) \in \Omega_0} (I(x, y) - I(x_0, y_0))^2 \tag{11}$$

where Ω_0 denotes a local window with a size of 3×3 that is entered at (x, y). (x_0, y_0) is a pixel point in the local window of Ω_0. In addition, the local SCM (LSCM) is expressed as Equation (12) to determinate the (x, y) neighborhood contrast.

$$LSCM(x, y) = \sum_{a=-M}^{M} \sum_{b=-N}^{N} SCM(x + a, y + b) \tag{12}$$

where $(2M + 1) \times (2N + 1)$ represents the neighborhood size.

Since PC and LSCM cannot completely reflect the local luminance information, the local energy (LE) is proposed as below.

$$LE(x, y) = \sum_{a=-M}^{M} \sum_{b=-N}^{N} (I(x + a, y + b))^2 \tag{13}$$

Therefore, according to the theory mentioned above, a new activity measure (NAM) is defined using PC, LSCM, and LE to measure various aspects of subgraph information.

$$NAM(x, y) = (PC(x, y))^{\alpha_1} \cdot (LSCM(x, y))^{\beta_1} \cdot (LE(x, y))^{\gamma_1} \tag{14}$$

where $\alpha_1, \beta_1, \gamma_1$ are set to 1, 2, and 2 respectively, which are used to adjust the value of PC, LSCM, and LE in NAM [24].

After the NAM is obtained, the fused high-frequency image can be determined by the Equation (15).

$$H_j^{fusion}(x, y) = \begin{cases} H_A(x, y) & if \, Lmap_A(x, y) = 1 \\ H_B(x, y) & otherwise \end{cases} \tag{15}$$

where the $H_j^{fusion}(x, y)$ represents the high-frequency fused subgraph of jth-level, $H_A(x, y)$ and $H_B(x, y)$ are high-frequency subgraphs of source image A and B. $Lmap_i(x, y)$ is a decision map for the fusion of high-frequency subgraph, which can be calculated by Equation (16).

$$Lmap_i(x, y) = \begin{cases} 1 & if \, \lceil S_i(x, y) \rceil > \frac{\widetilde{M} \times \widetilde{N}}{2} \\ 0 & otherwise \end{cases} \tag{16}$$

where the $S_i(x, y)$ is calculated by Equation (17).

$$S_i(x, y) = \{(x_0, y_0) \in \Omega_1 | NAM_i(x_0, y_0) \geq \max(NAM_1(x_0, y_0), \cdots, \\ NAM_{i-1}(x_0, y_0), NAM_{i+1}(x_0, y_0), \cdots, NAM_K(x_0, y_0))\} \tag{17}$$

where Ω_1 denotes a sliding window with a size of $\widetilde{M} \times \widetilde{N}$, and (x, y) is the center of it. K is the number of source images.

According to the low-frequency and high-frequency subgraph fusion method mentioned above, the fused low-frequency and high-frequency NSCT coefficients $L_J^{fusion}(x, y)$ and $H_j^{fusion}(x, y)$ can be obtained. Then the fused fusion image F can be reconstructed by inverse NSCT transformation.

4. Photoelectric Image Fusion PD Detection Based on Improved NSCT

4.1. Overall Detection Process

In order to improve the accuracy rate of the pattern recognition, we propose an improved NSCT image fusion algorithm for the problem of missing PD signals in optical and UHF detections, that is, NSCT is used to decompose the grayscale optical and UHF PRPD patterns into low-frequency subgraphs and high-frequency subgraphs accordingly. The above fusion method is used to fuse the optical pattern with the UHF pattern to obtain the photoelectric fusion PD pattern. This pattern is then subjected to a series of processing such as feature extraction, dimensionality reduction, and pattern recognition. The overall process of the test verification is shown in Figure 11.

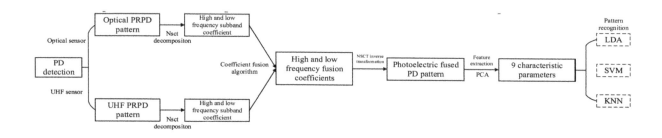

Figure 11. Experiment verification of the overall process.

4.2. Decomposition of PD Patterns Based on Improved NSCT

On the basis of the GIL PD experimental data mentioned above, the PD of the needle defect at 20 kV is taken as an example to perform NSCT decomposition and the fusion of the PD optical pattern and the PD UHF pattern.

In the NSCT decomposition of the PRPD pattern, one low-frequency subgraph and three high-frequency subgraphs can be obtained after the three-level NSPFB decomposition. In order to ensure the anisotropy of the image decomposition and preserve the information of the image in all directions more completely, the high-frequency subgraphs on the NSPFB decomposition of the fjirst-, second- and third-level are decomposed by the NSDFB directional decomposition with 2^1-, 2^2-, and 2^3-direction, respectively. Therefore, each high-frequency subgraph can be decomposed into 2^1-, 2^2-, and 2^3-direction subgraphs.

As a result, $1 + \sum_{k=1}^{3} 2^k$ subgraphs that are equal in size to the original image can be obtained after the optical pattern and the UHF pattern are decomposed by the NSCT, respectively, as shown in Figures 12 and 13. It can be seen that the PD pattern is decomposed into multiscale, multidirectional high-frequency subgraphs and low-frequency subgraphs, which can retain the contour and detailed information of the source image well.

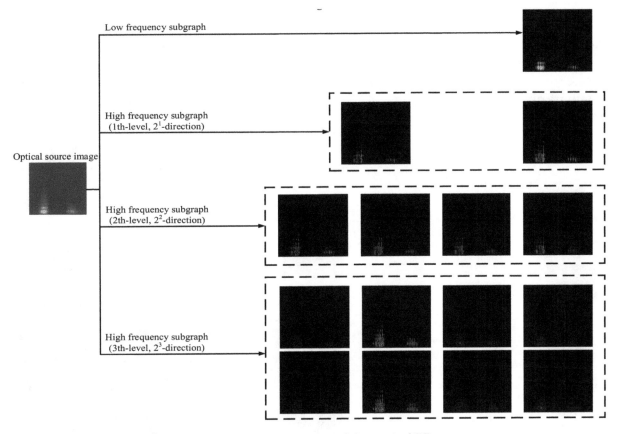

Figure 12. NSCT decomposition of the optical PD pattern.

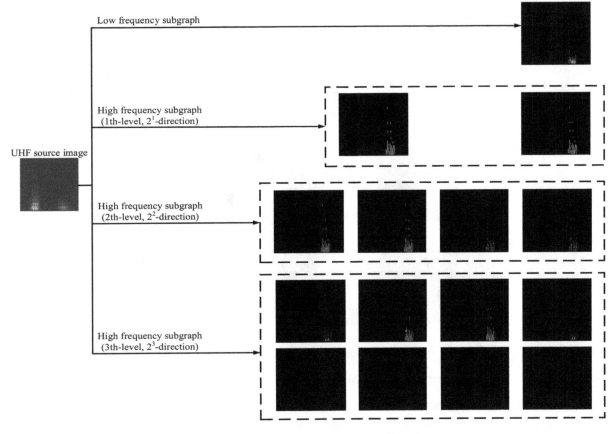

Figure 13. NSCT decomposition of the UHF PD pattern.

4.3. Fusion of the Photoelectric PD Pattern

According to the image fusion rule in Section 3.2., the corresponding fusion rule is performed on each of the low-frequency subgraph and the high-frequency subgraph, respectively, which can obtain one low-frequency photoelectric fusion subgraph and $\sum_{k=1}^{3} 2^k$ high-frequency photoelectric fusion subgraphs, as shown in Figure 14. The inverse NSCT transform is performed on the fused photoelectric subgraph to obtain a photoelectric fusion PD pattern, as shown in Figure 15.

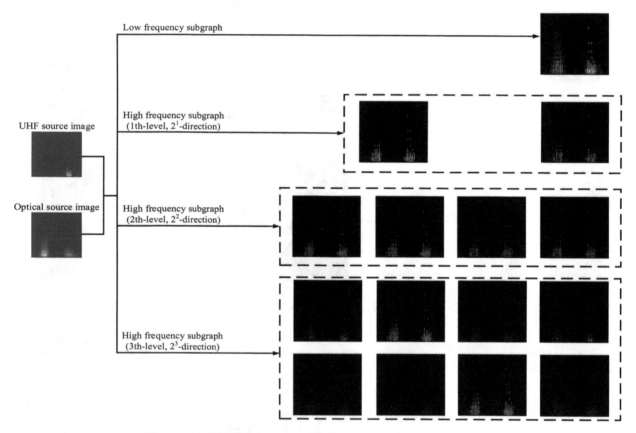

Figure 14. Fusion framework of optical and UHF patterns.

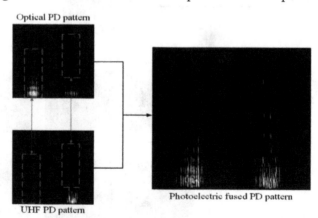

Figure 15. Photoelectric fusion PD pattern.

4.4. Feature Extraction and Dimension Reduction

In order to effectively reflect the image features, the feature parameters of the photoelectric fusion PD pattern are extracted. In this study, the eigenvector space is formed by Tamura texture features, gray-gradient symbiotic matrix, Hu invariant moment, and color moment of the image, a total of 28 characteristic parameters.

Among them, Tamura texture features theoretically include six components. But according to the Tamura feature extraction results of photoelectric fusion images, the characteristic parameters of Tamura texture features that are insensitive to photoelectric fusion images are ignored, and only the roughness, contrast, and directionality are used as features of pattern recognition. Then, 15 gray-gradient symbiotic matrix features, seven texture features, and three color moment features are, respectively, extracted to form the eigenvector space of the photoelectric fusion PD pattern [25–28].

However, there may be overlapping information between different features, resulting in multiple collinearity between the feature parameters. Meanwhile, too many dimensions of the feature vector can easily stress the training of the model, resulting in a lower recognition rate.

Therefore, in order to more fully characterize the characteristic information of the PD pattern and reduce the burden of the model, we use the principal component analysis (PCA) method to reduce the dimension of the eigenvector space. First, the factor correlation analysis of eigenvector space is carried out, gaining the value of KMO: 0.8356. The value of Bartlett spherical test is 132.96. It can be seen that there is strong partial correlation between feature vectors, which is suitable for dimensionality reduction by PCA [29]. In this study, according to the contribution rate of the feature factors, the first eight principal component factors with the cumulative contribution rate of 98% are selected as the input parameters of the recognition model. The PCA results are shown in Table 1.

Table 1. Contribution rate of principal component factors.

Factor Number	1	2	3	4	5	6	7	8	9	10	...	28
Contribution rate/%	65.00	11.46	7.78	6.01	3.02	2.21	1.58	0.86	0.64	0.58	...	0.01
Cumulated contribution rate/%	65.00	76.46	84.24	90.25	93.27	95.48	97.06	97.92	98.56	99.14	...	100

4.5. Pattern Recognition Results of Different PD Patterns

On the basis of the principal component factors described above, different classifiers are used to identify the PD experimental data. The classifiers used, in this study, were linear discriminant analysis (LDA) [30], k-nearest neighbor (KNN) [31], and support vector machine (SVM) algorithm [32]. LDA is a dimensionality reduction technology for supervised learning, which projects the sample onto a sorting line determining the category of the new sample based on the position of the projected point. KNN is a classification and regression method, which determines the classification by calculating which category of k-nearest samples in the feature space of a sample most belongs. The SVM algorithm maps points in low-dimensional space to high-dimensional spaces making them linearly separable, and then classifying them by the principle of linear partitioning.

In order to verify the applicability of the photoelectric fusion PD pattern proposed in this study, the above three classifiers were used to test the photoelectric fusion PD pattern samples, comparing with the recognition results of optical patterns and UHF patterns. At the same time, in order to verify the influence of different training sample numbers, four pattern recognition tests were carried out, respectively. In these four tests, the total number of training samples was 300, 240, 180, and 120, respectively, including all three types of defects. Therefore, the number of testing samples in these four tests were 60, 120, 180, and 240 correspondingly, including all three types of defects as well. It can be seen from the recognition results that different types of PD patterns have a significant influence on the recognition rate, as shown in Figure 16. The definition of recognition rate is:

$$Recognition\ rate\ =\ \frac{sample\ number\ of\ correct\ recognition}{number\ of\ testing\ sample}$$

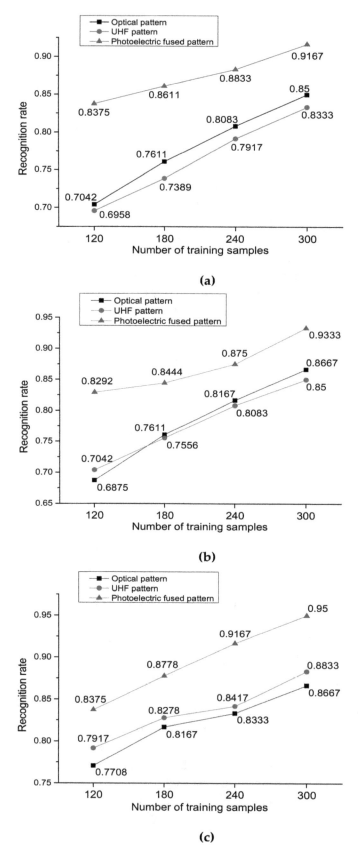

(a)

(b)

(c)

Figure 16. Recognition results of different classifiers. (**a**) linear discriminant analysis (LDA) classifier, (**b**) k-nearest neighbor (KNN) classifier, and (**c**) support vector machine (SVM) classifier.

According to Figure 16, the recognition rate of the photoelectric fusion pattern is the highest using any classifier. For all three types of PD patterns, the recognition rate increases with the increase of training samples. When using the SVM classifier with 300 training samples, the recognition rate of the photoelectric fusion pattern can reach up to 0.95. Moreover, the recognition rate of the photoelectric fusion pattern can still reach about 0.83 when the number of samples is only 120. Therefore, it can be concluded that the photoelectric fusion pattern can significantly improve the accuracy of PD pattern recognition.

In addition, when the number of training samples is the same, the accuracy of the recognition of the photoelectric fusion pattern by the three classifiers is higher than that of the optical pattern and the UHF pattern. Because photoelectric fusion pattern proposed in this study contains more abundant PD characteristic information, it can effectively improve the accuracy of the recognition. When the number of training samples is only 120, the average recognition rate of the three classifiers in each case is calculated in Figure 17, where three types of PD defects are, respectively, identified by different patterns. It can be seen that the recognition rate of the needle defect by the UHF pattern is especially low because of the loss of UHF signals. Moreover, the recognition rate of the free particle defect by the optical pattern is lower than the others because of the loss of optical signals. However, pattern recognition using photoelectric fusion pattern not only greatly increases the recognition rate of the needle defect and the free particle defect, but also slightly improves the recognition rate of the floating defect. Therefore, the proposed photoelectric fusion pattern has practicality.

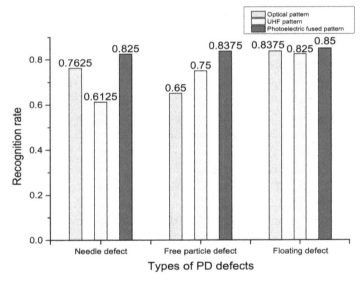

Figure 17. Average recognition rate of each defect under different kinds of patterns (with 120 training samples).

5. Conclusions

In this study, the optical-UHF integrated PD detection of GIL experimental platform is built, collecting PD patterns of three typical PD defects. Aiming at the PD pattern recognition problem of the GIL, an image fusion algorithm based on NSCT is proposed, which can fuse the optical PD pattern with the UHF PD pattern. Finally, three different classifiers are used to identify the photoelectric fusion PD pattern, which can verify the feasibility of this method. The conclusions are summarized as follows:

1) Due to the limitation of the PD detection principle and the influence of the GIL structure, the UHF pattern of the needle PD defect and the optical pattern of the free-particle PD defect have the loss of PD signals in the GIL. This phenomenon results in the reduction of the effective characteristic information in the PD pattern, which can reduce the pattern recognition accuracy of the PD.

2) The photoelectric fusion pattern can effectively avoid signals loss of UHF detection and optical detection in some situations, which can reduce the negative influence of false mode and pattern

aliasing on recognition. Through the photoelectric fusion pattern, the characteristic information of optical patterns and UHF patterns can complement each other, improving the accuracy and reliability of PD pattern recognition.

3) Compared with the optical pattern and the UHF pattern, the photoelectric fusion pattern can significantly improve the recognition rate of PD pattern recognition under the three kinds of classifiers, which can reach up to 0.95. In addition, when the number of training samples is small, the recognition rate can still reach about 0.83. Furthermore, the photoelectric fusion pattern not only greatly improves the recognition rate of the needle defect and the free particle defect, but the recognition accuracy of the floating defect can also be slightly improved. Therefore, the photoelectric fusion pattern has a good application effect.

Author Contributions: Data curation, Y.Z.; formal analysis, Y.Z.; funding acquisition, W.L.; methodology, Y.Z.; supervision, Y.Q., Y.X., G.S., and X.J.; writing—original draft, Y.Z.

References

1. Volcker, O.; Koch, H. Insulation co-ordination for gas-insulated transmission lines (GIL). *IEEE Trans. Power Deliv.* **2001**, *16*, 122–130. [CrossRef]
2. Magier, T.; Tenzer, M.; Koch, H. Direct Current Gas-Insulated Transmission Lines. *IEEE Trans. Power Deliv.* **2018**, *33*, 440–446. [CrossRef]
3. Miyazaki, A.; Takinami, N.; Kobayashi, S.; Yamauchi, T.; Hama, H.; Araki, T.; Nishima, H.; Hata, H.; Yamaguchi, H. Line constant measurements and loading current test in long-distance 275 kV GIL. *IEEE Trans. Power Deliv.* **2001**, *16*, 165–170. [CrossRef]
4. Schoffner, G. In A directional coupler system for the direction sensitive measurement of UHF-PD signals in GIS and GIL. In Proceedings of the 2000 Annual Report Conference on Electrical Insulation and Dielectric Phenomena (Cat. No.00CH37132), Victoria, BC, Canada, 15–18 October 2000; pp. 634–638.
5. Okubo, H.; Yoshida, M.; Takahashi, T.; Hoshino, T.; Hikita, M.; Miyazaki, A. Partial discharge measurement in a long distance SF/sub 6/gas insulated transmission line (GIL). *IEEE Trans. Power Deliv.* **1998**, *13*, 683–690. [CrossRef]
6. Li, J.; Han, X.; Liu, Z.; Yao, X. A Novel GIS Partial Discharge Detection Sensor With Integrated Optical and UHF Methods. *IEEE Trans. Power Deliv.* **2018**, *33*, 2047–2049. [CrossRef]
7. Han, X.; Li, J.; Zhang, L.; Pang, P.; Shen, S. A Novel PD Detection Technique for Use in GIS Based on a Combination of UHF and Optical Sensors. *IEEE Trans. Instrum. Meas.* **2019**, *68*, 2890–2897. [CrossRef]
8. Yongpeng, X.; Yong, Q.; Gehao, S.; Xiuchen, J.; Xiaoli, Z.; Zijie, W. Simulation analysis on the propagation of the optical partial discharge signal in I-shaped and L-shaped GILs. *IEEE Trans. Dielectr. Electr. Insul.* **2018**, *25*, 1421–1428. [CrossRef]
9. Cunha, A.L.D.; Zhou, J.; Do, M.N. The Nonsubsampled Contourlet Transform: Theory, Design, and Applications. *IEEE Trans. Image Process.* **2006**, *15*, 3089–3101. [CrossRef]
10. Wang, X.; Song, R.; Song, C.; Tao, J. The NSCT-HMT Model of Remote Sensing Image Based on Gaussian-Cauchy Mixture Distribution. *IEEE Access* **2018**, *6*, 66007–66019. [CrossRef]
11. Hanai, M.; Kojima, H.; Hayakawa, N.; Mizuno, R.; Okubo, H. Technique for discriminating the type of PD in SF6 gas using the UHF method and the PD current with a metallic particle. *IEEE Trans. Dielectr. Electr. Insul.* **2014**, *21*, 88–95. [CrossRef]
12. Ren, M.D.M.; Zhang, C.; Zhou, J. Partial Discharge Measurement under an Oscillating Switching Impulse: A Potential Supplement to the Conventional Insulation Examination in the Field. *Energies* **2016**, *9*, 623. [CrossRef]
13. Firuzi, K.; Vakilian, M.; Phung, B.T.; Blackburn, T.R. Partial Discharges Pattern Recognition of Transformer Defect Model by LBP & HOG Features. *IEEE Trans. Power Deliv.* **2019**, *34*, 542–550.
14. Mahyari, A.G.; Yazdi, M. Panchromatic and Multispectral Image Fusion Based on Maximization of Both Spectral and Spatial Similarities. *IEEE Trans. Geosci. Remote. Sens.* **2011**, *49*, 1976–1985. [CrossRef]
15. Yang, Y.; Tong, S.; Huang, S.; Lin, P. Multifocus Image Fusion Based on NSCT and Focused Area Detection. *IEEE Sens. J.* **2015**, *15*, 2824–2838.

16. Lei, T.; Feng, Z.; Zong-Gui, Z. In The nonsubsampled contourlet transform for image fusion. In Proceedings of the 2007 International Conference on Wavelet Analysis and Pattern Recognition, Beijing, China, 2–4 November 2007; pp. 305–310.

17. Bhatnagar, G.; Wu, Q.M.J.; Liu, Z. Directive Contrast Based Multimodal Medical Image Fusion in NSCT Domain. *IEEE Trans. Multimed.* **2013**, *15*, 1014–1024. [CrossRef]

18. Bhateja, V.; Patel, H.; Krishn, A.; Sahu, A.; Lay-Ekuakille, A. Multimodal Medical Image Sensor Fusion Framework Using Cascade of Wavelet and Contourlet Transform Domains. *IEEE Sens. J.* **2015**, *15*, 6783–6790. [CrossRef]

19. Shao, L.; Kirenko, I. Coding Artifact Reduction Based on Local Entropy Analysis. *IEEE Tran. Consum. Electron.* **2007**, *53*, 691–696. [CrossRef]

20. Jia-Shu, Z.; Cun-Jian, C. In Local variance projection log energy entropy features for illumination robust face recognition. In Proceedings of the 2008 International Symposium on Biometrics and Security Technologies, Islamabad, Pakistan, 23–24 April 2008; pp. 1–5.

21. Zhang, L.; Zhang, L.; Mou, X.; Zhang, D. FSIM: A Feature Similarity Index for Image Quality Assessment. *IEEE Trans. Image Process.* **2011**, *20*, 2378–2386. [CrossRef]

22. Li, H.; Qiu, H.; Yu, Z.; Zhang, Y. Infrared and visible image fusion scheme based on NSCT and low-level visual features. *Infrared Phys. Technol.* **2016**, *76*, 174–184. [CrossRef]

23. Zhu, Z.; Zheng, M.; Qi, G.; Wang, D.; Xiang, Y. A Phase Congruency and Local Laplacian Energy Based Multi-Modality Medical Image Fusion Method in NSCT Domain. *IEEE Access* **2019**, *7*, 20811–20824. [CrossRef]

24. Qu, X.-B.; Yan, J.-W.; Xiao, H.-Z.; Zhu, Z.-Q. Image Fusion Algorithm Based on Spatial Frequency-Motivated Pulse Coupled Neural Networks in Nonsubsampled Contourlet Transform Domain. *Acta Autom. Sin.* **2008**, *34*, 1508–1514. [CrossRef]

25. Amadasun, M.; King, R. Textural features corresponding to textural properties. *IEEE Trans. Syst. Man Cybern.* **1989**, *19*, 1264–1274. [CrossRef]

26. Tamura, H.; Mori, S.; Yamawaki, T. Textural Features Corresponding to Visual Perception. *IEEE Trans. Syst. Man Cybern.* **1978**, *8*, 460–473.

27. Weng, T.; Yuan, Y.; Shen, L.; Zhao, Y. In Clothing image retrieval using color moment. In Proceedings of the 2013 3rd International Conference on Computer Science and Network Technology, Dalian, China, 12–13 October 2013; pp. 1016–1020.

28. Soh, L.; Tsatsoulis, C. Texture analysis of SAR sea ice imagery using gray level co-occurrence matrices. *IEEE Trans. Geosci. Remote Sens.* **1999**, *37*, 780–795. [CrossRef]

29. Gonzalez-Audicana, M.; Saleta, J.L.; Catalan, R.G.; Garcia, R. Fusion of multispectral and panchromatic images using improved IHS and PCA mergers based on wavelet decomposition. *IEEE Trans. Geosci. Remote Sens.* **2004**, *4*, 1291–1299. [CrossRef]

30. Wei-Shi, Z.; Jian-Huang, L.; Yuen, P.C. GA-fisher: A new LDA-based face recognition algorithm with selection of principal components. *IEEE Trans. Syst. Man Cybern. Part B* **2005**, *35*, 1065–1078.

31. Xiong, J.; Zhang, Q.; Sun, G.; Zhu, X.; Liu, M.; Li, Z. An Information Fusion Fault Diagnosis Method Based on Dimensionless Indicators With Static Discounting Factor and KNN. *IEEE Sens. J.* **2016**, *16*, 2060–2069. [CrossRef]

32. Laufer, S.; Rubinsky, B. Tissue Characterization With an Electrical Spectroscopy SVM Classifier. *IEEE Trans. Biomed. Eng.* **2009**, *5*, 525–528. [CrossRef]

12

Electrical Detection of Creeping Discharges over Insulator Surfaces in Atmospheric Gases under AC Voltage Application

Michail Michelarakis [1,*], **Phillip Widger** [1], **Abderrahmane Beroual** [2] **and Abderrahmane (Manu) Haddad** [1]

[1]   Advanced High Voltage Engineering Research Centre, School of Engineering, Cardiff University, The Parade, Cardiff CF24 3AA, UK

[2]   École Centrale de Lyon, University of Lyon, Ampère CNRS UMR 5005, 36 Avenue Guy Collongue, 69134 Écully, France

*   Correspondence: MichelarakisM@cardiff.ac.uk

Abstract: Creeping discharges over insulator surfaces have been related to the presence of triple junctions in compressed gas insulated systems. The performance of dielectric materials frequently utilised in gaseous insulating high voltage applications, stressed under triple junction conditions, has been an interesting topic approached through many different physical perspectives. Presented research outcomes have contributed to the understanding of the mechanisms behind the related phenomena, macroscopically and microscopically. This paper deals with the electrical detection of creeping discharges over disc-shaped insulator samples of different dielectric materials (polytetrafluoroethylene (PTFE), epoxy resin and silicone rubber) using atmospheric gases (dry air, N_2 and CO_2) as insulation medium in a point-plane electrode arrangement and under AC voltage application. The entire approach implementation is described in detail, from the initial numerical field simulations of the electrode configuration to the sensing and recording devices specifications and applications. The obtained results demonstrate the dependence of the generated discharge activity on the geometrical and material properties of the dielectric and the solid/atmospheric gas interface. The current work will be further extended as part of a future extensive research programme.

Keywords: creeping discharge; AC voltage; point-plane; atmospheric gases; flashover voltage; polytetrafluoroethylene (PTFE); epoxy resin; silicone rubber

1. Introduction

In gaseous insulation applications, triple junctions are defined as points where the gaseous medium, dielectric material and electrode meet leading to local electric field enhancements. When a certain electric field level is reached, the initiation of discharge activity is possible. In the case of their appearance, such phenomena become present at voltages below the optimised rated operating and withstand levels of the affected apparatus, introducing additional concerns in the overall effort of designing effective and reliable insulating systems. Some significant examples of triple junctions are incorporated into the design of Gas Insulated Lines (GIL), Gas Insulating Switchgear (GIS), transformer bushings and Gas Circuit Breakers (GCB).

The most common discharge phenomenon linked with enhanced electric fields in the vicinity of solid dielectric-gas interface is the streamer creeping discharge. Over the years, major contributions have been published, approaching the development and propagation of streamers over dielectric surfaces from several different experimental and computational perspectives. Characteristics of single streamer events in homogenous electric field arrangements have been reported [1,2] where the dependence of the propagation velocity and associated electric field over the dielectric material is described. In [3], the effect of the dielectric permittivity on streamer propagation along insulating surfaces using electrical and optical techniques is reported. The morphology and propagation length of creeping discharges and their dependence on voltage waveform, voltage levels, dielectric material and gaseous medium have been extensively studied in more recent works [4–7]. Reported works using dust figure [8] and Pockels effect methods [9] always constitute very interesting optical approaches. Another very important aspect, related to the phenomena described above, is the surface charge accumulation and its impact on the flashover voltage [10,11] and degradation of solid insulators under tests [12].

This paper examines the development of creeping discharge over disc-shaped insulator samples of different dielectric materials, namely polytetrafluoroethylene (PTFE), epoxy resin and silicone rubber, which are frequently utilised in high voltage technology applications. These samples are also insulated by a gaseous medium of either dry air, nitrogen (N_2) or carbon dioxide (CO_2). A needle-plane electrode configuration is employed with a strongly non-uniform electric field at 50 Hz AC voltage applied to the needle electrode, in order to replicate triple junction conditions. The implementation of the test procedure is described from the early stages of the numerical field simulation process of the electrodes arrangement using a Finite Element Solver (FEM) software package. The aim of the accurate modeling of the electrode geometry is the optimisation of the electric field distribution along the insulator surface. Detection of surface discharges is performed by means of sensing associated currents using a high-sensitivity, high frequency current transformer (HFCT) ranging from a few kHz up to several MHz bandwidth. The current transformer sensing technique provides several performance advantages over other techniques implementing different physical principles [13], such as low power loss, high accuracy and no need for further amplification of the output. Additionally, the overall convenience of a HFCT installation, together with the provided electrical isolation between the system under test and the high-cost recording devices, make the technique a very practical choice for laboratory-based high voltage testing.

2. Design and Implementation of the Test Procedure

2.1. Electrodes Configuration Simulation and Electric Field Computation

As mentioned in the previous section, the aim of the test procedure is to replicate triple junction conditions in the vicinity of a dielectric insulator surface surrounded by a gaseous insulating medium. For that reason, a needle-plane electrode configuration is employed. A two-dimensional illustration along with an actual picture of the electrode configuration, showing the disc-shaped insulator sample under test conditions are shown in Figure 1.

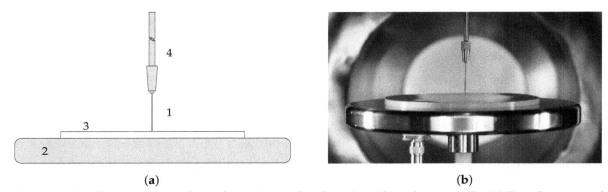

(a)   (b)

Figure 1. Needle-plane electrode configuration with a disc-shaped insulator sample: (**a**) Two-dimensional illustration of the configuration with the following numbered features: 1-needle electrode, 2-plane electrode, 3-disc-shaped insulator sample and 4-needle holder; (**b**) Actual picture of the practical test configuration as set-up inside the pressure vessel.

The needle is made of tungsten with a high-precision tip diameter of $20 \pm 0.5\,\mu m$, $0.51\,mm$ shaft diameter and a length of $32\,mm$. It is attached to a stainless-steel needle holder placed perpendicularly and in close proximity to the insulator sample disc center. The plane electrode incorporates a dull polished stainless-steel planar surface of $150\,mm$ diameter and $15\,mm$ thickness and is electrically separated from the rest of the mounting system with a nylon threaded rod. The entire configuration is placed vertically and centered inside a cylindrical 90 l dull polished stainless-steel pressure vessel rated up to 10 bars operational gauge pressure, which has a diameter of $480\,mm$ and a height of $500\,mm$.

The described design is further examined from the numerical electric field perspective using a Finite Element Method (FEM) simulation model. The detailed geometry is transferred to COMSOL Multiphysics® through its integrated three-dimensional geometry builder, with some geometry simplifications that can improve significantly the computational time without affecting the quality and reliability of the generated result. As the design is largely symmetrical around the central axis of the entire configuration, only half of the geometry needs to be computationally solved without affecting the accuracy of the computed results. It is expected that the maximum electric field stress will appear at the needle tip because of its small diameter and short distance from the zero potential plane electrode. Figure 2 shows the simulated full geometry of the test system, as designed within COMSOL Multiphysics®, together with a generated illustration of the three-dimensional equipotential surfaces (isosurfaces) after solving the simulation model. For the example shown in Figure 2b, the considered insulator sample is made of PTFE with a 4 mm thickness; the insulating gaseous medium being air at standard atmospheric conditions. The applied voltage is set to 1 V.

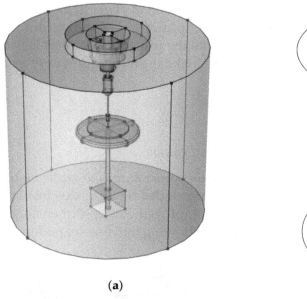
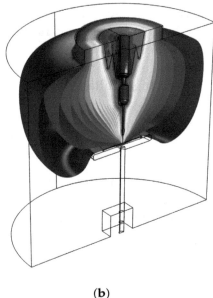

(a) (b)

Figure 2. Finite Element Method (FEM) simulation model: (**a**) Full geometry including electrodes arrangement and surrounding volume; (**b**) Three-dimensional equipotential surfaces (isosurfaces) within the half symmetrical volume of the initial geometry.

The distribution of the electric field on the surface of the insulator sample is of great significance as the possibility of large irregularities may lead to inaccurate observations. In a broad sense, in case of discharge activity appearance, it will initiate from the sharp needle tip and will have equal chance of propagating in any direction towards the edges of the disc insulator. Figure 3 shows the computed electric field distribution along the diameter of the insulator sample top surface. The presented electric field values are normalised to the maximum surface electric field that appears at the point with the shortest distance from the needle tip.

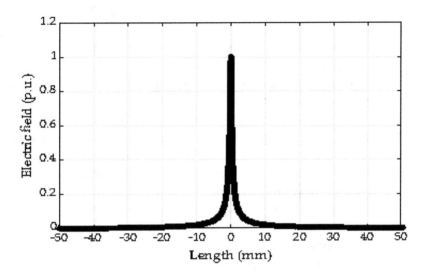

Figure 3. Electric field distribution along insulator surface. Here, 0 mm represents the center of the top surface of the insulator sample while ±50 mm the edges of it. The presented values are normalised based on the maximum computed value.

2.2. Experimental Set-Up

Figure 4 summarises the main parts of the experimental configuration used for the purposes of this work. For the AC voltage application, a 50 kV/3.75 kVA transformer is used, which is controlled through an isolating transformer and a voltage regulator with adjustment of the voltage level on the low-voltage side. An RC voltage divider of ratio 3750:1 is used to scale down the generated voltage to safe measurable levels. Voltage is applied to the pressure vessel through a bushing rated up to 39 kV AC rms. As described previously, the electrodes configuration, together with the insulator samples under test, are vertically and horizontally centered inside the cylindrical stainless-steel dull polished chamber. The side apertures are covered with stainless-steel blanking plates, preventing ambient light entering into the pressure vessel.

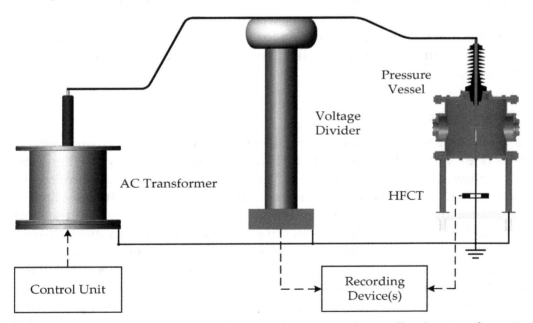

Figure 4. Experimental set-up for AC voltage application to the needle-plane configuration and insulator samples.

The insulator samples tested are disc-shaped with a diameter of 100 mm and thicknesses of 4 and 6 mm. Three different materials are considered namely polytetrafluoroethylene (PTFE), silicone rubber and epoxy resin with relative permittivities (ε_r) of 2.1, 2.9 and 3.5, respectively. The PTFE and epoxy resin samples surfaces were polished with sandpaper of 800 grit gradually increasing up to 7000 grit size. The silicone rubber samples were vacuum casted using a highly polished stainless-steel mold, followed by at least 24 h curing at 50 °C. For all cases, arithmetical mean height (R_a) is maintained between 0.6–0.8 μm. The insulator samples are cleaned with high-purity isopropyl alcohol and dried for at least 4 h at 50 °C. Between drying and placing the samples inside the final test set-up, a brief time interval of 30 min is kept allowing the samples to return to room temperature levels. After each test series, each insulator is examined for traces, indicating degradation/damage, and if any is observed on the surface, it is replaced with a new unused one. As described previously, the needles are made of pure tungsten, a material well-known for its high melting temperature, allowing for a sufficiently large number of tests without degradation of the needle tip quality. During the creeping discharge detection tests, a maximum of twenty voltage applications are made before the needle is replaced. For the case of flashover tests, the needle is used for a maximum of ten attempts prior to being replaced, assuming that no abrupt deviations in the readings of successive attempts occur which would indicate a damaged needle. For the purposes of the presented work, the test vessel housing the test electrode configuration is filled and tested using three different atmospheric gases, respectively: dry air, nitrogen (N_2) and carbon dioxide (CO_2) at 1 bar absolute pressure. Preceding gas injection, the pressure vessel is vacuumed for 30 min after a vacuum level of -1000 mbar gauge pressure is reached. This additional vacuum is held in an attempt to maintain low humidity levels inside the test chamber.

Two different, although similar, patterns for the AC voltage application are followed. For flashover tests, the applied voltage is manually increased with a rate of 1 kV/s until flashover occurs. A lower sensitivity, 0.1 V/A, current transformer is used for triggering the recording device. The last full AC-cycle recorded, preceding the flashover event, corresponds to the measured rms flashover voltage. Ten flashover tests are performed for each insulator sample/gaseous medium combination. The arithmetic mean value and standard deviation for the flashover voltage for each case are specified as implied from the relevant standards [14]. For creeping discharge detection tests, test voltages are specified as a percentage of the flashover voltage for each case study. Following that, the test voltage is gradually reached with an increment of 2% per second resulting in a ramp duration of 50 s for each case. After that, the voltage level is maintained for a maximum of 10 s in order to avoid overstress of the insulator sample and the utilised needle. Between two successive voltage applications, a time interval of at least 2 min is maintained.

2.3. Current Sensing & Recording

Several methods have been reported for the detection of electrical phenomena within the high-frequency (HF), very-high-frequency (VHF) and ultra-high-frequency (UHF) regions, including acoustic, optical, chemical and electrical methods. These methods find wide application in the detection of partial discharges [15–19]. In this work, direct measurement of the currents associated with creeping discharge is performed using a high-sensitivity 5 V/A, high frequency current transformer (HFCT) ranging from 4.8 kHz up to 400 MHz. Considering that the test configuration is installed inside a grounded metal fabricated pressure vessel, it is important that the generating signals will exit the test chamber and, consequently, reach the sensing device undistorted and unattenuated. For that purpose, a flange mount SMA connector is installed at the bottom side of the plane electrode of Figure 1. A 50 Ω RG-405 coaxial cable connects the plane electrode with a pre-installed high-pressure rated coaxial feedthrough. A coaxial cable connected to the low-pressure side of the feedthrough is then connected to the configuration shown in Figure 5, similar implementations of which were presented in [20,21].

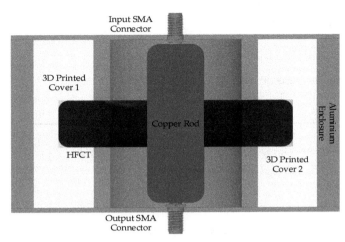

Figure 5. HFCT enclosure configuration.

Two SMA connectors are mounted on the top and bottom of a C101 alloy copper rod. The diameter of this central conductor was calculated based on the $50\,\Omega$ coaxial characteristic impedance principle using Equation (1) [15],

$$Z_0 = \frac{138}{\sqrt{\varepsilon_{\text{air}}}}\log\left(\frac{d_1}{d_2}\right) \tag{1}$$

where, Z_0 is the characteristic impedance, d_1 the diameter of the HFCT aperture, d_2 the diameter of the copper conductor and ε_{air} the relative permittivity of atmospheric air. The entire configuration is enclosed in an EMI/RFI shielded aluminium enclosure. The scattering parameters of the configuration were measured using an R&S® ZVL Vector Network Analyzer $(9\,\text{kHz} - 6\,\text{GHz})$, following a $50\,\Omega$ two-port calibration, and the results are presented in Figure 6. As shown, the upper cut-off frequency is measured at $359.46\,\text{MHz}$, which is quite close to the bandwidth rating of the HFCT. The output side is terminated through a short-circuited $20\,\text{dB}/50\,\text{W}$, DC-8.5 GHz, high-power fixed attenuator, preventing possible reflections reaching the sensing configuration.

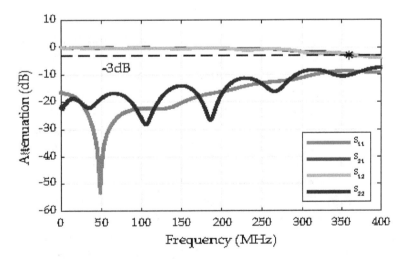

Figure 6. S-parameters of the configuration shown in Figure 5, for the frequency range corresponding to the rating of the used HFCT.

The recording of the voltage waveform and associated currents was performed in two different stages: high-resolution single trigger recordings were made using a Teledyne LeCroy HDO6104, 1 GHz, $2.5\,\text{GS/s}$, 12-bit oscilloscope while, for multiple successive trigger events, a large buffer PicoScope® 5000 series, 200 MHz, 250 MS/s, 12-bit was used.

3. Experimental Results and Discussion

3.1. Flashover Tests

A series of flashover tests were performed using all the solid dielectric-insulating gas combinations considered in this work. Ten flashover events were recorded for each case. These tests were implemented in order to quantify the equivalent threshold voltage level for creeping discharges as a percentage of the corresponding mean flashover voltages of each case. The obtained results are summarised in Figure 7. For the case of the 6 mm thick PTFE sample in dry air, no flashover events were recorded within the allowed application voltage levels up to 39 kV rms, hence, it is not included in the graph.

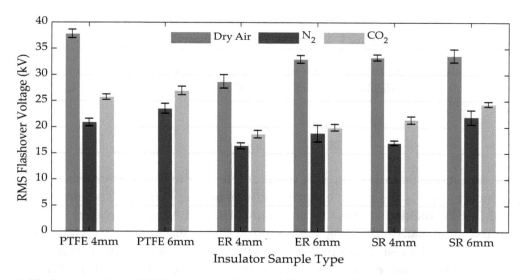

Figure 7. Flashover voltage (FOV) test results for six different insulator samples in dry air, nitrogen (N_2) and carbon dioxide (CO_2), all at 1 bar absolute pressure. Here, arithmetic mean values are depicted together with the corresponding standard deviations after 10 voltage applications. ER stands for epoxy resin while SR for silicone rubber.

For all the different sample types, dry air showed the best insulating performance as compared to the two other gaseous mediums. Additionally, pure CO_2 showed a better performance compared to pure nitrogen, with the difference in FOVs being larger for 4 mm samples compared to those of 6 mm thickness. Focusing on the calculated standard deviations, N_2 showed a more stable insulating behaviour for thinner samples, for all the materials compared to thicker ones insulated with the same gas, while CO_2 seemed to have on average the lowest standard deviations. The electrode system with PTFE ($\varepsilon_{PTFE} = 2.1$) had the highest resistance to flashover for all gaseous mediums and sample thicknesses considered. Despite the results of the other two materials being close, the configuration with silicone rubber ($\varepsilon_{SR} = 2.9$) seemed to have a slightly higher withstand voltage compared to that with epoxy resin ($\varepsilon_{ER} = 3.5$). Figure 8 shows the dependence of the computed electric field and measured flashover voltages on the different values of dielectric permittivity. The values of each curve are normalised to the, computed or measured, corresponding value for PTFE. It is obvious that, as the dielectric permittivity increases, the electric field on the center of the disc insulator sample also increases, while, the flashover voltage levels decrease for all the cases considered. Such an observation can be correlated with results reported in research works [5] where the stopping length of creeping discharges is examined, showing that insulator discs made of higher dielectric permittivity materials are responsible for the propagation of longer streamer channels on their surface for both N_2 and CO_2, but also for SF_6. AC breakdown test results of point-plane arrangements for different gap distances, presented in [7], also showed that the insulating performance of CO_2 is better when compared to N_2, without a dielectric material between the electrodes.

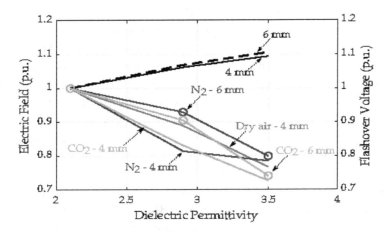

Figure 8. Computed electric field (black curves) and measured flashover voltages over dielectric permittivity variation. All the curves are normalised over the corresponding value of $\varepsilon_r = 2.1$ for PTFE.

3.2. High-Resolution Recordings

High-resolution recordings were obtained through single trigger events of the recording device at dual-channel, sampling at 1.25 GS/s rate. Two signals of 40 ms duration each were captured, corresponding to two full AC-cycles. The applied voltage levels considered for each case correspond to 85% of the flashover levels presented in previous section. That way, it is possible to compare between the different cases, where the variations in the FOVs do not allow the application of the same test voltage levels. Additionally, a 15% safety margin from flashover events was maintained, avoiding unwanted stress of the test configuration in case of flashover. Figure 9 shows the obtained recordings for PTFE of 4 mm thickness in dry air, nitrogen (N_2) and carbon dioxide (CO_2) at 1 bar absolute pressure. Because similar patterns were observed for the remaining insulator samples types, only the above selected results were shown here.

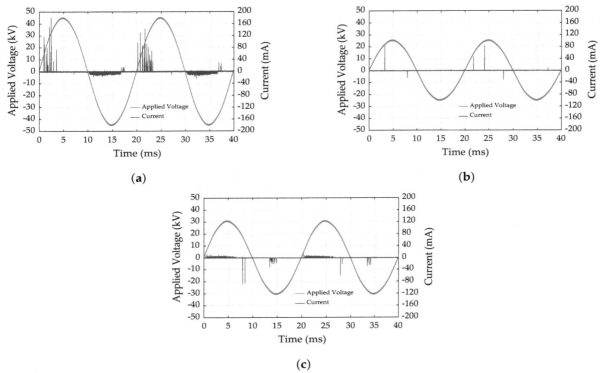

Figure 9. PTFE of 4 mm thickness in (**a**) dry air at 1 bar absolute pressure and 31.74 kV rms applied voltage, (**b**) N_2 at 1 bar absolute pressure and 17.74 kV rms applied voltage and (**c**) CO_2 at 1 bar absolute pressure and 21.61 kV rms applied voltage.

For the case of dry air in Figure 9a, a high density of positive and negative polarity current pulses is observed. It is obvious that positive polarity current peaks amplitudes dominate over the negative spikes. However, significant differences in the current pulse characteristics were observed. For nitrogen (N_2), as shown in Figure 9b, a very different behaviour is observed when compared to dry air. During the positive AC half cycle of the applied voltage, a small number of positive current pulses, although quite high in amplitude, were detected while absence of activity during the negative half cycle is obvious. Finally in Figure 9c, the behaviour of carbon dioxide (CO_2) differs significantly from both dry air and N_2. Low amplitude positive current pulses were detected during the positive half cycle of the applied voltage while, during the negative half cycle, higher amplitude but less dense negative polarity current pulses were observed.

Wide bandwidth sensing and recording devices combined with high sampling rates allow the capturing of current pulses corresponding to creeping discharge events. For the presented test results, applied voltages were very close to the FOVs and most of the recordings consist of superimposed pulses, indicating multiple discharge events occurring in very fast time frames on the surface of the insulator sample. Single streamer current pulses are also present however, less frequent. Current pulses are classified into two different categories: those having the same polarity as the applied voltage and those with opposite polarity. Positive polarity pulses during the positive half cycle of the AC voltage application are present for all the cases with varying amplitudes and characteristics. Examples of positive polarity pulses exported from the recordings in Figure 9 are shown in Figure 10 with their time resolved characteristics being summarised in Table 1. Negative polarity current pulses are also detected, the recordings of which are shown in Figure 11, with their corresponding characteristics summarised in Table 2. Similar pulse shapes were observed for all the insulator sample materials. The apparent charge, which is also depicted, was calculated by integrating the recorded pulse over the corresponding time domain.

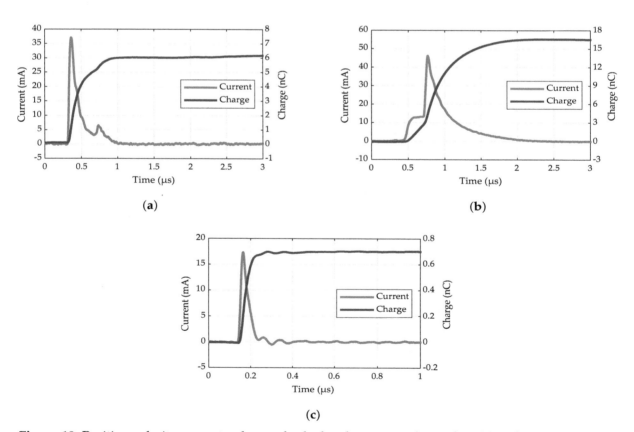

Figure 10. Positive polarity current pulses and calculated apparent charge for PTFE of 4 mm thickness in: (**a**) dry air, (**b**) nitrogen (N_2) and (**c**) carbon dioxide (CO_2).

Table 1. Positive polarity current pulses characteristics.

Gaseous Medium	Current Peak (mA)	Rise Time (ns)	Fall Time (ns)	Width (ns)	Duration (ns)
Dry air	37.07	28.49	263.42	110.26	326.21
N_2	46.14	274.25	662.93	192.94	975.02
CO_2	17.35	13.77	49.06	35.54	73.08

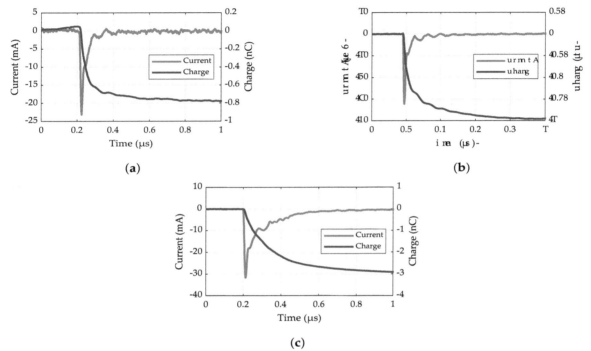

(a) **(b)**

(c)

Figure 11. Negative polarity current pulses and calculated apparent charge for PTFE of 4 mm thickness in: (**a**) dry air, (**b**) nitrogen (N_2) and (**c**) carbon dioxide (CO_2).

Table 2. Negative polarity current pulses characteristics.

Gaseous Medium	Current Peak (mA)	Rise Time (ns)	Fall Time (ns)	Width (ns)	Duration (ns)
Dry air	−23.19	51.81	7.95	16.14	65.00
N_2	−32.23	41.63	5.76	12.94	52.04
CO_2	−31.63	232.38	8.53	38.58	247.65

Significant differences in the measured rise and fall times are observed. For all the tests performed for the purposes of the presented work, the fastest rise and fall times, for these kinds of pulses, were observed in carbon dioxide (CO_2) while the slowest were seen in nitrogen (N_2). The calculated apparent charge appears to be dependent and proportional to these values, with the highest charge being observed for the case of N_2 in Figure 10b. In [22], the authors describe the correlation between rise and fall times with the electron avalanche process during partial discharge (PD) and electron drift after the full extension of the discharge activity. For the single double-exponential shaped pulses, such an approach can provide valuable information. Figure 11 includes captures of negative polarity current pulses during the negative half cycle of the AC applied voltage for dry air and CO_2, while for nitrogen the negative polarity current pulse appears during the positive AC half cycle. Pulses of opposite polarity to that of the applied voltage were observed in all measurements. Similar findings, using the Pockels effect optical method, were reported in [9] and identified as back-discharges. Following the provided description for the negative discharges during positive polarity of the applied voltage, the negative charge expands uniformly on the positively charged surface neutralizing the positive charges. The opposite process occurs for positive discharges during the negative half cycles.

3.3. Current Peaks Density Recordings

High resolution recordings, within the GS/s sampling rate levels, usually do not allow capturing of multiple trigger events which are able to demonstrate the repeatability of the generated phenomena, during the same applied voltage test. For that purpose, a lower sampling rate, larger buffer size, fast-triggering device was employed. In this way, the quick recording of multiple full AC-cycle trigger events for the same applied voltage attempt was possible without overstressing the test object. Post-processing of these recordings involves isolating the peak values of the current pulses within the time domain and incorporate them in a concatenating plot where colour grading was applied based on the density of these peaks. The colour identification was normalised over the total number of detected peaks based on their polarity. Figures 12–14 illustrate part of the results using the described technique for the cases of 4 mm thickness insulator samples while Figure 15 includes selected datasets for 6 mm thick samples. Here, 25 trigger events of 20 ms duration each are considered, resulting in a total duration of 500 ms for each case. For all the presented measurements, the applied voltage is approximately equal to the 85% of the corresponding FOV.

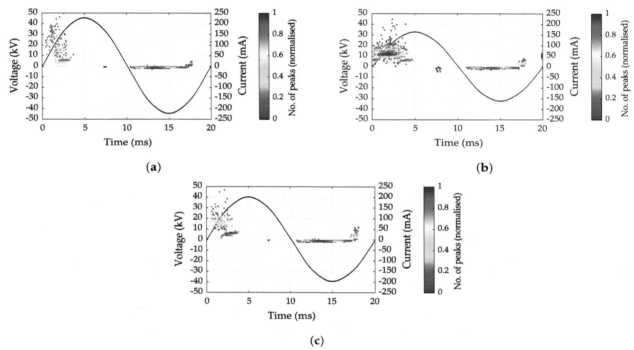

Figure 12. Current peaks densities for 4 mm thickness insulator samples in dry air at 1 bar absolute pressure: (**a**) PTFE, (**b**) epoxy resin and (**c**) silicone rubber.

In Figure 12, for samples in dry air, 4 mm thickness samples of PTFE and silicone rubber have fairly similar behaviour while epoxy resin shows a higher density of discharge activity during the positive half cycle of the applied voltage. For samples of the same thickness in N_2 in Figure 13, negative polarity pulses during the negative half cycle were not detected. For silicone rubber, a high density of low amplitude positive polarity discharges was found during the positive half cycles while, for epoxy resin samples, very high positive peaks were captured, possibly indicating that any increase in the applied voltage could have resulted in a flashover. PTFE shows a well distributed activity within the first quarter of the applied voltage waveform. It could be said that CO_2 in Figure 14 shows the most stable behaviour. PTFE and silicone rubber once again are very similar while PTFE shows less dense activity during the negative polarity of the applied voltage. For tests with the epoxy resin sample, no detectable pulses were recorded during the second half of the AC waveform for that specific applied voltage level. Tests with thicker samples in Figure 15 show that PTFE in CO_2 shows decreased peak amplitudes during the negative half cycle, while epoxy resin in N_2 shows considerably lower peaks and high repeatability for the 6 mm samples compared to the 4 mm. Silicone rubber of 6 mm thickness in dry air shows increased resistance to creeping discharge activity in comparison with the 4 mm sample of the same material, especially when the polarity of the applied voltage is positive. Overall, the dependence of the insulator sample thickness is visible for the presented cases. For tests with epoxy resin of 4 mm thickness in CO_2, the applied voltage margin for occurrence of the discharge activity in the negative half cycle until flashover is small. Results of N_2 combined with 4 mm thickness samples need further investigation, especially for the case of epoxy resin.

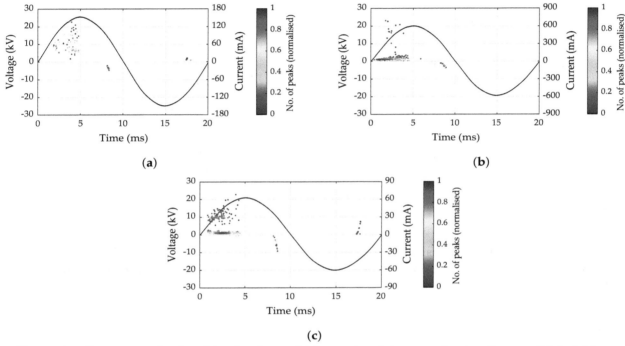

Figure 13. Current peaks densities for 4 mm thickness insulator samples in nitrogen (N_2) at 1 bar absolute pressure: (**a**) PTFE, (**b**) epoxy resin and (**c**) silicone rubber.

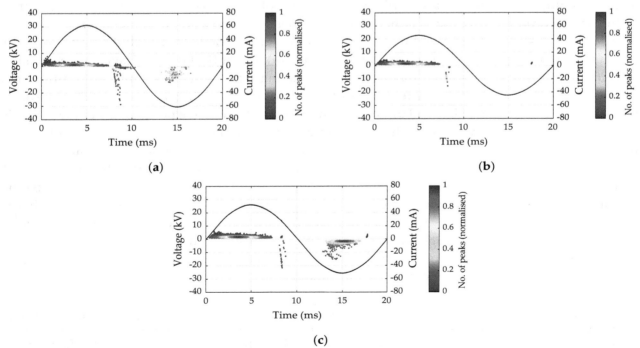

Figure 14. Current peaks densities for 4 mm thickness insulator samples in carbon dioxide (CO_2) at 1 bar absolute pressure: (**a**) PTFE, (**b**) epoxy resin and (**c**) silicone rubber.

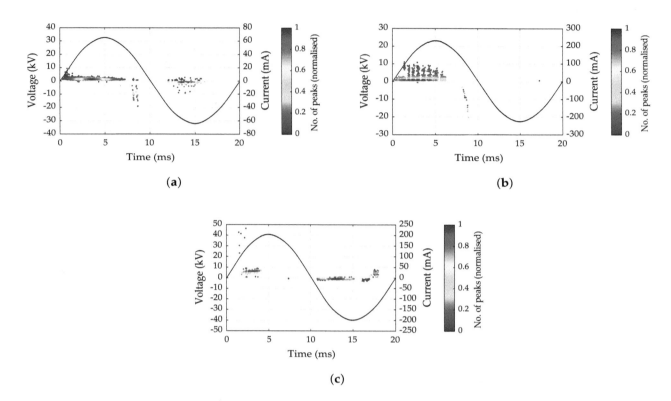

Figure 15. Current peaks densities for 6 mm thickness insulator samples: (**a**) PTFE in CO_2 at 1 bar absolute pressure, (**b**) epoxy resin in N_2 at 1 bar absolute pressure and (**c**) silicone rubber in dry air at 1 bar absolute pressure.

4. Conclusions

In this paper, electrical detection of creeping discharge under AC voltage application was presented. The complete procedure, starting from the numerical simulation of the electrode configuration until the final test implementation was described in detail. Emphasis was given to the selection of current sensing and recording devices so that their cut-off frequency is kept as high as possible.

Flashover tests results showed that, for the examined electric field distribution and for all the examined insulating gaseous media (dry air, CO_2 and N_2), lower flashover voltages were measured for electrode configurations with insulating materials having higher relative permittivity. Furthermore, high-frequency recordings of the current pulses associated with creeping discharges clearly indicated that there is more significant electrical discharge activity when the relative permittivity of the insulating material is higher. For lower permittivity insulator sample materials, fewer current pulses were detected however, the activity was spread greatly within a range corresponding to relatively higher amplitudes. It could be said that, the higher the difference between the relative permittivities of the insulator sample material and the insulating gaseous medium, the stronger the detected creeping discharge events are, especially for the cases of dry air and N_2. For the case of CO_2, the higher the relative permittivity of the insulator material the narrower the margin between creeping discharge and flashover is. Back-discharges were detected for all the cases tested in this investigation.

Future research will extend the current work and further explore the presented scenario using different physical perspectives and detection techniques.

Author Contributions: Conceptualization, M.M., P.W., A.B. and A.H.; methodology, M.M.; software, M.M.; validation, P.W., A.B. and A.H.; formal analysis, M.M.; investigation, M.M.; data curation, M.M.; writing—original draft preparation, M.M.; writing—review and editing, M.M., P.W., A.B. and A.H.; visualization, M.M.; supervision, P.W., A.B. and A.H.

References

1. Allen, N.L.; Mikropoulos, P.N. Streamer propagation along insulating surfaces. *IEEE Trans. Dielectr. Electr. Insul.* **1999**, *6*, 357–362, doi:10.1109/94.775623. [CrossRef]

2. Allen, N.L.; Mikropoulos, P.N. Dynamics of streamer propagation in air. *J. Phys. D Appl. Phys.* **1999**, *32*, 913–919, doi:10.1088/0022-3727/32/8/012. [CrossRef]

3. Akyuz, M.; Gao, L.; Cooray, V.; Gustavsson, T.G.; Gubanski, S.M.; Larsson, A. Positive streamer discharges along insulating surfaces. *IEEE Trans. Dielectr. Electr. Insul.* **2001**, *8*, 902–910, doi:10.1109/94.971444. [CrossRef]

4. Sadaoui, F.; Beroual, A. DC creeping discharges over insulating surfaces in different gases and mixtures. *IEEE Trans. Dielectr. Electr. Insul.* **2014**, *21*, 2088–2094, doi:10.1109/TDEI.2014.004486. [CrossRef]

5. Sadaoui, F.; Beroual, A. AC creeping discharges propagating over solid—gas interfaces. *IET Sci. Meas. Technol.* **2014**, *8*, 595–600, doi:10.1049/iet-smt.2014.0050. [CrossRef]

6. Beroual, A.; Coulibaly, M.L.; Aitken, O.; Girodet, A. Investigation on creeping discharges propagating over epoxy resin and glass insulators in the presence of different gases and mixtures. *Eur. Phys. J. Appl. Phys.* **2011**, *56*, 30802, doi:10.1051/epjap/2011110122. [CrossRef]

7. Beroual, A.; Khaled, U.; Coulibaly, M.L. Experimental Investigation of the Breakdown Voltage of CO_2, N_2, and SF_6 Gases, and CO_2–SF_6 and N_2–SF_6 Mixtures under Different Voltage Waveforms. *Energies* **2018**, *11*, doi:10.3390/en11040902. [CrossRef]

8. Murooka, Y.; Takada, T.; Hiddaka, K. Nanosecond surface discharge and charge density evaluation Part I: review and experiments. *IEEE Electr. Insul. Mag.* **2001**, *17*, 6–16, doi:10.1109/57.917527. [CrossRef]

9. Zhu, Y.; Takada, T.; Inoue, Y.; Tu, D. Dynamic observation of needle-plane surface-discharge using the electro-optical Pockels effect. *IEEE Trans. Dielectr. Electr. Insul.* **1996**, *3*, 460–468, doi:10.1109/94.506221. [CrossRef]

10. Kumara, S.; Alam, S.; Hoque, I.R.; Serdyuk, Y.V.; Gubanski, S.M. DC flashover characteristics of a polymeric insulator in presence of surface charges. *IEEE Trans. Dielectr. Electr. Insul.* **2012**, *19*, 1084–1090, doi:10.1109/TDEI.2012.6215116. [CrossRef]

11. Winter, A.; Kindersberger, J. Surface charge accumulation on insulating plates in SF_6 and the effect on DC and AC breakdown voltage of electrode arrangements. In Proceedings of the Annual Report Conference on Electrical Insulation and Dielectric Phenomena, Cancun, Mexico, 20–24 October 2002; pp. 757–761.

12. Nakanishi, K.; Yoshioka, A.; Arahata, Y.; Shibuya, Y. Surface Charging On Epoxy Spacer At DC Stress In Compressed SF_6 Gas. *IEEE Trans. Power Appar. Syst.* **1983**, *PAS-102*, 3919–3927, doi:10.1109/TPAS.1983.317931. [CrossRef]

13. Ziegler, S.; Woodward, R.C.; Iu, H.H.; Borle, L.J. Current Sensing Techniques: A Review. *IEEE Sens. J.* **2009**, *9*, 354–376, doi:10.1109/JSEN.2009.2013914. [CrossRef]

14. British Standard, B.S. *High-Voltage Test Techniques—Part 1: General Definitions and Test Requirements*; BSI; BS EN 60060-1:2010; 28-02-2011.

15. Reid, A.J.; Judd, M.D.; Stewart, B.G.; Fouracre, R.A. Partial discharge current pulses in SF_6 and the effect of superposition of their radiometric measurement. *J. Phys. D Appl. Phys.* **2006**, *39*, 4167–4177, doi:10.1088/0022-3727/39/19/008. [CrossRef]

16. Saitoh, H.; Morita, K.; Kikkawa, T.; Hayakawa, N.; Okubo, H. Impulse partial discharge and breakdown characteristics of rod-plane gaps in N_2/SF_6 gas mixtures. *IEEE Trans. Dielectr. Electr. Insul.* **2002**, *9*, 544–550, doi:10.1109/TDEI.2002.1024431. [CrossRef]

17. Judd, M.D.; Farish, O. High bandwidth measurement of partial discharge current pulses. In Proceedings of the Conference Record of the 1998 IEEE International Symposium on Electrical Insulation (Cat. No.98CH36239), Washington, DC, USA, 7–10 June 1998.

18. Mansour, D.A.; Kojima, H.; Hayakawa, N.; Hanai, M.; Okubo, H. Physical mechanisms of partial discharges at nitrogen filled delamination in epoxy cast resin power apparatus. *IEEE Trans. Dielectr. Electr. Insul.* **2013**, *20*, 454–461, doi:10.1109/TDEI.2013.6508747. [CrossRef]

19. Rodrigo Mor, A.; Castro Heredia, L.C.; Muñoz, F.A. A Novel Approach for Partial Discharge Measurements on GIS Using HFCT Sensors. *Sensors* **2018**, *18*, 4482, doi:10.3390/s18124482. [CrossRef] [PubMed]

20. Zachariades, C.; Shuttleworth, R.; Giussani, R.; MacKinlay, R. Optimization of a High-Frequency Current Transformer Sensor for Partial Discharge Detection Using Finite-Element Analysis. *IEEE Sens. J.* **2016**, *16*, 7526–7533, doi:10.1109/JSEN.2016.2600272. [CrossRef]

21. Hu, X.; Siew, W.H.; Judd, M.D.; Peng, X. Transfer function characterization for HFCTs used in partial discharge detection. *IEEE Trans. Dielectr. Electr. Insul.* **2017**, *24*, 1088–1096, doi:10.1109/TDEI.2017.006115. [CrossRef]
22. Okubo, H.; Hayakawa, N.; Matsushita, A. The relationship between partial discharge current pulse waveforms and physical mechanisms. *IEEE Electr. Insul. Mag.* **2002**, *18*, 38–45, doi:10.1109/MEI.2002.1014966. [CrossRef]

13

Seasonal Influences on the Impulse Characteristics of Grounding Systems for Tropical Countries

Muhd Shahirad Reffin [1], Abdul Wali Abdul Ali [1], Normiza Mohamad Nor [1,*],
Nurul Nadia Ahmad [1], Syarifah Amanina Syed Abdullah [1], Azwan Mahmud [1]
and Farhan Hanaffi [2]

[1] Faculty of Engineering, Multimedia University, 63100 Cyberjaya, Malaysia;
muhd_shahirad@yahoo.com (M.S.R.); walikdr17@gmail.com (A.W.A.A.);
nurulnadia.ahmad@mmu.edu.my (N.N.A.); synina@gmail.com (S.A.S.A.);
azwan.mahmud@mmu.edu.my (A.M.)

[2] Faculty of Electrical Engineering, Universiti Teknikal Malaysia Melaka, 76100 Durian Tunggal, Malaysia;
farhan@utem.edu.my

* Correspondence: normiza.nor@mmu.edu.my

Abstract: One of the most important parameters of the performance of grounding systems is the soil resistivity. As generally known, the soil resistivity changes seasonally, hence the performance of grounding systems, at DC and under high impulse conditions. This paper presents the performance of grounding systems with two different configurations. Field experiments were set up to study the characteristics of the grounding systems seasonally at power frequency and under high impulse conditions. A review of field testing on practical grounding systems was also presented. It was found that the soil resistivity, RDC and impulse characteristics of grounding systems were improved over time, and the improvement was higher for electrodes that have more contact with the soils.

Keywords: grounding; grounding electrodes; high impulse conditions; seasonal; soil resistivity

1. Introduction

Grounding systems are necessary to discharge high fault currents to ground and ensure the safe operation of power systems at all time. It was reported in IEEE Standard 81 [1] that due to the soil settling process and compactness of the soil, the earth impedance of ground electrode decreases slightly over a year or more after installation, due to the soil settling process and improved compaction in the soil. A few research investigations have also been reported on the influence of soil resistivity on practical grounding systems at power frequency and high impulse current [2,3]. As generally known, soil composition, inhomogeneity, hydrological, geological process can vary seasonally, and some studies have analysed the seasonal influence on grounding systems in terms of resistance values at low voltage, and low currents by field experiments [4–7]. However, limited studies have been published on the seasonal influence on the performance of grounding systems under high impulse conditions. He et al. [8] observed the seasonal influence on grounding systems under high impulse conditions for two seasons: winter and summer. They observed that the soil resistivity at the top layer was stable, whereas a higher soil resistivity of the bottom layer was observed during winter than during summer [8]. This resulted in higher resistance at the power frequency in winter than during summer. When the impulse factor was measured as the ratio of $R_{impulse}$ to RDC, a close impulse factor was obtained for both seasons. However, the investigations on the impulse characteristics of grounding systems were completed for two seasons, and no continuous measurement was made in between the seasons.

Further, the time period required for the checking and maintenance of grounding systems for both at power frequency and transient conditions has not been intensively studied or suggested. IEEE Standard 80:2013 [9] suggests the ground resistance be checked periodically after completion of construction, however, no specific time period is mentioned. Similarly, IEEE Standard 142: [10] and IEEE Standard 81 [1] state that power frequency tests should be conducted periodically, however, again no specific time period is suggested. Due to the lack of study on the seasonal performance of grounding systems and suggested time periods that can be found in literature, this paper therefore aims to address this shortfall.

In this study, field tests were used to investigate the characteristics of two practical grounding systems under high magnitude current surges throughout the year. Seasonal influences on the steady-state and impulse resistances were investigated, which represents the condition when the grounding systems were left over a period of time in real practice. These measurements allow a better understanding of the seasonal performances of grounding systems and provide information on how frequently grounding systems need to be checked and maintained after installation.

2. Experimental Arrangement

2.1. Review of Field Testing and Measurements

Different test arrangements may result in different and unreliable results when measuring the impulse characteristics of practical grounding systems. A review of field testing and measurement of impulse tests on grounding systems was firstly performed, to present various test set-ups and arrangements adopted in previously published works [2,8,11–27]. The study of soil behavior under impulse condition by means of field testing showed that the resistance decreased with increasing current [2,12–27]. It was noted from these studies [2,11–27] that the main concerns in the field measurements are the costs, logistics and time challenges. The guidelines were also found to be limited, due to the limited standards for field testing and measurements of grounding systems under high impulse conditions.

There are a few well established methods suggested in the standards on the measurements of earth resistance of earthing systems at low voltage and low frequency currents, namely the two-point method, three point method, ratio method, staged fault tests and fall-of-potential method [1,9,10]. However, it is now well accepted that soil characteristics under high magnitude impulse currents would become 'non-linear' and different than measured at low voltage and low frequency currents [2,8,11–27]. Due to the different characteristics of earthing systems under high impulse currents than when under low voltage, low frequency currents, there is a need to assess the practical earthing systems under high impulse conditions.

Field measurements undoubtedly can provide important results concerning the impulse characteristics of earthing systems since they represent the closest scenario to when high currents are practically discharged to grounding. With the improvement of impulse voltage/current generators which can be mobilized to the sites, impulse characterisations of earthing systems under high impulse conditions by field measurements have now become popular. Since then, a lot more studies have been directed towards impulse tests on earthing systems using field measurements. However, the measurement and testing methods found in these papers [11,13–27] differ from one another, which could be due to the lack of standards emphasising the required guidelines, as well as the great dependence on the available configurations and test site.

Since it is now well accepted that the impulse characteristics of earthing systems are different from those at low voltage low frequency currents, it is equally important to assess the performance of earthing systems under high impulse conditions. IEEE Standard 81 [1] provides some guidance and recommendations for measurements of earthing systems under high impulse conditions by field measurements, however they do not address it quantitatively, probably due to the little research work that has been carried out on grounding assessment under high impulse currents by field measurements. Other than the costs, logistics and time challenges in the field measurements, technical challenges are the main issues faced by researchers in performing field measurements of the impulse characteristics of earthing systems. This is due to the limited testing standards for field testing and measurements of earthing systems under high impulse conditions. Another technical challenge is due to the limited number of impulse generator manufacturers who are willing to tailor their designs to allow them to be mobilized to the field sites. So far, to the authors' knowledge, standards on impulse tests on practical earthing systems available in the literature are limited.

2.1.1. Distance of Remote Earth from the Electrode under Tests

In IEEE Std 81 [1], brief guidelines are provided for measurements at field sites using a mobile impulse generator. The work presented in IEEE Std 81 [1] is based on the tests conducted at the Georgia Institute of Technology. It was suggested in IEEE Standard 81 [1] to have the same leads and reference ground arrangement as that used for low-frequency Fall-of-Potential (FOP) tests. Figure 1 shows the test arrangement for the experimental test set-up of earthing systems under high impulse conditions. The distance of the electrode under test to the earth probe is 62% of the distance between the electrode under test to the remote earth. As for the ground mat, it was recommended in IEEE Std. 80 [9] that the dimensions of the electrode under test to the remote earth be extended by 3 to 4 times the diagonal dimensions of the ground mat. However, the IEEE standard guidelines are brief, and many other authors [11,13–22] have adopted other test set-up arrangements in their measurements, where all these distances were not as specific as those highlighted in the IEEE Standard [1].

2.1.2. Impulse Generator

As for the mobile impulse generator, no generally accepted standard has been presented so far. Some studies [18] used laboratory facilities, without any special design changes to the impulse generator for the testing of earthing systems by field measurements. On the other hand, some studies [13,15,16,19] have used impulse generators purposely built for their tests of earthing systems under high impulse conditions by field measurements. Marimoto et al. [16] have exclusively developed a weatherproof mobile impulse voltage generator in an effort to test the grounding systems of power substations.

No detailed information has been published in the standards on the specific methods, procedures and precautions of the measurements for testing the earthing systems at field sites. It was briefly highlighted in [1] that the current and voltage leads should be isolated from earth to avoid any interference, which can be done by hanging the leads over polyvinyl chloride (PVC) conduits. Different methods of hanging these leads to isolate the earth have also been found in some other studies [16,19]. However, other studies have not mentioned any consideration of the isolation of the leads [17–21]. This shows that there is a need to clarify the proper measurement methods.

2.1.3. Remote Earth/Auxiliary Earth

Another important parameter that needs to be considered in the experimental arrangement for the testing of earthing systems by field measurements is the auxiliary ground, which is needed to carry the return current to the impulse generator via the ground under tests. So far, to the authors' knowledge, no specific standard has discussed the design requirements of this auxiliary ground.

Some authors have constructed auxiliary grounds around the electrode under tests. In this arrangement, a bigger remote earth size is expected since it is installed around the electrode under test. Thus, the earth resistance values of the ring electrode may be expected to have lower earth resistance values than the electrode under test. References [20,21] have used a remote earth installed around the electrode under tests. A few clarifications are needed when having the remote earth around the electrode under tests such as the appropriate radius of remote earth that should be constructed around the electrode under tests.

On the other hand, some authors [12,13,15–19] used separate earthing systems, which are constructed away from the electrode under tests as the auxiliary ground. It is stated in IEEE Standard 80 [1] that the earth resistance of remote earth should be lower than that the electrode under tests.

However, many times the steady state earth resistance values of the remote earth have not been mentioned or addressed by some authors. Though the remote earth constructions look bigger in the published work [12,13,15–19], the earth resistance values of the remote earth were not specifically mentioned in the papers. Yunus et. al. [17] studied the effects of earth resistance values of remote earths on the electrodes under test for when the remote earth was placed away from the electrode under test. Some significant observations were that the results were found to be different than the findings in most literatures, with higher earth resistance values of the remote earth than that the electrode under tests [17], and where the impulse earth resistance value was found to be higher than that measured at low voltage and low frequency currents (DC earth resistance value) when higher remote earth resistance values were used. They also found that the earth resistance values of electrodes under test increased with increasing currents in the higher earth resistance values of the remote earth [17].

Further, Abdullah et al. [22] compared both arrangements (remote earth around the electrode under test and remote earth placed at some distance away from the electrode under tests). They [22] found close agreement between the results of these two arrangements. This shows that both test arrangements are acceptable, as long as the condition that the earth resistance value of the remote earth be smaller than that of the electrode under testing is met.

2.1.4. Placement of Current Transducer

Other concerns for the field measurements and testing of the earthing systems under high impulse conditions are the placement of current transducers during the field measurements. In some published papers [23–26], the experimental arrangement was not shown at all, though some parameters of the generators and test circuit are described. Due to the different possible arrangements of the remote earth, surrounding the electrode under tests and away from the electrode under test, as highlighted in Section 3, the placement of the current transducer (CT) would expectedly be different too.

Here, the placement of the CT for the remote earth placed at a distance away from the electrode under test is discussed first. In IEEE Standard 81 [1], the current transducer was placed at the electrode under test, at the same point where the cable of voltage divider was also connected. Some studies [15,21,24,25,28] also positioned the CT in series to the electrode under tests.

On the other hand, for the arrangement where the remote earth is placed at some distance away from the electrode under tests, Chen and Chowdhuri [27] measured the current at the remote earth, and not in series to the electrode under tests. So far, to the authors' knowledge, such an arrangement was only found in their paper [27]. As for other papers, i.e., [13,15,16], they did not clearly show the CT placement, though there were some current measurements. Due to a lack of test set-up arrangements, the authors feel that there is a need for a proper standard for impulse tests on earthing systems by field measurements. As for other kinds of arrangement where the remote earth is installed around the electrode under tests, the CT is more commonly placed at the remote earth [12,20,22].

2.2. Test Set up for This Study

In the investigation described in this paper, a mobile impulse generator which is capable of generating high voltages up to 300 kV and high impulse currents up to 10 kA was used (see Figure 1). The impulse generator was powered by a diesel generator. Current measurements were achieved with a current transformer of sensitivity of 0.01V/A and with an attenuation of the probe of ×10. A resistive divider with a ratio of 3890:1 was used for voltage measurements. Two commercially available digital storage oscilloscopes (DSOs), powered by batteries, were used to capture the voltage and current signals separately. In order to avoid any interferences and flashover in the circuit, all equipment and leads were placed above ground. For the diesel generator and impulse generator, an epoxy was used as a frame to separate them from the ground. For the DSOs, the batteries are placed on an insulation table, made of epoxy. All the leads and mesh cables were isolated from the ground by hanging the leads on epoxy insulation rods. Figure 2 shows the experimental arrangement of the impulse tests of grounding systems under high impulse conditions.

Figure 1. Equipment used for field measurements.

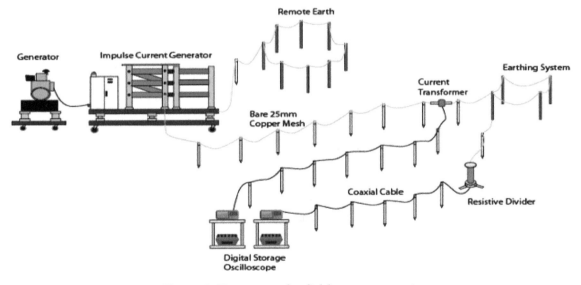

Figure 2. Test set-up for field measurements.

2.3. Electrodes under Tests

In this study, two configurations were adopted: a grounding device with spike rods (configuration 1) and one with four rod electrodes (configuration 2), shown in Figures 3 and 4, respectively. Configuration 2 represents a conventional electrode, where the design does not emphasize any enhancement of the ionization process. Configuration 1 was developed based on the evidence from the work by Petropolous [28] that when spikes or needles were attached to the spherical electrode, lower impulse resistance values were noted for the same voltage due to the higher field intensities at the spikes. This indicates the effectiveness of the earth electrodes with spikes as compared to electrodes without spikes. In this paper, a new grounding electrode with spike rods is introduced. By having spike rods at the electrodes, a high electric field intensity will be concentrated at the spikes, which encourages soil ionisation and breakdown processes to take place, hence allowing more current to be discharged to the ground. This will result in a more effective grounding system. For this reason, and to provide more contact between the electrodes and surrounding soil, a grounding device with spike rods (consisting of two rods (inner shaft (120) and outer shaft electrodes (110))) was adopted. The grounding device with spike rods is 1.5 m in length. The diameter of the inner rod is 3 cm, and outer rod is 5 cm, with a gap between inner and outer rods of 1 cm. There are five spikes (123), each one of 20 cm length. During the installation, the spike rods (123) are kept closed from the surface. For the installation, due to its large diameter, a pre-bore with a diameter of 4 cm was firstly performed using an auger. Upon completing a pre-bore hole with a depth of 1.5 m, the grounding device with spike rods (100) is positioned into the hole, in a generally vertical configuration, with all the spike rods (123) concealed and protected from damage within the shaft. The outer shaft (110) is subjected to impact force while it is driven through hammering into the ground. During driving, the top end is protected by a dolly or capping to protect the top end of the shaft against any damage from driving or hammering the electrodes. Once the grounding device with spike rods (100) reached the required distance, 1.5 m, the inner shaft (120) was turned using the provided winch (121) in such a way that the grounding spike rods (123) protruded out and pierced into the soil mass. An indication that all the spike rods (123) protrude out is when the winch stops, and is not able to turn the inner shaft (120) anymore. Another type of ground electrode used in this study consists of four ground rod electrodes, with each rod of 20 mm diameter, and a depth of 1.5 m. Copper mesh of 25 mm width and thickness of 2 mm was used to connect the rod electrode at all four points, as shown in Figure 4.

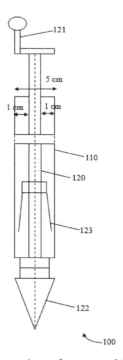

Figure 3. Configuration 1, consists of a grounding device with spike rods.

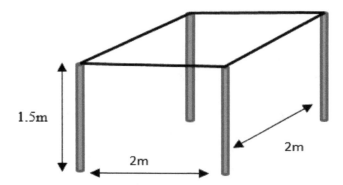

Figure 4. Configuration 2 for the tests, consisting of a 2 m × 2 m grounding grid.

In this study, the Finite Element Method (FEM) was utilized to obtain the voltage profiles for both configurations. Figures 5 and 6 show the point of electric profile is taken for configurations 1 and 2, respectively. The soil resistivity was taken at the beginning of the tests using a Wenner Method, which was computed as a two-layer medium model, using a Current Distribution, Electromagnetic Fields, Grounding and Soil Structure Analysis (CDEGS) software. It was found that the resistivity at the top layer is 119.5 Ωm with a depth of 7.18 m, and the bottom of a two-layered soil is 391 Ωm, with an infinite depth. Both configurations were injected at 5 kA. The corresponding trends of the electric field for configurations 1 and 2 are shown in Figures 7 and 8, respectively. As can be noted, the shapes of the electric fields for both configurations are different when a similar soil resistivity layer was used. A higher and more non-uniform electric field was noted for configuration 1, where the maximum surface potential is 820 MV (see Figure 7). For the case of configuration 2, the potential is rather uniform, with 650 MV at the rod, and 180 MV at the copper strips (see Figure 8). The FEM simulation indicates that configuration 1 is preferred, since it can increase the electric field significantly, and has non-uniform electric field, thus reducing the earth resistance value more significantly under transient conditions, compared to configuration 2.

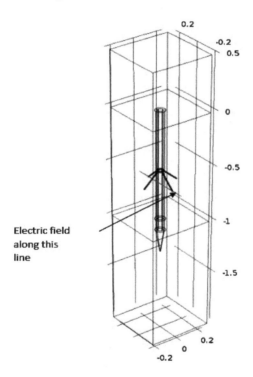

Figure 5. Electric profile is taken along the spike rods.

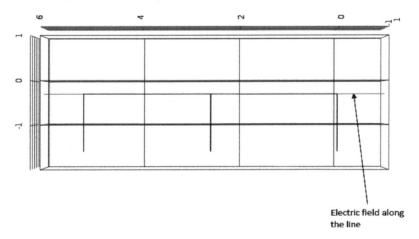

Figure 6. Electric profile taken along the conventional rods.

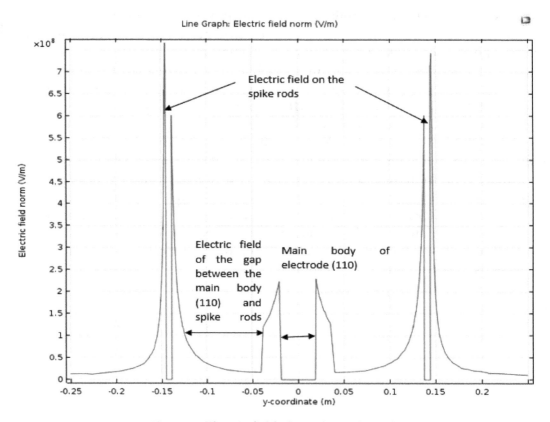

Figure 7. Electric field along the spike rods.

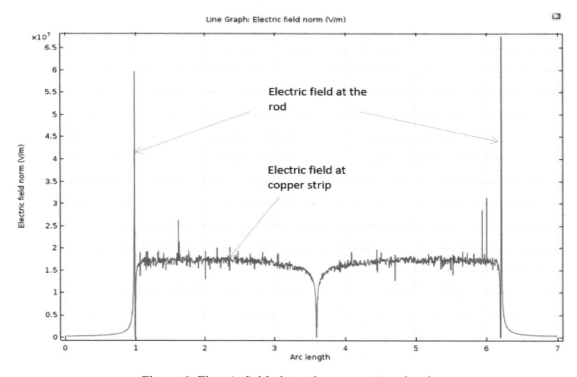

Figure 8. Electric field along the conventional rods.

2.4. Auxillary or Remote Electrodes

A larger size of auxillary or remote earth grounding system was used, so that a lower earth resistance value was achieved, in comparison to the test electrodes, presented earlier in Section 2.2. In this study, the remote earth was installed 50 m away from the electrode under tests. Figure 9 shows the remote earth used in this study, which consists of a 20 m × 30 m grounding grid. The spacing between copper strips is 5 m apart for the 20 m side and 15 m apart for the 30 m side. Hard copper strips of 30 mm width and 2 mm thick were used, which were buried 300 mm below the earth's surface, and welded exothermically to 12-rod electrodes, where each one is 20 mm in diameter, and 1.8 m long. Using a Fall-of-Potential (FOP) method, the earth resistance values of the remote earth were measured, and found to range between 8.5 Ω to 10 Ω throughout the year, which is always lower than the electrodes under tests.

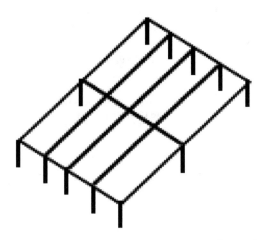

Figure 9. A 20 m × 30 m remote earth mesh.

3. Results and Analysis

Using the Wenner method, the soil resistivity was measured in March 2018 and March 2019 to see variations in soil profile over the year. Configurations 1 and 2 are 30 m apart. The soil resistivity was measured at the site over 150 m-long lines within the test site, and these configurations were installed within this 150 m-long line. The results were then modeled into 2-layer soil model using Current Distribution, Electromagnetic Interference, Grounding and Soil Structure Analysis (CDEGS), which results are summarised in Table 1. As can be seen, the soil resistivity slightly varied over the year, with the height of the first layer also being reduced after a year. FOP, as outlined in IEEE Standard 81 [1] was applied to measure the DC and low-current resistances of configurations 1 and 2 throughout the year, which are shown in Table 2. The resistances of both configurations were found to reduce after the installation, where a higher reduction was seen for configuration 1, with a decrease by more than 20% throughout the year. This could be due the presence of five spike rods, which provide more contact within the soil, as compared to configuration 2. It was also noted that for the measurement in September 2018, high RDC values were noted for both configurations, with respect to any other months. It was experienced by the authors that during the measurement, the weather was hotter than the rest of the months.

Table 1. Soil resistivity profile.

Month of Measurement	Soil Resistivity of Layer 1, ϱ_1 (Ωm)	Soil Resistivity of Layer 2, ϱ_2 (Ωm)	Height of Layer 1 (m)
March 2018	119.5	391.0	7.2
March 2019	111.4	454.2	5.2

Right after RDC measurements, impulse tests were performed on both test electrode configurations on the same day. Impulse currents and the corresponding voltages were measured and shown in Figures 10 and 11 for configuration 1, at charging voltage of 30 kV and 210 kV, respectively, for test no. 1. As can be seen in the figures, faster times to discharge to zero for both voltage and current traces was seen at higher voltage magnitudes. Similar voltage and current traces were seen for configuration 2, and for other test nos., where faster times to discharge to zero for both voltage and current traces at higher currents were observed. The time to discharge to zero for voltage and current was plotted against the applied voltage for all the tests, the time to discharge to zero decreased with applied voltage, as shown in Figures 12 and 13 for configuration 1 and 2, respectively. Both configurations were found to have the slowest time to discharge to zero for test no.4, which has high RDC. Both configurations with the lowest RDC (in Dec. 2018) were found to have the fastest discharge time to zero, indicating a good conductivity of the grounding systems. Time to discharge to zero was also found to be higher for configuration 1, in comparison to configuration 2 (see Figures 14 and 15) for test no. 1 to 3, and no. 3–7 respectively). The results also indicated that the lower the RDC, the faster the time for current and voltage to discharge to zero.

Table 2. Measured resistance of electrodes at power frequency, low-current tests.

Test No.	Date of Measurement	DC Resistance, RDC (Ω)			
		Conf. 1	Percentage Difference from the First Reading (%)	Conf. 2	Percentage Difference from the First Reading (%)
1	21/03/2018	91	0	60.9	0
2	07/05/2018	69.1	24.1	55.2	9.36
3	02/08/2018	69.7	23.4	57.5	5.58
4	03/09/2018	84.2	7.25	80.3	−31.9
5	17/10/2018	71.9	20.99	57.1	6.24
6	21/11/2018	70.7	22.3	55.9	8.2
7	17/12/2018	70.1	23	57.1	6.24
8	25/2/2019	69	24.2	55.5	8.9

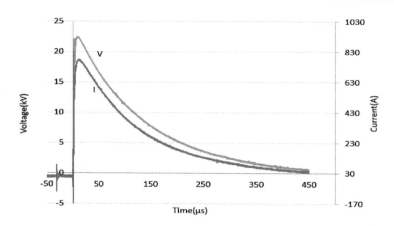

Figure 10. Voltage and current traces for configuration 1 at a charging voltage of 30 kV.

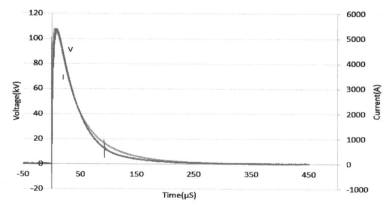

Figure 11. Voltage and current traces for configuration 1 at a charging voltage of 210 kV.

Impulse resistance values were measured as the ratio of the voltage at current peak to the current peak, and plotted versus the peak current, as shown in Figures 16 and 17 for configurations 1 and 2, respectively. The resistance values obtained from these tests were found to decrease with current magnitude, indicating the effect of impulse currents on the characteristics of test electrodes for configurations 1 and 2. It was noted that the impulse resistance are lower a few months after installation, showing improvement in the grounding systems for both configurations. In order to determine the effectiveness of the test electrodes in comparison to its RDC, the percentage of reduction of impulse resistance from its corresponding RDC was measured as (1), where the $R_{impulse}$ was taken as the average impulse resistance measured from varying the charging voltage of 30 kV until 210 kV:

$$\text{Percentage of resistance reduction} = \left(\frac{R_{DC} - R_{impulse}}{R_{DC}} \right) \times 100\% \tag{1}$$

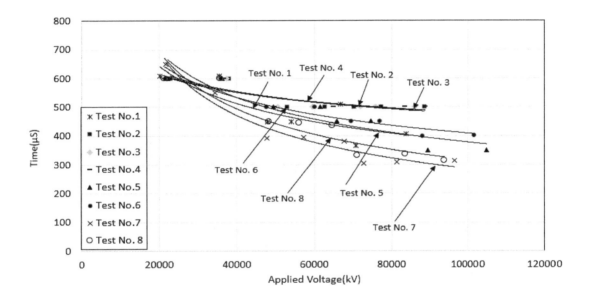

Figure 12. Time to discharge to zero for voltage and current traces for configuration 1 throughout the year.

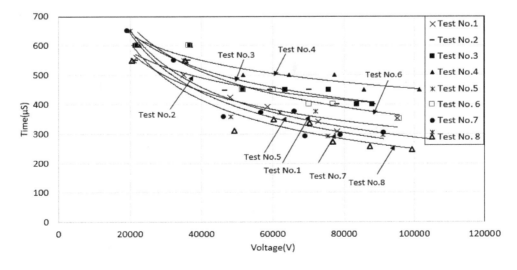

Figure 13. Time to discharge to zero for voltage and current traces for configuration 2 throughout the year.

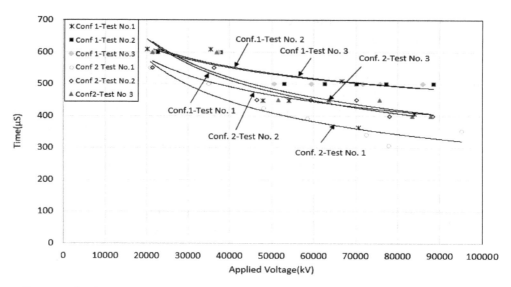

Figure 14. Time to discharge to zero for voltage and current traces for configuration 1 and 2 for tests no. 1 to 3.

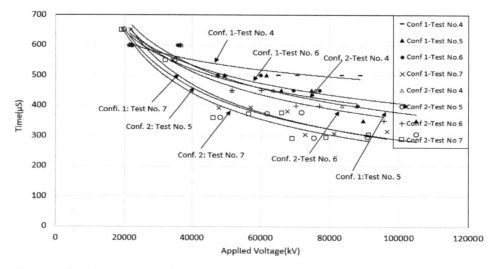

Figure 15. Time to discharge to zero for voltage and current traces for configuration 1 and 2 for tests no. 4 to 7.

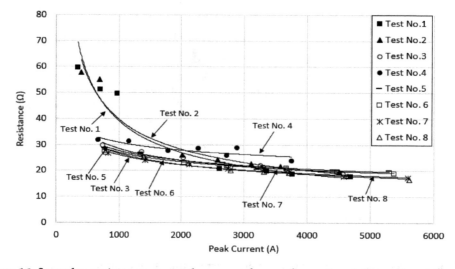

Figure 16. Impulse resistance vs. peak current for configuration 1 throughout the year.

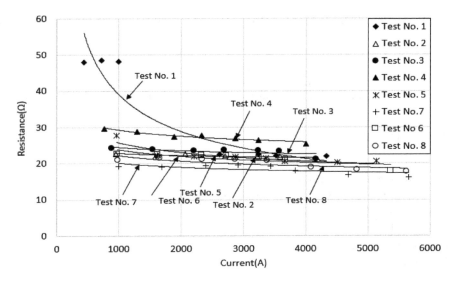

Figure 17. Impulse resistance vs. peak current for configuration 2 throughout the year.

Table 3 summarises the impulse factor for both configurations, measured throughout the year. As can be seen from the table, percentage of resistance reduction is higher for configuration 1, in comparison to configuration 2. A higher percentage of resistance reduction for configuration 1 after a months of installation was also noted, showing the effectiveness of the earth electrode design with spike rods.

Table 3. Percentage of impulse resistance reduction from RDC throughout the year.

Test No.	Configuration 1			Configuration 2		
	R_{DC}	Average $R_{impulse}$ (Ω)	Resistance Reduction from R_{DC} (%)	R_{DC}	Average $R_{impulse}$ (Ω)	Resistance Reduction from R_{DC} (%)
1	91	33.94	62.7	60.9	33.4	45.1
2	69.1	32.4	53.1	55.2	22.0	60.1
3	69.7	23.5	66.3	57.5	23.3	59.5
4	84.2	28.2	66.5	80.3	27.6	65.6
5	71.9	23	68	57.1	22.2	61.1
6	70.7	22.5	68.2	55.9	21.4	61.7
7	70.1	21.1	69.9	57.1	18.4	67.8

4. Conclusions

Power frequency and impulse tests by field measurements on a grounding device with spike rods and a small grid were performed throughout the year. It was found that under high impulse conditions, the RDC of both configurations became lower after months of installation. The percentage of reduction of RDC from the first installation was found to be more pronounced for configuration 1 than configuration 2, suggesting the effectiveness of the new grounding electrode with configuration 1. This could be due to the spike rods in the grounding electrode, providing more and better contact of the electrodes with the soil, hence reducing the RDC. The characteristics of both configurations were also investigated under high impulse currents. The percentage of earth resistance reduction of $R_{impulse}$ from its RDC was found to be higher for configuration 1. In addition, a higher percentage of earth resistance reduction was exhibited for configuration 1 after months of installation, indicating an improvement in the grounding systems for electrodes with spikes. The results from the study provide an important guideline for power engineers concerning the time period that needs to be considered for maintenance of grounding systems, where it is desirable to perform the maintenance of grounding systems after one year of installation. The results indicate that it may be desirable to consider using electrode rods with spike rods, so that earth resistance values can be reduced under transient conditions. This scenario gives a more effective grounding system, which enhances the ionisation process and increases the dissipating current.

Author Contributions: M.S.R., A.W.A.A., N.M.N., S.A.S.A. and N.N.A. involved in setting up the experiment, M.S.R. and A.W.A.A. sorted out the data, M.S.R. and A.W.A.A. prepared the draft and N.M.N. helped with validation of analysis. A.M., N.M.N. and N.N.A. helped in securing funding. F.H. helped with the FEM simulation.

Subscripts and Abbreviations

DC	Direct current
$R_{impulse}$	Resistance values measured under impulse condition
RDC	Resistance values measured at low voltage, low frequency currents
DSO	Digital Storage Oscilloscope
FEM	Finite Element Method
CDEGS	Current Distribution, Electromagnetic Fields, Grounding and Soil Structure Analysis
FOP	Fall-of-Potential

References

1. ANSI/IEEE Std 81-2012. *IEEE Guide for Measuring Earth Resistivity, Ground Impedance, and Earth Surface Potentials of a Ground System*; IEEE: Piscataway, NJ, USA, 2012.
2. Reffin, M.S.; Nor, N.M.; Ahmad, N.N.; Abdullah, S.A. Performance of Practical Grounding Systems under High Impulse Conditions. *Energies* **2018**, *11*, 3187. [CrossRef]
3. Tu, Y.; He, J.; Zeng, R. Lightning Impulse Performances of Grounding Devices Covered with Low-Resistivity Materials. *IEEE Trans. Power Deliv.* **2006**, *21*, 1701–1706. [CrossRef]
4. Gustafson, R.J.; Pursley, R.; Albertson, V.D. Seasonal Grounding Resistance Variation on Distribution Systems. *IEEE Trans. Power Deliv.* **1990**, *5*, 1013–1018. [CrossRef]
5. Abdullah, N.; Marican, A.; Osman, M.; Abdul, N. Rahman Case Study on Impact of Seasonal Variations of Soil Resistivities on Substation Grounding Systems Safety in Tropical Country. In Proceedings of the 7th Asia-Pacific International Conference on Lightning, Chengdu, China, 1–4 November 2011; pp. 150–154.
6. Gonos, I.; Gonos, I.F.; Moronis, A.X.; Stathopulos, I.A. Variation of Soil Resistivity and Ground Resistance during the Year. In Proceedings of the 28th International Conference on Lightning Protection (ICLP), Kanazawa, Japan, 17–21 September 2006; pp. 740–744.
7. Androvitsaneas, V.P.; Gonos, I.; Stathopulos, I.A. Performance of Ground Enhancing Compounds during the Year. In Proceedings of the 34th International Conference on Lightning Protection (ICLP), Rzeszow, Poland, 2–7 September 2018; pp. 1–5.
8. He, J.; Wu, J.; Zhang, B.; Yu, S. Field Testing for Observation of Seasonal Influence on Grounding Device at Impulse Condition. In Proceedings of the Asia-Pacific Symposium on Electromagnetic Compatibility, Singapore, 21–24 May 2012; pp. 445–448.
9. *ANSI/IEEE Std 80-2013: IEEE Guide for Safety in AC Substation Grounding*; IEEE: Piscataway, NJ, USA, 2013.
10. *IEEE 142-2007: IEEE Recommended Practice for Grounding of Industrial and Commercial Power Systems*; IEEE: Piscataway, NJ, USA, 2007.
11. Bellaschi, P.L. Impulse and 60-Cycle Characteristics of Driven Grounds. *IEE Trans. Power Appar. Syst.* **1941**, *60*, 123–128.
12. Kosztaluk, R.; Loboda, M.; Mukhedkar, D. Experimental Study of Transient Ground Impedances. *IEEE Trans. Power Appar. Syst.* **1981**, *100*, 4653–4660. [CrossRef]
13. Sekioka, S.; Hara, T.; Ametani, A. Development of a Nonlinear Model of a Concrete Pole Grounding Resistance. In Proceedings of the International Conference on Power Systems Transients, Lisbon, Portugal, 3–7 September 1995; pp. 463–468.
14. Dick, W.K.; Holliday, H.R. Impulse and Alternating Current Tests on Grounding Electrodes in Soil Environment. *IEEE Trans. Power Appar. Syst.* **1978**, *PAS-97*, 102–108. [CrossRef]
15. Sekioka, S.; Sonoda, T.; Ametani, A. Experimental Study of Current Dependent Grounding Resistance of Rod Electrode. *IEEE Trans. Power Deliv.* **2005**, *20*, 1569–1576. [CrossRef]
16. Morimoto, A.; Hayashida, H.; Sekioka, S.; Isokawa, M.; Hiyama, T.; Mori, H. Development of Weatherproof

Mobile Impulse Voltage Generator and Its Application to Experiments on Nonlinearity of Grounding Resistance. *Trans. Inst. Electr. Eng. Jpn.* **1997**, *117*, 22–33. (In English) [CrossRef]

17. Yunus, S.; Nor, N.M.; Agbor, N.; Abdullah, S.; Ramar, K. Performance of Earthing Systems for Different Earth Electrode Configurations. *IEEE Trans. Ind. Appl.* **2015**, *51*, 5335–5342. [CrossRef]

18. Ramamoorty, M.; Narayanan, M.M.B.; Parameswaran, S.; Mukhedkar, D. Transient Performance of Grounding Grids. *IEEE Trans. Power Deliv.* **1989**, *4*, 2053–2059. [CrossRef]

19. Yang, S.; Zhou, W.; Huang, J.; Yu, J. Investigation on Impulse Characteristic of Full-Scale Grounding Grid in Substation. *IEEE Trans. Electromagn. Compat.* **2017**, *60*, 1993–2001. [CrossRef]

20. Clark, D.; Guo, D.; Lathi, D.; Harid, N.; Griffiths, H.; Ainsley, A.; Haddad, A. Controlled Large-Scale Tests of Practical Grounding Electrodes- Part II: Comparison of Analytical and Numerical Predictions with Experimental Results. *IEEE Trans. Power Deliv.* **2014**, *29*, 1240–1248. [CrossRef]

21. Harid, N.; Griffiths, H.; Haddad, A. Effect of Ground Return Path on Impulse Characteristics of Earth Electrodes. In Proceedings of the 7th Asia-Pacific International Conference on Lightning, Chengdu, China, 1–4 November 2011; pp. 686–689.

22. Abdullah, S.; Nor, N.M.; Etopi, N.; Reffin, M.; Othman, M. Influence of Remote Earth and Impulse Polarity on Earthing Systems by Field Measurements. *IET Sci. Meas. Technol.* **2017**, *12*, 308–313. [CrossRef]

23. Towne, H.M. Impulse Characteristics of Driven Grounds. *Gen. Electr. Rev.* **1928**, *31*, 605–609.

24. Vainer, A.L. Impulse Characteristics of Complex Earth Grids. *Elektrichestvo* **1965**, *3*, 107–117.

25. Vainer, A.L.; Floru, V.N. Experimental Study and Method of Calculating of the Impulse Characteristics of Deep Earthing. *Electical Technol. Ussr (Gb)* **1971**, *2*, 18–22.

26. Liew, A.C.; Darveniza, M. Dynamic Model of Impulse Characteristics of Concentrated Earths. *IEE Proc.* **1974**, *121*, 123–135. [CrossRef]

27. Chen, Y.; Chowdhuri, P. Correlation between Laboratory and Field Tests on the Impulse Impedance of Rod-type Ground Electrodes. *IEE Proc. Gener. Transm. Distrib.* **2003**, *150*, 420–426. [CrossRef]

28. Petropoulos, G.M. The High-Voltage Characteristics of Earth Resistances. *J. IEE* **1948**, *95*, 172–174.

14

Statistical Study on Space Charge Effects and Stage Characteristics of Needle-Plate Corona Discharge under DC Voltage

Disheng Wang, Lin Du * and Chenguo Yao

State Key Laboratory of Power Transmission Equipment and System Security and New Technology, Chongqing University, Chongqing 400030, China
* Correspondence: dulin@cqu.edu.cn

Abstract: The air's partial discharges (PD) under DC voltage are obviously affected by space charges. Discharge pulse parameters have statistical regularity, which can be applied to analyze the space charge effects and discharge characteristics during the discharge process. Paper studies air corona discharge under DC voltage with needle-plate model. Statistical rules of repetition rate (n), amplitude (V) and interval time (Δt) are extracted, and corresponding space charge effects and electric field distributions in PD process are analyzed. The discharge stages of corona discharge under DC voltage are divided. Furthermore, reflected space charge effects, electric field distributions and discharge characteristics of each stages are summarized to better explain the stage discharge mechanism. This research verifies that microcosmic process of PD under DC voltage can be described based on statistical method. It contributes to the microcosmic illustration of gas PD with space charges.

Keywords: partial discharge; needle-plate model; statistical rule; discharge stage; space charge

1. Introduction

Nowadays, the insulation requirements of DC equipment are further improved, with the wide application of DC system in high voltage direct current (HVDC) transmission and other aspects. As a reflection of insulation faults in DC system, PD under DC is also concerned in many researches.

Gas partial discharge (PD) under DC voltage is a process closely related to the generation, migration and dissipation of space charges. The main forms of gas PD in DC system are the surface discharge on dielectric materials and the corona discharges under non-uniform electric field [1,2]. Researches on gas corona discharges under DC mainly includes three directions. One is researching on microcosmic discharge theory and discharge mechanism explanation [3–5]. Another one is considering influencing factors of discharge [6–9], such as electrode structure [6,7], air pressure [8] and air flow [9]. The rest one is studying discharge parameters characteristics, and apply them to pattern recognition and stage division of discharges [10–14].

Different from the periodic generation and dissipation of space charges under alternate current (AC), many space charges will be retained in discharge region under DC because the polarity of electric field does not change. Therefore, one of the main points in the gas corona discharge researches under DC is the space charge effect and the electric field distortion caused by it. As yet, due to the lack of space charge measurement methods in gas, it is difficult to describe the space charge and the electric field distributions through testing. Reference [15] studies calibration of field-mill to measure DC electric field with space charges. However, the measurement scope cannot reach the distance of molecular level, so it still cannot precisely describe the distribution of space charge and electric field. Some researches apply simulation methods to describe the discharge mechanism [16,17]. In [17], applying corona discharge fluid model, space charge intensity and electric field variation during a single discharge process are described. But the simulation starts with none space charge situation and only lasts for the short time of a single discharge, which does not take the retained space charges into account. Space charge distribution are also studied in some experimental methods [18,19]. References [19] describes the corona layer morphology by gray value and thickness on the luminescent image. In this method, only space discharges distribution of ionization region can be obtained. In addition, the space charge behaviors and electric field variation of a single discharge cannot be described. Therefore, proposing new method to reflect the space charge effect and the electric field distortion is of significance.

It is obvious that the variation of electric field will affect PD phenomenon, and reflects in PD parameters. In theory, through the deep analysis of discharge parameter characteristics, the space charge and electric field distribution can be reflected. While existing researches on the discharge parameters characteristics are either focused on the differences of characteristics from other discharge stages and patterns [10–12], or focused on the characteristic extraction [13,14]. Few researches relate the statistical rules to the space charge effect and electric field distribution.

So in this paper, it is proposed to apply mathematical statistics methods and microcosmic discharge and space charge theories, to analyze space charge effect and electric field distribution on the basis of characteristic parameter rules of PD pulses.

In this paper, the air discharge of needle-plate model under DC is researched. Applying mathematical statistics, the stage characteristics of repetition rate (n), amplitude (V) and interval time (Δt) characteristic parameters are figured out to divide the discharge stages. The variation rule of characteristic parameters of each stage and transition process are explained by the space charge effects and electric field distribution. Furthermore, deduced space charge effects, electric field distributions and discharge characteristics of each stages are summarized to better explain the stage discharge mechanism.

2. Principle

2.1. Experiment Platform and Method

The schematic diagram of the experiment platform is shown in Figure 1. DC voltages are generated by the high voltage (HV) DC power supply Matsusada AU-30R2, whose output voltage range is ±30 kV. PD pulse acquisition adopts pulse current method. Oscilloscope is adapted to measure the voltage pulse signals on the 50 Ω non-inductive resistance. The needle-plate model is set in a cylindrical organic glass box (radius of 0.1 m), which is filled with air and blocks air flow. The curvature radius of needle tip r is 0.5 mm, and the distance between the needle and plate d is 15 mm. d/r is greater than 4, and the needle-plate electrical field is extremely uneven [20].

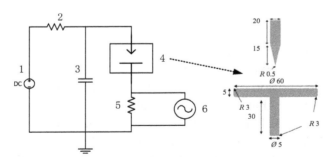

1- DC power supply 2- protective resistance 3- coupling capacitor
4- needle-plate model 5- non inductive resistance 6- oscilloscope

Figure 1. Schematic diagram of the experimental platform.

During the experiment process, applied voltage is raised with step-rise method. After initial discharge voltage are found, the voltage is increased to a certain voltage level with step of 0.25 kV (0.1 kV for some continuous voltage levels with marked PD signal change), and then sustains. Under each voltage level, when PD are relatively stabilized, PD signals in window width of 0.1 s are acquired with the sampling frequency of 500 MHz, which is repeated with 2 min interval. 5 groups of signals are recorded under each voltage level. To avoid occasionality, whole experiment process is repeated more than three times under each polarity. The temperature range is 293 K to 295 K, and humidity range is 75% to 79% during the experiment processes.

2.2. Relevant Space Charge Theory

According to the streamer theory, in a long gap, the PD will develop in form of streamer. When electric field intensity (E) on the head of streamer reaches a certain level, photons excited in strong ionizations induce photoelectric ionization, which produces new electrons and forms the secondary electron avalanches. Then the streamer can self-sustaining develop. The self-sustaining process of streamer is described in [21] under negative polarity needle-plate model. The ions on the head of streamer are considered to aggregate as a sphere, and self-sustaining condition can be expressed as:

$$\int_r^R \frac{2}{3} p f \mu e^{-\mu\rho} \frac{\sqrt{\rho}r}{(d-x_1)^{1/2}} \alpha e^{\int_\rho^r \alpha d\rho'} d\rho = 1, \tag{1}$$

where r is the radius of the sphere. ρ denotes radius of the photons from the center. R is the radius of the region where photons can induce photoionization, which is related to E. p denotes the probability of photoionization. d is the distance between the electrodes, and x_1 is the distance between the trigged electrons and the plate. The collision ionization coefficient α is related to E and pressure P, which can be expressed as:

$$\alpha/P = Ae^{-BP/E}, \tag{2}$$

where A, B are coefficients. According to Equation (1) and Equation (2), E has obvious influence on the self-sustaining development of the streamer. When other variables are confirmed, the minima electric field intensity which meets the self-sustaining condition (E_0) can be determined.

During the formation process of electron avalanches, ionizations produce electrons and cations. Migration velocity v of ions is connected with the E. The charge–mass ratio of electron is much lower than ions, thus its migration rate is about two magnitudes higher. For needle-plate module, E in tiny region near the needle tip is significantly higher. Under high E, produced electrons mainly have migrated away from the discharge region when self-sustaining discharge stops. Meanwhile, high-speed electrons can hardly be trapped by molecules, thus merely small proportion of the electrons form anions in ionization region. Only in low E regions can more anions be formed gradually. Therefore, in the development path of streamers, the densities of retained electrons and anions are relatively low. Low-speed cations almost retain and play a major role.

2.3. PD Pulse Characteristic Parameters

The acquired PD pulses reflect the electric field variation of the gap in a discharge process, which is caused by the instantaneous change of electric charge quantity (q). Characteristic parameters amplitude (V), interval time (Δt) and repetition rate (n) of the PD pulses are selected for statistical study.

V can reflect the maximum change of q in a single discharge. From the beginning to the end of a streamer, space charges produced before are not completely dissipated by migration and recombination. Thus space charges retain and q rises. The maximum q should occur at the moment the self-sustaining discharge stops. Combined with the discussions above, amplitude of discharge pulse is related to E_0.

Δt is defined as the time interval from the end of a PD pulse to the beginning of the next. It is mainly composed of the electric field recovery time and the discharge delay of the next pulse. Discharge delay refers to the time delay of a pulse after applied voltage reaches the initial discharge voltage value. For streamer-type PD, discharge delay is about 10^{-8} s and can be ignored. The recovery time of electric field is related to the dissipation of space charges, which is influenced by q of the retained space charges and E of the space after a streamer. According to the analysis above, the maximum q in a discharge determines V. Therefore, Δt should have relationship with V if E is similar.

n is defined as the discharge repeating times per millisecond, which is related to Δt and mainly used for auxiliary analysis of discharge rules when Δt is too small to figure out its variation.

Acquired PD pulses under positive or negative polarity DC voltage are shown in Figure 2. In this research, the first peak value of a pulse is recorded as V. For positive polarity, the trigger voltage level is 0.2 V. For negative polarity, the trigger level is −0.1 V.

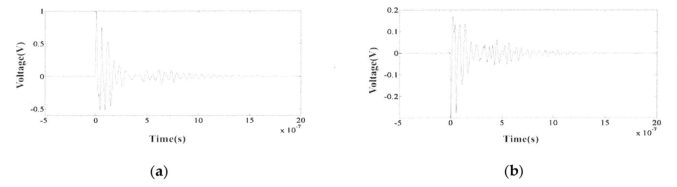

(a) (b)

Figure 2. Single pulse under positive or negative polarity DC voltage. (**a**) Positive pulse; (**b**) Negative pulse.

3. Statistic Rules and Stage Characteristics of Positive PD

3.1. Formation of Positive PD Pulse

First of all, how space charge behaviors and electric field variation contribute to the formation of a PD pulse is deduced.

When E of the tiny region near the needle tip reaches the initial discharge value, electron avalanches occur here. When electron avalanche develops forward, produced ions will reduce E in the region where ionizations just happen. Thus ionizations mainly happen on the head regions. When ionizations on the head stop, so does the whole electron avalanche.

For positive polarity voltage, cations are concentrated in the head of electron avalanches. When the self-sustaining condition is satisfied, electron avalanche discharge transfers to streamer discharge. The cations produced by secondary electron avalanches further increase the density of cations on the head of the streamer. Cations can be approximately considered as motionless in the discharge path during the short time of discharge. While electrons generated by secondary electron avalanches are injected into the body of the streamer, and most are transformed to anions because of the low E. Then, the anions and the cations form plasma, which is approximately electrically neutral inside. Thus the streamer shows positive polarity on the whole affected by the high-density cations on the head during the discharge process. Because the charge–mass ratio of cations is relatively high, the interaction effect between cations is low. The paths of streamers under positive polarity DC voltage are relatively concentrated. An instantaneous, enhanced positive polarity electric field will be generated by the cations near the needle tip. The E near the plate will also increase. A growing number of electrons flow toward the gap from the plate, thus forming the rising edge of the pulse current on the sampling resistance.

The streamer continues to develop towards the plate until E is insufficient to maintain self-sustaining discharges on the head. Then under the action of electric field, retained cations migrate toward the plate and anions migrate toward the needle. The electric field intensity gradually returns to the initial value, and the enhanced electric field decreases. The number of electrons flowing from the plate decreases gradually, and the falling edge of PD pulse occurs. Because the E on the discharge path at the falling stage is lower than before, the falling time is longer than the rising time. When the electric field intensity of the region resumes to the initial discharge value, a new PD can be generated.

3.2. Discharge Stages

Air corona discharge of the needle-plate module under DC has obvious space charge effect. When E of discharge paths resumes to the initial discharge value, part of ions still retain. After multiple discharges, ion densities reaches an equilibrium state. The cation density near the needle tip should be higher than the average. When the density reaches a certain level, the ionic electric field will distort to the original electric field.

The minimum applied voltage under which the PD pulse can be detected is defined as the initial discharge voltage. First, the average value of parameters n, V and Δt are extract. The expression of the average value is:

$$\bar{y} = \sum_{1}^{N} y_i / N,\qquad(3)$$

where N denotes the number of data, and y_i denotes the i th data.

In this research, for each voltage level, five or more groups of acquired PD data (0.1 s each) are chosen from different measurements, ensuring that inside which are at least one hundred pulses. Then characteristic parameters are extracted from these PD pulse data.

Figure 3 shows the average value of PD parameters under each voltages. According to rules of $\bar{n}$, $\bar{V}$ and $\overline{\Delta t}$, the discharge stages are divided. The discharge process can be divided into three stages: initial streamer stage, glow-like discharge stage and breakdown streamer stage.

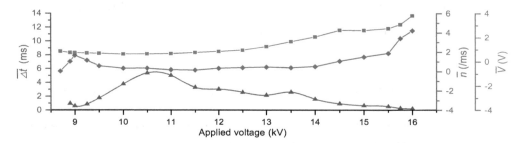

Figure 3. Average value diagram of n, V and Δt on positive polarity.

The initial streamer stage is 8.8 kV to 9 kV. The density of retained cations at this stage is relatively low and is close to the needle tip, so the effect of cation electric field is weak. E of outer regions is below E_0. Thus the development degrees of streamers are similar, and the generated space charge amounts of each discharge are close. Thus, $\bar{V}$ basically retains unchanged. Under initial discharge voltage, the high-intensity electric field region is tiny. The discharges is random, thus $\overline{\Delta t}$ is large (84.44 ms, not shown in Figure 3). With the increase of applied voltage, amount of retained cations increases. Because the cations group is close to the tip, thus their main effect is enhancing original E near the needle tip. The amount of produced cations and anions is relatively small, so when E increases, the electric field recovers quickly, thus $\bar{n}$ rises.

Nine kV to 9.5 kV is the transition stage, where the cation group gradually becomes intensive and moves away from the needle tip. For the electric field between cations and the needle tip, the cation group generates an opposite electric field and weakens the original one. In some regions, E is weakened to the value below E_0, so some streamers are weakened and stopped inside cation group. While in some regions, the lowest E in cation group is higher than E_0, and even stronger more outside, streamers are enhanced. So on average, $\bar{V}$ changes little. Weakened E inside results in the increasing of electric field recovery time. So $\overline{\Delta t}$ increases.

The glow-like discharge stage is about 9.5 kV to 13.5 kV. At this stage, the development of streamer is significantly inhibited by the cations group. The cations group stays at a certain distance from the tip and the opposite electric field is more remarkable. Majority of streamers are stopped inside the cation group. In this situation, the equipotential line of E where E is equal to the E_0 is basically unchanged, so the development degrees of the streamers at the initial stage are similar. $\bar{V}$ is basically unchanged. Then at the later transition stage, the distance of the cation group from the needle expands under the action of the needle-tip electric field. The cation electric field cannot offset the original electric field increases. So E gradually increases and streamers develop further. $\bar{V}$ rises slightly. At the initial stage, due to the increasing amount of generated space charges and the decreasing E, the discharge recovery time increases and $\overline{\Delta t}$ is large. In 10 kV to 11.5 kV, E inside decreases significantly and $\overline{\Delta t}$ reaches the maximum value. Then with the expansion of the cation group and the slightly increase of E, recovery time slightly decreases, and so does $\overline{\Delta t}$.

As E between the needle tip and the cation group is weakened, electrons are more likely to form anions and accumulate near the needle tip under the action of the electric field. If the curvature radius of the positive polarity electrode is relatively large, the density of accumulated anions can reach a certain level and anion group can be formed. The electric field between the anion group and the needle tip will be enhanced, while the electric field between anion group and the cation group will be further weakened. Then, the stable glow discharge may occur between the needle tip and the cations. There is no obvious glow discharge phenomenon under positive polarity voltage in this needle–plate gap.

At the breakdown streamer stage (13.5 kV to 16 kV), the cations group is far away from the needle tip and its density is also relatively reduced. Its weakening effect to the electric field is reduced, and the streamers can develop through the cation region. As a result, the streamers are stronger than the stage before. Raising the applied voltage, more cations are generated and $\overline{V}$ rises. E in the discharge regions increases. The recovery time decreases significantly and so does $\overline{\Delta t}$.

When it close to breakdown (after 15.5 kV), the negative polarity plate begins to affect the discharge process. At this time, the head of streamers will be close to the negative polarity plate. Due to the high conductivity of the plasma, the electric field between the streamer head and the plate electrode is significantly enhanced, which is conducive to ionization. The streamers will be strongly enhanced here. Therefore, V increases with obviously rising trend. After a discharge, E of the discharge path is still in an enhanced state. So Δt decreases a lot (about 0.23 ms at 16 kV). Continue to increase the applied voltage, streamers link the gap and breakdown occurs quickly. Due to the randomness of the breakdown streamer, the breakdown may occur under a certain applied voltage within the scope of 16.5 kV to 17.5 kV.

3.3. Derived Parameter Rule

The equivalent charge quantity q is defined as the charge quantity of electrons passing through the sampling resistance in unit time, which can indirectly reflect the changes of the space electric field and discharge energy. Regarding single PD pulse as a triangular wave approximately, the expression of q is:

$$q = \overline{n} \cdot \frac{\overline{V}t_w}{2R}. \tag{4}$$

where t_w denotes single PD pulse width, which is set as 50 ns. R denotes the sampling resistance value.

Equivalent charge quantities of positive PD are shown in Figure 4. At the initial streamer stage, there is an obvious q, while at the glow-like discharge stage q is close to 0. Then in the breakdown streamer stage, q increases obviously, especially near breakdown. It indicates that the discharges of glow-like stage are obviously inhibited under the action of space charges, which also means that the electric field variation is little, and the discharges consume less energy. q is obvious at breakdown streamer stage.

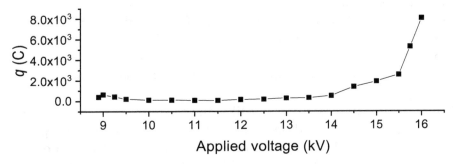

Figure 4. Equivalent charge quantity diagram of positive PD.

The variation coefficient (VC) is adopted to reflect the fluctuation of PD parameters under different voltages, furthermore, to judge the unevenness of electric field distribution. The expression of variation coefficient is:

$$VC = \frac{\sqrt{\sum\limits_{1}^{N}(y_i - \overline{y})^2 / N}}{\overline{y}}. \tag{5}$$

The VC of V and Δt under different voltages are shown in Figure 5. VC for V basically remains unchanged at the initial streamer stage. Then VC decreases at the early stage of glow-like discharge and reaches the lowest value. It rises at the end of glow-like discharge stage, and reaches a higher value at the breakdown streamer stage and then retains. When approaching breakdown, VC has a small increase. Amplitude fluctuation in the glow-like discharge stage are relatively smaller, indicating that the streamers are relatively regular, and the electric field distribution of discharge regions is even at this stage because of the cation group. Especially at the initial stage of glow-like discharge, the space charge electric field plays an obvious role.

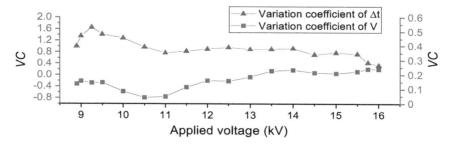

Figure 5. Variation coefficients diagram of positive polarity.

The VC for Δt increases at the initial streamer stage where space charges are few, and decreases at the transition stage with the increasing density of cations. Then VC sustains in glow-like discharge stage. At last, it decreases again near breakdown. It can be seen that the fluctuation of Δt will be lower with certain cation density, because space charges retained after a single discharge have less effect on the electric field. The cation group contributes to a relatively stable electric field in discharge regions, thus ions migrate regularly and the fluctuation of discharge recovery time is low.

3.4. Δt–V Distribution Rules

Discharges of initial streamer stage are very few and relatively random, whose Δt–V distribution is not researched. Instead, typical Δt–V distribution of transition stage after initial streamer stage is described. In Figure 6, the Δt–V points is concentrated in two regions, which forms two "sharp peaks". As discussed before, the points in high-amplitude peak represent the pulses generated by the streamers passing through the cation group, while points in low-amplitude peak represent the pulses generated by the streamers stopped inside. Figure 7 shows V and Δt proportion distribution respectively. Each Δt proportion curves has two peaks. While in each Δt proportion curve, there is only one peak, and low-value Δt makes up the majority. Combined with Figure 6, it can be concluded that the difference of the two forms of discharges is mainly reflected in the amplitude. Δt distributions of this two forms of discharges are basically the same. It may because the streamers corresponding to the high-amplitude pulses generate more ions, but after the discharge, the recovery electric field for ion migration has a higher E.

The typical Δt–V scatter diagram of glow-like discharge stage is shown in Figure 8a,b. At the initial stage, the discharges are restrained on weaken electric field inside action group. V changes little, so Δt–V scatters basically form a horizontal line. Notice that the central value of V in 10 kV is close to that of the low-value peaks in 9 kV (about 0.65 V), which demonstrates the two discharge forms are similar. Then at the later stage of glow-like discharge, the range of V enhances. Δt–V points still distribute in a horizontal ribbon basically. Figure 9a,b show the proportion distributions of V and Δt. At the beginning, V is relatively concentrated (10 kV–11 kV). But at later stage, V gradually distributes dispersedly. The abscissa of the peak point is larger. For Δt, the abscissa of the highest proportion point and the abscissa ranges of each proportion curve are basically the same. It can be observed that the distribution curves of Δt can be divided into two parts. One part is of peak shape in abscissa of 0 to 0.005 s. Another part is of flat shape in 0.005 s to 0.01 s. The flat part may represent the Δt after the streamers that pass though the cation group region, since the ions take a long time to migrate because of the low E in this region. For Δt, their distribution are dispersed at the early stage. The abscissa of the highest proportion point decreases with applied voltage rises, reaching the lowest value under 11 kV. The Δt points then become more concentrated at the later stage, and the highest proportion value rises.

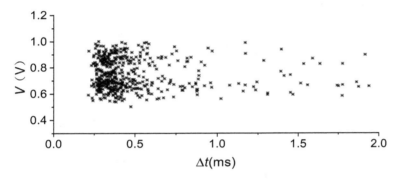

Figure 6. Δt–V distribution of 9 kV at the transition stage.

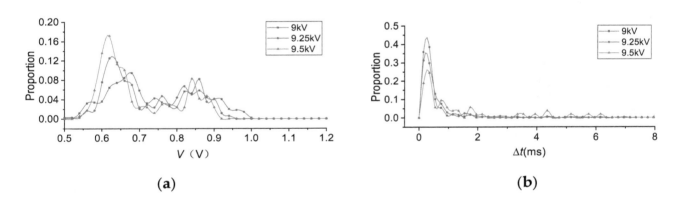

(a)   (b)

Figure 7. Proportion distribution diagrams at the transition stage. (**a**) Proportion distribution of V; (**b**) Proportion distribution of Δt.

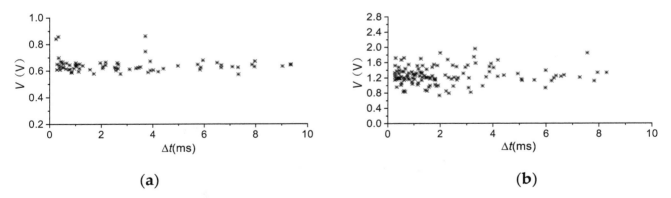

(a)   (b)

Figure 8. Δt–V distribution of glow-like discharge stage. (**a**) Δt–V distribution of 10 kV; (**b**) Δt–V distribution of 13 kV.

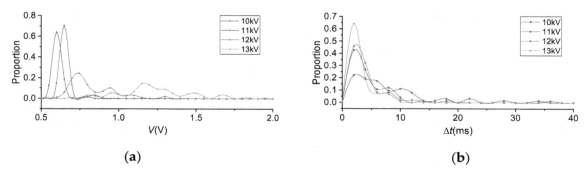

Figure 9. Proportion distribution diagrams at the glow-like discharge stage. (**a**) Proportion distribution of V; (**b**) Proportion distribution of Δt.

At breakdown streamer stage, E in the cation group region can satisfy self-sustaining condition and majority of streamers can pass through the cation region. As shown in Figure 10a, the main Δt–V points form a triangular. When Δt abscissa is small, the density of points is large and V range is wide. With the increase of Δt, density of point and V range are decreased. Compared Figure 10a to Figure 10b, with the increase of applied voltage, the range of V increases, and the Δt decreases. Figure 11a,b show the V and Δt distribution proportions. For V, the shapes of the proportion curves are similar in the beginning, and the abscissa of the highest proportion point gradually increases. While V overall increases, and V range expands significantly near breakdown. For Δt, with the increase of applied voltage, the concentration degree of Δt is significantly increased. The highest proposition of the curve increases, and the abscissa of highest proposition point decreases. Similarly, the variations are especially obvious near the breakdown. It can be concluded that breakdown streamers in positive polarity are strong and intensive.

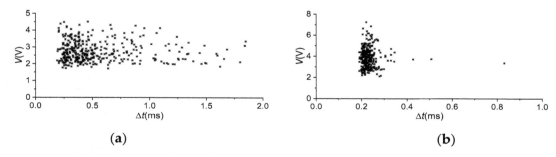

Figure 10. Δt–V distribution of breakdown streamer stage. (**a**) Δt–V distribution of 15 kV; (**b**) Δt–V distribution of 16 kV.

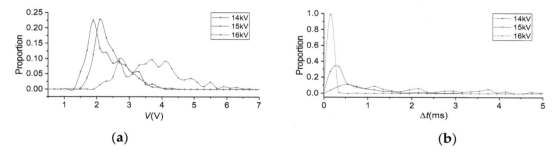

Figure 11. Proportion distribution diagrams at breakdown streamer. (**a**) Proportion distribution of V; (**b**) Proportion distribution of Δt.

4. Statistic Rules and Stage Characteristics of Negative PD

4.1. Formation of Negative PD Pulse

For negative polarity PD, there are many electrons in the head of streamers, which is different from the positive polarity situation. Due to the large charge–mass ratio, interaction forces between electrons play a role. As electrons migrate away from the needle tip dispersively, the discharge paths of the streamers are also dispersive. Retained space charges will distribute with a wider radius.

When streamers stop, the generated electrons are repelled by the electric field and leave the discharge regions quickly. Cations are regarded as immobile during the discharge process.

When the electron avalanches are transformed to streamers, the secondary collapses produce more cations, which gather in the head of the streamer. The ion density in the head of the streamer is high, and E in this region is low. Many electrons are transformed to anions. Cations main form plasma with the anions in the body of the streamer. Since the plasma is electrical neutrality, the main electrical property displayed by the streamer is negative polarity during the discharge process, which is caused by the electrons and anions in the front of streamer head. Thus an instantaneous negative polarity electric field toward the plate electrode will be generated, and E near the plate also increases instantaneously. Growing number of electrons flow towards plate, forming the rising edge of pulse. When streamer stops, the ions migrate to the electrodes, and E gradually recovers. The enhanced negative polarity electric field decreases, forming the falling edge of pulse. When the space charge dissipates, E near the needle tip returns to initial discharge value, and a new PD may be generated.

4.2. Discharge Stages

Due to the slow migration velocity, cations may retain in discharge regions. Meanwhile, the density of anions outside will increase. Under the influence of ion electric field, different forms of discharge are also formed under negative polarity voltage.

Similarly, average values of n, V and Δt are extracted and are shown in Figure 12. According to the rules of n, V and Δt, process of negative polarity corona discharge can be distinguished as four stages: initial streamer stage, Trichel discharge stage, glow-discharge stage and breakdown streamer stage.

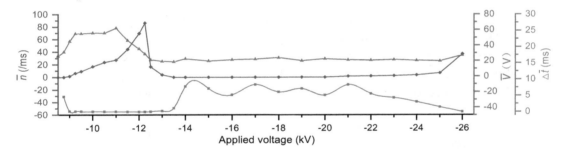

Figure 12. Average value diagram of n, V and Δt on negative polarity.

The initial streamer stage is −8.5 kV to −9.0 kV. At the initial discharge voltage, E is low and the streamers are weak. V is low and Δt is large. With the increase of the applied voltage, the amount of cations increases, and E between needle tip and cations increased slightly. Recovery time of the electric field becomes shorter; thus n rises. Streamers are strengthened, and the space charges generated by a single discharge are also slightly increased, so V increased slightly.

The Trichel discharge stage is −9.0 kV to −12.25 kV. At this stage, the density of cations near the needle tip reaches a certain level, forming a cation layer because of the dispersive development streamers. Cation layer greatly enhances the electric field between the needle tip and the cation layer. Outside the cation layer, E toward the plate is weakened, and anions in space also play a weaken role.

The Trichel discharge stage can be divided into initial and later stages. At initial stage, the streamers stop in the weakened periphery region of cation region. Since the weaken effect of cations layer basically offsets the increase of E produced by of applied voltage, streamers almost stop at same distance from the needle tip, and the development degrees of the stream are approximately the same. V is almost unchanged. E between the cation layer and the needle tip is flat and steady increases, so that n basically presents a linear growth trend.

At the later stage, electrons begin to form anions very close to the outer edge of cation layer, and plasma layer is formed when cations and anions are both very intensive. Variation of E in the plasma layer is very low. As the anions and cations get closer and closer, plasma layer gradually takes place the peripheral cation layer. E inside plasma layer increases. It is easier for E to recover to E_0 after a streamer, and ion variation generated by a single discharge is relatively less. Therefore, the recovery time of electric field decreases. n increases significantly, reaching the maximum value under -12.25 kV. Ion variation generated by a single streamer is less and density of original ions grows, thus single streamer affect the electric field less and less, instantaneous variation of space charge intensity is low. So V decreases.

After that, with the increase of applied voltage, E between the needle electrode and the cation layer is so high that it will not drop blow E_0 even after a streamer. So stable glow discharges occur (-12.5 kV to -24 kV). At this time, the generation and dissipation of ions basically reach a dynamic balance. Almost no instantaneous space charge changes occur, so there is no PD pulse. E is very low on the outer edge of the plasma layer, where streamers can hardly be produced. PD pulses will be generated only in a few of the peripheral regions of the cation layer when the E is weakened and may drop below E_0 after a streamer. These streamers are soon stopped by the plasma layer, so V is small. n is also small at this stage. Within a certain voltage range, the increase of applied voltage only causes the expanding of cation layer. The periphery weakened electric field still cannot supports streamers, so the negative polarity glow discharge stage lasts long.

The breakdown streamer stage is -24 kV to breakdown voltage. Glow discharges still happen inside. E on the outer edge of the plasma layer has risen high enough. So streamers are generated from the plasma region and continues to develop outward. Streamers are stronger and E in discharge regions increases. So both n and V increases. When breakdown streamer occurs, breakdown will happen soon when applied voltage increases a little. The breakdown voltage is approximately -27 kV to -28 kV.

4.3. Derived Parameter Rule

The q diagram of negative polarity discharge pulse is shown in Figure 13. At initial streamer stage, q is small. Then, in the Trichel discharge stage, q increases linearly throughout the stage, which is verified in Figure 14b. The linear increase of q in the Trichel discharge stage also indicates that the electric field inside the cations layer is relatively flat. Subsequently, in glow discharge stage, q is approximately 0. When it comes to the breakdown streamer stage, q rises again.

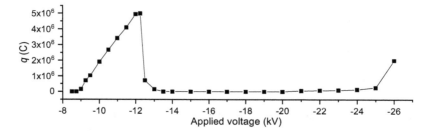

Figure 13. Equivalent charge quantity diagram of negative PD.

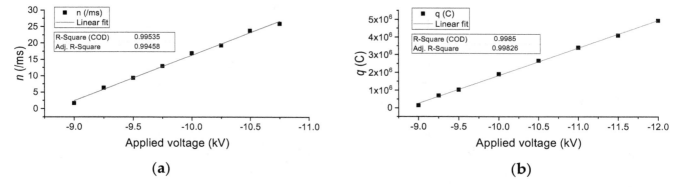

Figure 14. Linear fit analysis of n, q at Trichel discharge stage. (**a**) Linear fit of n; (**b**) Linear fit of q.

PD fluctuation is researched by VC. As shown in Figure 15, for V, VC increases at initial streamer stage, has a sudden rise in -9 kV and then stable at the initial stage of Trichel discharge. It indicates that the gradually formed cation layer makes the electric field distribution uniform. At the later stage of Trichel discharge, VC remains stable after rising to a certain extent. It is because with the increase of plasma, E between the needle tip and the plasma layer is quicker strengthened. The electric field intensity in a certain region is all higher than E_0, so the streamer may develop in many paths within the region. The development degree of the streamer varies, and the fluctuation of V is relatively large. At the glow discharge stage, only few PD happen in peripheral regions of plasma. Few PD pulses with small V are achieved in tiny peripheral regions. So VC basically remains stable at glow discharge stage, and it slightly decreases at the streamer breakdown stage.

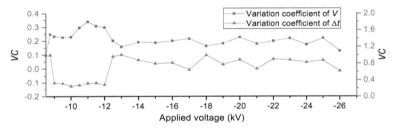

Figure 15. Variation coefficients diagram of negative polarity.

For Δt, due to the generation of the cation layer and the plasma layer, the fluctuation is relatively small. The remarkable characteristic is that VC remains low throughout the Trichel discharge stage. It indicates that the space charge migration after a single discharge is under a relatively stable electric field, and the recovery times of the electric field are similar. Subsequently, VC in glow discharge stage maintains in a larger value, and decreases a little in the breakdown streamer stage. It means discharges in the breakdown streamer stage are more regular than that in the glow-discharge stage.

4.4. Δt–V Distribution Rules

The typical Δt–V scatter diagram of the initial streamer stage is shown in Figure 16. When discharges are relatively regular, the scatter points of Δt–V basically form a horizontal line. It indicates that the discharge region is tiny. The development degree of streamers are similar, so V of PD pulses are unchanged. Figure 17 shows proportion distribution of V. When applied voltage increases, abscissas of V become bigger and the abscissa range is enhanced. Highest proportion value decreases. It indicates that the effect of ion electric field is weak, and the electric field generated by the needle tip mainly plays a role. So the change of applied voltage has a greater impact on the V. At this stage, Δt is large and decreases sharply with the increase of applied voltage. The scatter diagram of Δt is not listed.

The typical Δt–V scatter diagram of the Trichel discharge stage is shown in Figure 18. Some Δt–V points form an inclined spindle. It indicates there is an approximate proportional relationship between V and Δt, which confirms the analysis in Section 2.3. In Figure 18a, at the initial stage, the points in high Δt abscissas are relatively intensive. In Figure 18b, the points in low Δt abscissas are intensive. The density of the points outside the spindle rises because plasma layers form in some regions. The weaken effect of cation layer for electric field outside decrease because of the plasma. So regions where E is higher than E_0 were enhanced in some directions. Some Δt–V points in low-Δt abscissas have higher V, and their distribution does not follow the original proportional relationship.

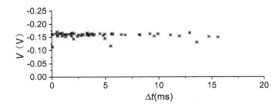

Figure 16. Δt–V distribution diagram of -8.75 kV at initial streamer stage.

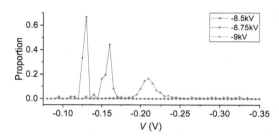

Figure 17. Proportion distribution diagrams of V at the initial streamer stage.

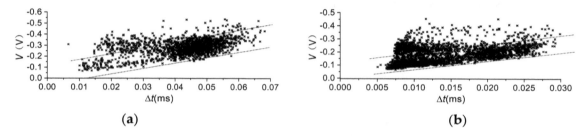

(a)   **(b)**

Figure 18. Δt–V distribution diagrams at Trichel discharge stage. (**a**) Δt–V distribution of -10 kV; (**b**) Δt–V distribution of -12 kV.

At the initial stage, the proportion distribution of V under different applied voltage are approximately similar, as shown in Figure 19a. At the later stage, the distribution curve of V overall moves to lower V abscissas gradually, and the highest proportion gradually increases, as shown in Figure 20a.

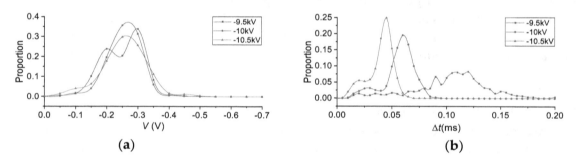

(a)   **(b)**

Figure 19. Proportion distribution diagrams at the initial stage of Trichel discharge. (**a**) Proportion distribution of V; (**b**) Proportion distribution of Δt.

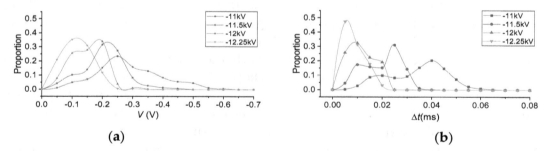

(a)   **(b)**

Figure 20. Proportion distribution diagrams at the last stage of Trichel discharge. (**a**) Proportion distribution of V; (**b**) Proportion distribution of Δt.

As shown in Figure 19b, the range of Δt decreases overall when applied voltage rises. Notice that Δt curves can be divided into two parts. One part is of flat shape and low proportion in low Δt abscissas, the other one is of peak shape and high proportion in high Δt abscissas. The flat shape part is caused by the plasma in some regions, because part of cation layer is transformed to plasma layer, and high E regions expand. At the later stage, as shown Figure 20b, with the increase of applied voltage, the proportion of the flat shape part increases, at last flat shape part becomes the main body. It indicates that diffuse plasma layer is formed, which almost covers all discharge regions.

Since PD in the glow-discharge stage are very few, only Δt–V distribution of the transition stages before and after glow-discharge stage are analyzed. The typical Δt–V scatter diagrams are shown in Figure 21a,b. The Δt–V points basically distribute in a horizontal band, which means V changes little, because PD can be measured are only generated in tiny regions. Δt–V distribution is similar to that of initial streamer stage.

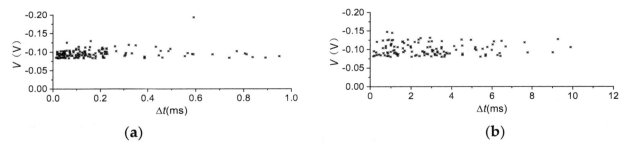

(a)   (b)

Figure 21. Δt–V distribution diagrams of glow discharge stage. (**a**) Δt–V distribution of −13 kV; (**b**) Δt–V distribution of −24 kV.

Proportion distributions of V and Δt at glow discharge stage are shown in Figure 22. In Figure 22a, it can be found that the V proportion curves at transition stage after glow discharge have double peak. The points in low-V abscissas peak should link to PD in few peripheral regions of cation layer where E drops below E_0 after a streamer. So the low V abscissas peak in proportion curves of transition stage after the glow discharge, are similar to the single peak in proportion curves of transition stage before the glow discharge. The points in high V peak link to PD starting from the outer edge of plasma layer. The two forms of discharge streamers have obvious difference, thus form the double peaks.

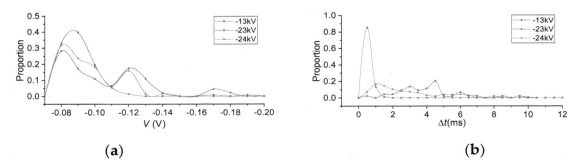

(a)   (b)

Figure 22. Proportion distribution diagrams at glow discharge stage. (**a**) Proportion distribution of V; (**b**) Proportion distribution of Δt.

At breakdown streamer stage, Δt significantly decreases. The typical Δt–V scatter diagram is shown in Figure 23. The distribution of Δt–V points form a sharp cone like shape. In this situation, the discharge region is close to the plate, and the axial electric field is stronger than other directions. In axial direction streamers are continuously produced. Therefore, low Δt and high V discharge pulses account for the main proportion. Proportion distribution characteristics of V and Δt at breakdown discharge stage are also shown in Figure 24. In Figure 24b, it can be observed that the concentration degree of Δt is significantly increased as it approaches breakdown.

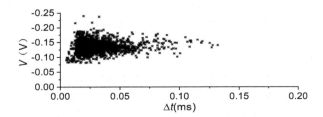

Figure 23. Δt–V distribution of −26 kV at breakdown streamer stage.

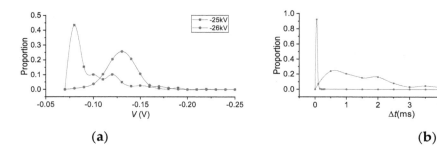

Figure 24. Proportion distribution diagrams at breakdown streamer stage. (**a**) Proportion distribution of V; (**b**) Proportion distribution of Δt.

5. Discussion

The deduced space charge effect in the corona discharge development process of both polarities are concluded and compared, to further discuss the PD mechanisms.

For initial streamer stage, space charge effect is weak. Streamers develops when E of tiny region reaches E_0. The development degree of streamers is low, so V is low and Δt is short. Discharges are relatively random.

Then space charges begin to retain, and actions play a main role. At transition stage of positive PD, the action group weaken the internal electric field while strengthen the external one. But E of the weaken regions are not even. In central axial region, electric field is the strongest and still higher than E_0 after being weakened. While E of regions around is not, so streams develop toward these regions are stopped earlier. Thus, two forms of streamer occur.

At Trichel discharge stage of negative PD, the action layer strengthen the internal electric field while weaken the external one. Actions are more dispersive and intensive, and are closer to the needle tip. Because of the strengthen effect, from needle tip to action layer, electric field decreases little, or even increase. While the E's outside cation layers are basically lower than E_0, which stops streamers in a similar degree and causes unchanged V. The electric field inside cation layer is high, flat, and linear, and grows with the increase of applied voltage, so some Δt–V has an approximate proportional relationship. Recovery times are short and stable, so n rises in a linear fashion. Discharges inside the cation layer are strong. More anions are transformed and gradually become close to cations at the later stage.

Because actions are more intensive and closer to the needle tip on negative polarity, its effects are more obvious, which reflects on the variations of parameters, especially n, q.

Next, retained space charges increase to a certain level, electric effects are obvious. At glow-like discharge stage of positive PD, the effect of actions are more obvious because of higher density. The streamers are basically limited inside action group and cause unchanged V. Notice that PD parameter rules of this stage are similar to that of Trichel discharges in negative polarity. So it can be concluded that the electric field distribution of discharge regions are similar at two stages. The phenomenon delay and are not so obvious on negative polarity because of the different effects of actions. At the later transition stage, because cation group expands and becomes sparse, its effect is weakened and streamers in some regions are not limited, discharge form is transforming to breakdown streamer.

At the glow-discharge stage of negative PD, intensive anions occur close to the outer edge of cation layer, and plasma layer is formed. Stable glow discharges occur inside plasma layer, which does not produce a PD pulse without instantaneous space charge variations. In a few regions of the cation layer and plasma layer, some intermittent streamers occur. Because the regions are tiny and E is relatively weak, the PD is weak. Within a certain voltage range, the increase of applied voltage only causes the expanding of cation layer, so negative polarity glow discharge stage lasts long.

At last, streamers with few limitation occur gradually and cause breakdown. At breakdown streamer stage of positive PD, the cations group is far from the needle tip and its density decreases. Its effects are reduced, and the streamers are almost not limited and become strong. While at breakdown streamer stage of negative PD, cation layer and plasma layer still work. When the plasma layer is close to the plate, the high E regions in peripheral regions of plasma expand, the negative streamers 'restart' and develop without much limitation. As such, the negative streamers are relatively weak, and cause a breakdown sooner because they start near the plate.

The statistic rules of PD parameters in this paper and others support each other in some aspects [10–14]. The mechanism explanations are based on the classical theories and well-founded. The deduced result can be interpreted in perspective of previous studies, for example, space charge and plasma distribution at some stages can be verified by luminescent image [19]. In follow-up research, we are hopeful the deduced space charge effects and electric field distributions will be verified by a simulation on the basis of the corona discharge fluid model mentioned previously [17].

The statistical analysis method and some deduced rules may also be effective in other gas PD under an uneven electric field, with DC voltage and AC voltage with low frequency, where space charges have enough time to retain.

6. Conclusions

In this paper, a 15-mm needle-plate model is adopted to study the air corona discharge under positive and negative polarity DC voltage. The statistical rules of characteristic parameters of each stage and transition process are explained by the space charge effects and electric field distributions. Furthermore, the discharge characteristics, the reflected space charge effects of each stage are summarized and compared. Discharge mechanisms are explained in microcosmic angle. Microcosmic process of PD under DC voltage can be described based on statistical methods, and several conclusions can be drawn.

Space charges obviously distort electric fields, and the most effective influence to PD development is decreasing E down to E_0 somewhere. For positive corona discharge, the space charge effects are most obvious at the glow-like discharge stage, where cations play a main role. The internal electric field of cation group is weakened but its distribution is relatively flat. The cation group inhibits the development of streamer, but inhibition effect is weakened later. At the breakdown flow stage, the influence of the action group is almost not reflected. For negative corona discharge, at Trichel discharge stage, cations work obviously, producing an electric field distribution with high and flat E inside. Then anions take part in and plasma layer forms in peripheral regions of cation layer, and further weakens the periphery electric field. At the glow discharge stage, E inside is high enough to sustain stable glow discharges inside. While E outside plasma layer is low, until in the breakdown streamer stage, E in outer edge of plasma layer reaches E_0 and streamers 'restart'.

Space charge effects are connected with PD parameter by electric field, and some electrical field distributions contribute to obviously PD parameter rules. When discharge regions are tiny and E is relatively weak, it may cause unchanged V, and Δt–V points distribute horizontally, which are shown at initial discharge stages of both polarities and transition stages of negative-glow discharge. When the electric field is high and relatively flat, Δt–V may have an approximate proportional relationship, and Δt–V points distribute inclined, which is shown at Trichel discharge stage of negative polarity.

Different forms of discharges may happen under a certain voltage, especially at transition stages, and different forms of discharges can be distinguished on PD parameters. For example, the V proportion distribution at glow-like stage of positive polarity, and Δt proportion distribution at Trichel discharge stage of negative polarity. The phenomenon arises in that E varies significantly in different discharge regions, which, to a large extent, are caused by the existence of cations and plasmas.

Author Contributions: Conceptualization, D.W. and L.D.; Methodology, D.W.; Software, D.W.; Validation, D.W.; Formal Analysis, D.W.; Investigation, D.W.; Resources, L.D. and C.Y.; Writing-Original Draft Preparation, D.W.; Writing-Review & Editing, L.D. and C.Y.; Visualization, D.W.; Project Administration, L.D.; Funding Acquisition, L.D.

Acknowledgments: Many thanks to Deming Zhan and Han Yan for their contributions in experiments and investigations.

References

1. Lutz, B.; Kindersberger, J. Surface charge accumulation on cylindrical polymeric model insulators in air: Simulation and measurement. *IEEE Trans. Electr. Insul.* **2011**, *18*, 2040–2048. [CrossRef]
2. Zhang, Z.J.; Zhang, D.D.; Zhang, W.; Yang, C.; Jiang, X.L.; Hu, J.L. DC flashover performance of insulator string with fan-shaped non-uniform pollution. *IEEE Trans. Electr. Insul.* **2015**, *22*, 177–184. [CrossRef]
3. Thomson, J.J.; Thomson, G.P. *Conduction of Electricity through Gases*; Cambridge University: London, UK, 1933.
4. Trichel, G.W. Mechanism of the negative point-to-plate corona near onset. *Phys. Rev.* **1938**, *54*, 1078–1084. [CrossRef]
5. Morshuis, P.H.F.; Smit, J.J. Partial discharge at DC voltage: their mechanism, detection and analysis. *IEEE Trans. Dielectr. Electr. Insul.* **2005**, *12*, 328–340. [CrossRef]
6. Kachi, M.; Nadjem, A.; Moussaoui, A. Corona discharge as affected by the presence of various dielectric materials on the surface of a grounded electrode. *IEEE Trans. Dielectr. Electr. Insul.* **2018**, *25*, 390–395. [CrossRef]
7. Piccin, R.; Mor, A.R.; Morshuis, P. Partial discharge analysis of gas insulated systems at high voltage AC and DC. *IEEE Trans. Dielectr. Electr. Insul.* **2015**, *22*, 218–228. [CrossRef]
8. Marek, F.; Barbara, F.; Pawel, Z. Partial discharge forms for DC insulating systems at higher air pressure. *IET Sci. Meas. Technol.* **2016**, *10*, 150–157.
9. Stephan, V.; Joachim, H. Experimental evaluation of discharge characteristics in inhomogeneous fields under air flow. *IEEE Trans. Dielectr. Electr. Insul.* **2018**, *25*, 721–728.
10. Si, W.R.; Li, J.H.; Yuan, P.; Li, Y.M. Digital detection, grouping and classification of partial discharge signals at DC voltage. *IEEE Trans. Electr. Insul.* **2008**, *15*, 1663–1674.
11. Tang, J.; Liu, F.; Zhang, X.X.; Meng, Q.H.; Zhou, J.B. Partial discharge recognition through an analysis of SF6 decomposition product part 1 decomposition characteristics of SF6 under four different partial discharges. *IEEE Trans. Electr. Insul.* **2012**, *19*, 19–36. [CrossRef]
12. Liu, M.; Tang, J.; Pan, C. Development processes of positive and negative DC corona under needle-plate electrode in the air. *High Voltage Eng.* **2016**, *42*, 1018–1027.
13. Wei, G.; Tang, J.; Zhang, X.X.; Lin, J.Y. Gray intensity image feature extraction of partial discharge in high-voltage cross-linked polyethylene power cable joint. *IEEE Trans. Electr. Insul.* **2016**, *23*, 1076–1087. [CrossRef]
14. Zhang, S.Q.; Li, C.R.; Wang, K.; Li, J.Z.; Liao, R.J.; Zhou, T.C.; Zhang, Y.Y. Improving recognition accuracy of partial discharge patterns by image-oriented feature extraction and selection technique. *IEEE Trans. Dielectr. Electr. Insul.* **2016**, *23*, 1076–1087. [CrossRef]
15. Zhang, B.; Wang, W.; He, J. Impact factors in calibration and application of field mill for measurement of DC electric field with space charges. *CSEE J. Power Energy Syst.* **2015**, *1*, 31–36. [CrossRef]
16. Abdel-Salam, M.; Wiitanen, D. Calculation of corona onset voltage for duct-type precipitators. *Ind. Appl.* **1993**, *29*, 274–280. [CrossRef]
17. Wu, F. Numerical analysis on microcosmic process of corona discharge and ionized filed of HVDC transmission lines. Ph.D. Thesis, Chongqing University, Chongqing, China, 2014.
18. Liu, K. Study on the effect of corona discharge space charge background on gap discharge. Master's Thesis, Huazhong University of Science and Technology, Wuhan, China, 2013.
19. Liu, Z.; Liu, T.; Miao, X.; Guo, W. Research on corona layer of needle-plate discharge in atmosphere. *Sci. Technol. Eng.* **2013**, *13*, 1553–1556.
20. Yang, J. *Gas Discharge*, 1st ed.; Science Press: Beijing, China, 1983; pp. 154–196.
21. Yan, Z.; Zhu, D. *High Voltage Insulation Technology*, 2nd ed.; China electric power press: Beijing, China, 2007; pp. 47–76.

15

Modeling of Dry Band Formation and Arcing Processes on the Polluted Composite Insulator Surface

Jiahong He *, Kang He and Bingtuan Gao

School of Electrical Engineering, Southeast University, Nanjing 210096, China; 220192809@seu.edu.cn (K.H.); gaobingtuan@seu.edu.cn (B.G.)
* Correspondence: hejiahong@seu.edu.cn

Abstract: This paper modeled the dry band formation and arcing processes on the composite insulator surface to investigate the mechanism of dry band arcing and optimize the insulator geometry. The model calculates the instantaneous electric and thermal fields before and after arc initialization by a generalized finite difference time domain (GFDTD) method. This method improves the field calculation accuracy at a high precision requirement area and reduces the computational complexity at a low precision requirement area. Heat transfer on the insulator surface is evaluated by a thermal energy balance equation to simulate a dry band formation process. Flashover experiments were conducted under contaminated conditions to verify the theoretical model. Both simulation and experiments results show that dry bands were initially formed close to high voltage (HV) and ground electrodes because the electric field and leakage current density around electrode are higher when compared to other locations along the insulator creepage distance. Three geometry factors (creepage factor, shed angle, and alternative shed ratio) were optimized when the insulator creepage distances remained the same. Fifty percent flashover voltage and average duration time from dry band generation moment to flashover were calculated to evaluate the insulator performance under contaminated conditions. This model analyzes the dry band arcing process on the insulator surface and provides detailed information for engineers in composite insulator design.

Keywords: composite insulator; dry band formation; heat transfer model; generalized finite difference time domain

1. Introduction

Composite insulators have been extensively used to provide electrical insulation and mechanical support for high voltage (HV) transmission lines [1–4]. The shank of the composite insulator is made of fiberglass or epoxy, and the sheds of the composite insulator are made of composite materials. The hydrophobic nature of composite materials discretizes water into small droplets on the insulator surface and ensures good performance of insulators under contaminated conditions [5,6]. However, the humid pollution could form a layer under the severely contaminated environment and increase the leakage current density [7,8]. The leakage current generates heat and evaporates water in the pollutant layer to create the dry band [9,10]. The arc initializes due to the significantly increased electric field close to the dry band [11–14]. Theoretical models for dry band formation and arcing processes are valuable because they contribute to the investigation of composite insulator flashover mechanism [15,16] and provide detailed information for engineers to optimize insulator geometry.

B.F. Hampton first studied the formation of dry bands in 1964 [17]. E.C. Salthouse and J.O. Löberg introduced the specific process of dry band formation in 1971. In terms of surface resistivity and electric field, E. C. Salthouse pointed out that dry band formation is caused by energy dissipation [18,19].

J.O. Löberg concluded that the width and speed of dry band formation are related to the surface temperature [20]. The distorted distribution of the field strength of dry bands also plays an important role in dry band expansion [21,22]. By analyzing a 3-D insulator model with the finite element method (FEM), J. Zhou et al. gave the opinion that the distortion field strength increases the length of the dry band. The number of dry bands also influences the electric field distribution [21]. A. Das et al. summarized that the dry band position has a significant influence on the maximum electric field strength [22]. These studies present the characteristics of the dry band and analyze the effects of dry bands on electric field distribution and flashover phenomena. However, the process of dry band formation caused by leakage current and electric arc has not been investigated in detail. This paper proposes a model to analyze instantaneous electric and thermal field variation during the dry band formation and arcing processes. The fields were calculated using a generalized finite difference time domain (GFDTD) method.

The GFDTD method consists of the generalized finite difference method (GFDM) and the finite difference time domain (FDTD) method. The GFDM is a method improved from the finite difference method (FDM). The traditional FDM depends on mesh-dividing, which is not suitable for fields with complicated boundaries, while the GFDM is a meshless method to compute the relationship of any discrete point in the field of the boundary conditions. The GFDM has an advantage over traditional FDM in the sense that the density of the calculation points could be different according to the boundary conditions and precision required in the field domain. The concept of the GFDM was first put forward by J.J. Benito in 2001 [23]. L. Gavate et al. compared the GFDM with other methods and reviewed its application in fluid and force fields [24]. J. Chen et al. then calculated electromagnetic field using the GFDM to reduce the computation time [25]. Currently, GFDM has been used in field calculation problems such as heat transfer and fluid mechanics to increase the calculation accuracy of a relatively small area in a large field domain [26,27].

This paper analyzed instantaneous electric and thermal field distributions close to composite insulators and arcs. Finite difference time domain (FDTD) was utilized to investigate the characteristics of continuously changing fields. In 1966, K.S. Yee dispersed Maxwell's equations with time variables using the method of discretization later-called Yee cell [28], which was gradually developed into FDTD. This paper investigated electric and thermal fields variation by combining the GFDM with FDTD. GFDTD is capable of increasing the calculation accuracy in a high precision requirement area and reducing the computational complexity in a low precision requirement area. The arc propagation and heat transfer processes are modeled based on the electric and thermal field distribution.

Many theories and laboratory experiments have demonstrated that the elongation of the insulator creepage distance is an effective way to increase flashover voltage, but it also increases the weight and reduces the mechanical stress endurance of the composite insulator. Therefore, recent studies have focused on insulator parameter optimization when creepage distances remain the same [29–31]. The simulation models proved that the flashover probability slightly increases with the insulator shank diameter and decreases with shed spacing [32–34]. The creepage factor (CF), shed angle, and the ratio of overhangs between alternating sheds are the factors that impact dry band formation and arcing processes on the composite insulator surface. These parameters were optimized in the paper to analyze the 50% flashover voltage and the average duration time from pollutant layer formation to flashover.

This paper investigated the mechanism of dry band arcing by simulating the processes of dry band formation and arcing under the influence of a heat transfer model and an arc propagation model. The simulation results were compared with the results of the experiment to verify the model. This paper optimized the composite insulator geometry when the creepage distances remain the same.

2. Model Schematic and Method

2.1. Insulator Model Schematic

The composite insulator dimension and geometry were selected according to IEC 60815. Due to the symmetric geometry of composite insulators, a two-dimension model was applied to simulate the dry band formation and arc propagation processes and reduce the computational complexity. The composite insulators were designed for a 110 kV transmission line with 15 large sheds and 14 small sheds. The insulator shed radius and the dimensions of the electrodes are show in Figure 1a. The environment temperature and air pressure were 293 K and 101.325 kPa, respectively.

In Figure 1a, the pollution distribution on the top and bottom surface of the insulator was defined as $ESDD_T$ and $ESDD_B$. The flashover voltage reduces with the increase of the ratio of $ESDD_T$ to $ESDD_B$. The range of the ratio was 0.1 to 1 [35]. In this paper, the ratio was set as 1 to simulate the dry band formation and arcing phenomena the under a severe polluted scenario with relatively low flashover voltage. Therefore, the ESDD value was 0.1 mg/cm^2 and the surface resistivity is 8.3×10^5 Ω·m under the influence of environment temperature and air humidity, and water particles in the air were not considered in the model [36,37].

In Figure 1b, θ is the shed angle. CF is defined as the ratio of the insulator creepage distance to the arcing distance.

$$CF = \frac{l_1 + l_2}{d} \tag{1}$$

where the sum of l_1 and l_2 is the total nominal creepage distance of the insulator. d is the arcing distance of the insulator.

r_1 and r_2 are the radius of large and small sheds respectively. k_{shed} is defined as the ratio of r_2 to r_1.

$$k_{shed} = \frac{r_2}{r_1}. \tag{2}$$

In order to reduce the probability of dry band arcing and arc propagation, the geometry structure of insulator was optimized under the premise that creepage distances remain the same. The optimization variables of insulator geometry were CF, k_{shed}, and θ.

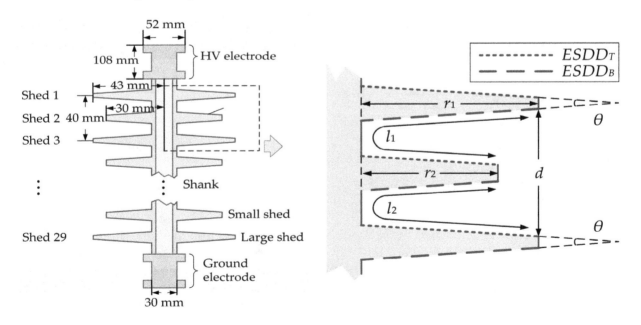

(a) Composite insulator schematic   (b) Geometry parameters

Figure 1. Composite insulator model schematic.

2.2. Dry Band Formation and Arc Propagation Models

2.2.1. Electric Field and Arc Propagation Model

In the electric field close to the insulator, the Poisson is shown below:

$$
\begin{cases}
\nabla^2\varphi = \dfrac{\partial^2\varphi}{\partial x^2} + \dfrac{\partial^2\varphi}{\partial y^2} = -\dfrac{\rho_c}{\varepsilon} & \text{Possion equation} \\[2mm]
\varphi(x,y)\big|_\Gamma = f_1(\Gamma) & \text{Dirichlet boundary condition} \\[2mm]
\dfrac{\partial\varphi}{\partial n}\Big|_\Gamma = f_2(\Gamma) & \text{Neumann boundary condition}
\end{cases}
\tag{3}
$$

where φ is the electric potential, ρ_c is bulk charge density and ε is permittivity.

Before the arc ignition, the electric field calculation model computed the electric field distribution to determine the arc ignition and obtain the leakage current density on the insulator surface. After the arc ignition, the electric field and arc propagation model computed the instantaneous electric field strength around the arc leader during the propagation. The random theory was utilized to determine the arc propagation directions based on the instantaneous electric field.

The instantaneous electric field close to the composite insulator was calculated by the GFDTD method shown in Appendix A. The advantage of GFDTD is that the density of discrete calculation points could be different in the field domain according to the precision requirement and boundary conditions. To focus on the field close to the arc and reduce the computational complexity in the low precision requirement area, the distribution of points close to the arc is denser than the points distribution in other parts of the field (Figure 2a).

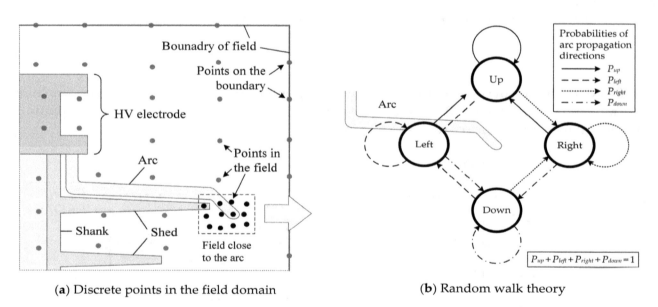

(a) Discrete points in the field domain   (b) Random walk theory

Figure 2. Electric field calculation of the discretized points and arc propagation model based on the random walk theory.

Random walk theory calculated the probabilities of arc propagation in all the directions (Figure 2b). The random number was generated at each step of arc propagation to determine the exact direction of the next step. Therefore, the arc growth direction could be different even when the electric field distribution remains the same, which describes the stochastic characteristics of arc propagation [38].

$$
P = \frac{E^2}{\sum E^2} a(E - E_c)
\tag{4}
$$

where E is the electric field strength summation of all possible directions with $E > E_c$, and E_c (2.1 kV/mm) is the RMS value of the threshold field. a is the step function. The arc propagation velocity is in proportion to the magnetite of the electric field strength.

2.2.2. Heat Transfer Model

The heat transfer model simulates the energy balance of the evaporation process, including the leakage current injection energy, heat conduction and convection energies on the insulator surface, heat radiation energy of the arc, and the water evaporation energy of phase changing.

Before the arc ignition, the source of the thermal field was the accumulated energy on the insulator surface generated by the leakage current density. After the arc ignition, the heat transfer model included the heat radiation of the arc as the dominant factor to affect the dry band formation during arc propagation.

The leakage current injection energy is calculated as follows:

$$W_{leakage} = \sum_{t=0}^{t_0} \sum_{i=1}^{l} \frac{E_i^{t_n 2}}{\rho_r}. \tag{5}$$

Heat conduction and convection are the main forms of heat dissipation on the composite insulator surface before arc initializes. Heat conduction partial differential equation (PDE) and boundary conditions are given as Equation (6).

$$\begin{cases} \rho c \frac{\partial T}{\partial t} = \lambda \left(\frac{\partial^2 T}{\partial x^2} + \frac{\partial^2 T}{\partial y^2} \right) + \Phi & \text{Heat conduction PDE} \\ T(x,y)|_\Gamma = f_1(\Gamma) & \text{Dirichlet boundary condition} \\ \frac{\partial T}{\partial n}|_\Gamma = f_2(\Gamma) & \text{Neumann boundary condition} \end{cases} \tag{6}$$

where T is the thermal temperature, t is time, ρ, c and λ are the density, specific heat capacity and thermal conductivity of different insulating materials, respectively. Φ is the internal heat sources caused by dry band arcing and the leakage current density of the insulator surface.

The GFDTD method in the heat conduction calculation is similar to the electric field computation. The discretized heat conduction PDE is shown in Equation (7)

$$\rho_i c_i \left(\frac{T_i^{t_{n+1}} - T_i^{t_n}}{\Delta t} \right) = \lambda_i d_{1,1} T_i^{t_{n+1}} + \sum_{j=1}^{n} \lambda_j d_{1,(j+1)} T_j^{t_{n+1}} + \lambda_i d_{2,1} T_i^{t_{n+1}} + \sum_{j=1}^{n} \lambda_j d_{2,(j+1)} T_j^{t_{n+1}} \tag{7}$$

where the superscript "t_{n+1}" represents the next stage in the discrete time domain.

Φ is calculated below:

$$\Phi_i = E_i^{t_n} J_i^{t_n} = \frac{\left(E_i^{t_n} \right)^2}{\rho_r} \tag{8}$$

where E is the electric field strength, J is the leakage current density and ρ_r is the resistivity of the insulator surface.

Thermal conduction and convection energies on the insulator surface are calculated below:

$$W_{conduction} = \sum_{t=0}^{t_0} \sum_{i=0}^{l} \lambda \Delta T_i^{t_n} \tag{9}$$

$$W_{convention} = \sum_{t=0}^{t_0} \sum_{i=0}^{l} h(T_i^{t_n} - T_0) \tag{10}$$

where l is the length of the insulator creepage distance, t_0 is the time duration, λ is the thermal conductivity of the insulating material, T_i^{tn} is the thermal temperature on the insulator surface, ΔT is the temperature difference as a function of distance and time. T_0 is the environment temperature. h is the heat transfer coefficient of convection.

Heat radiation becomes the dominant factor to cause heat transfer on the insulator surface after arc initialization. Heat radiation is the process of arc generating radiant energy. Arc radiation energy $W_{arc\_radiation}$ is calculated below:

$$W_{arc\_radiation} = \sum_{t=0}^{t_0} \sum_{i=0}^{l} \varepsilon_{emit} \sigma (T_i^{tn})^4 \tag{11}$$

where ε_{emit} is the emissivity of actual objects, $\sigma = 5.67 \times 10^{-8}$ is the Stefan–Boltzmann constant, and t_0 is the time period from the radiation start to the moment of field calculation.

Water in the pollutant layer evaporates during the heat transfer process. The Clausius–Clapeyron equation describes enthalpy variation based on air pressure and thermal temperature.

$$\ln \frac{P_2}{P_1} = \frac{\Delta H_{water}^{steam}}{R} \left(\frac{1}{T_1} - \frac{1}{T_2} \right) \tag{12}$$

where ΔH_{water}^{steam} is the phase-changing enthalpy of water, $R = 8.314$ is the universal gas constant, P_1 and P_2 remain the same as the standard atmospheric pressure (101.325 kPa), and T_1 and T_2 are the thermal temperature change before and after arc initialization. Therefore, ΔH is a function of thermal temperature during the dry band formation and the arc propagation processes. The evaporation energy is calculated in Equations (13) and (14).

$$\Delta H_{water}^{steam} = \frac{RT_1}{T_2} \tag{13}$$

$$W_{water\_steam} = \Delta H_{water}^{steam} V_{water}. \tag{14}$$

The thermal balance equation of dry band formation on the insulator surface is shown below:

$$W_{water\_steam} + W_{conduction} + W_{convection} = W_{arc\_radiation} + W_{leakage}. \tag{15}$$

3. Simulation Results

The dry band formation process from the moment of the insulator energization ($t = 0$ s) to the moment of arc initialization was simulated, in the first place, to analyze the effects of leakage current density on dry band formation. Then, the arc propagation process was simulated after arc initialization to investigate the effects of arc energy dissipation on further dry band formation and flashover.

3.1. Dry Band Formation and Arcing Simulations

The three stages of dry band formation before arc initialization are shown in Figure 3a–c respectively, when time t equals 0 s, 0.9 s, and 1.6 s.

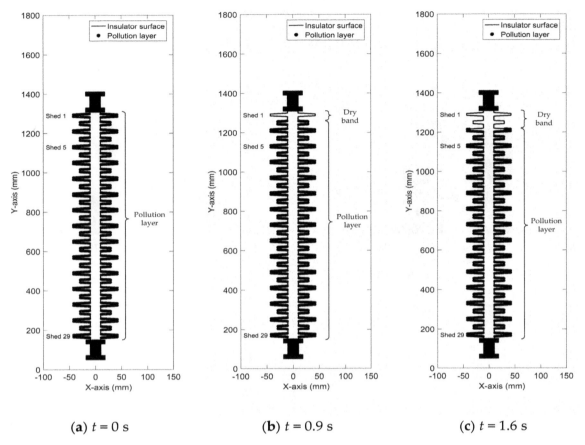

(**a**) $t = 0$ s   (**b**) $t = 0.9$ s   (**c**) $t = 1.6$ s

Figure 3. Dry band formation process on the insulator surface at different time nodes.

Figure 4 shows the electric and thermal field distributions at the initial state ($t = 0$ s) in Figure 3a before dry band formation. From Figures 3a and 4b, it is evident that the dry band was first generated at the location with the maximum thermal field.

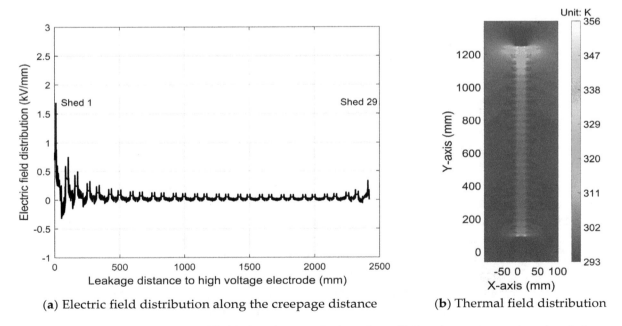

(**a**) Electric field distribution along the creepage distance   (**b**) Thermal field distribution

Figure 4. Electric and thermal field distributions before the pollutant layer generation. ($t = 0$ s).

The dry band area expands as the water in the pollutant layer continues evaporating. The thermal field distributions on the insulator surface close to the HV electrode in Figure 3b,c are shown in Figure 5. Figure 5 indicates the mutual positive effects on thermal temperature and dry band length.

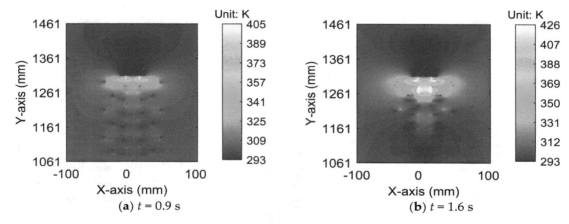

Figure 5. Thermal field distributions close to the high voltage (HV) electrode when the length of the dry band increases.

The electric field distributions on the insulator surface close to the HV electrode in Figure 3a–c are compared in Figure 6. Figure 6 shows that the maximum electric field with the dry band was higher than the maximum electric field without the dry band. The maximum electric field reduced when the dry band expanded.

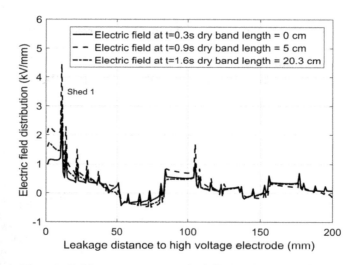

Figure 6. Electric field comparison with different lengths of the dry band.

The arc initializes when the maximum electric field exceeds the dielectric strength of air. However, the arc did not ignite immediately in Figure 3b because the water evaporation consumed the energy so that the maximum electric field could not maintain above the dielectric strength of the air. The arc initialized at $t = 1.6$ s, even though the electric field reduced slightly due to the expansion of the dry band. ($t = 5.42$ s). The first arc initialization and the thermal field distribution are shown in Figure 7. Arc initializes at the location on the insulator surface with the maximum electric field. It is observed that the thermal temperature significantly increases when the arc ignites, the arc thermal radiation dissipates energy from arc to the air and insulator surface. The dominant factor of dry band formation becomes arc energy radiation during the propagation process. However, the arc extinguishes when the length and number of dry bands increase because the leakage current reduces as the surface resistivity at the dry band is dramatically higher than the resistivity at the pollutant layer.

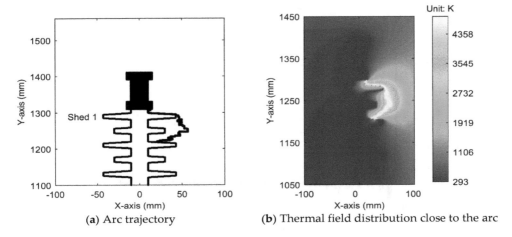

(a) Arc trajectory   **(b)** Thermal field distribution close to the arc

Figure 7. Arc trajectory and thermal field distributions close to the HV electrode when the first arc ignites.

Two dry bands were generated during the arc propagation process. The dry band and thermal field distributions are shown in Figure 8a,b. The electric field distribution along the creepage distance is shown in Figure 9. Figure 9 shows that the maximum electric field was lower than the maximum field with one dry band in Figure 6. However, the electric field distortion along the creepage distance was more severe than the field distribution with fewer dry bands. The electric field at more than one location on the insulator surface exceeded the dielectric strength of air. Therefore, multiple arcs reignited at different places on the composite insulator surface.

The distorted electric field led to the same distribution of the leakage current density. Therefore, temperature increased more significantly close to the dry band than the other locations on the insulator surface (Figure 8b).

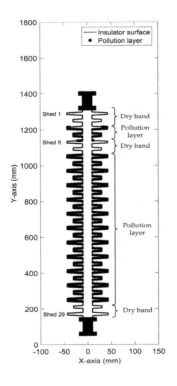

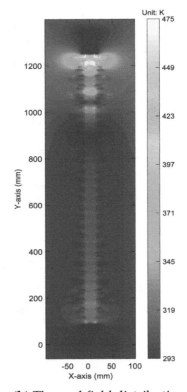

(a) Dry bands distribution   **(b)** Thermal field distribution

Figure 8. Dry bands and thermal field distributions close to the HV electrode.

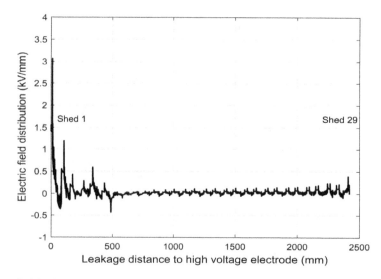

Figure 9. Electric field distribution along the creepage distance with three dry bands on the insulator surface.

Arcs ignited at different locations on the insulator surfaces after multiple dry bands generation when t was equal to 9.89 s (Figure 10a). Arcs distinguished with the expansion of dry band and ignited due to the distorted electric field close to dry bands. The iterations were repeated six times, and the number of arcs significantly increased after each iteration. The separated arcs were finally connected to a conductive path from the HV electrode to the ground electrode of the composite insulator and caused flashover (Figure 10b) when t was equal to 14.64 s.

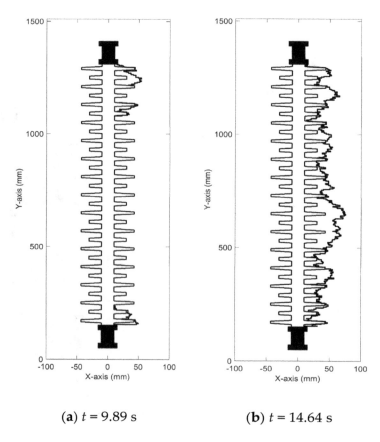

(a) $t = 9.89$ s   (b) $t = 14.64$ s

Figure 10. Arc trajectory when multiple arcs occur and connect into a conductive path.

It was observed that the arc can jump between insulator sheds rather than traveling along the creepage distance. The arc trajectories could be slightly different due to the random walk theory. The arc jumping between sheds and stochastic characteristics in the model are consistent with the physical phenomena of the arc.

3.2. Experiment Results

The scheme of the experiment system is shown in Figure 11. Due to the hydrophobicity of the composite material, the insulator samples were coated with dry kaolin powder and NaCl to form the contamination layer [39]. The ESDD value of the contamination layer was 0.1 mg/cm^2 to evaluate the dry band formation and arcing processes under a severely polluted scenario. The samples were firstly wetted by the clean fog and then energized with 110 kV at rated voltage to observe the dry band formation and arcing processes. The surface resistivity was 8.3×10^5 $\Omega \cdot$m. The HV and ground electrodes had a diameter of 52 mm and a length of 108 mm as shown in Figure 1a. The high-speed camera (2F01) recorded 500 video frames per second at 800 pixels $\times$ 600 pixels from the start of the experiment to flashover. The videos were transmitted to the computer with 400 MB/s Ethernet.

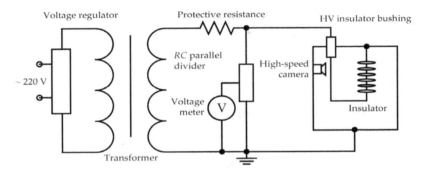

Figure 11. The schematic of the dry band formation and arcing experiment system.

Figure 12 shows the experiment results of the dry band formation and arc propagation processes. The arc propagation at different time frames was compared to analyze the dry band location and arcing phenomena. Figure 12a shows the dry band formation process before arc ignition. Figure 12b shows the first arc ignition due to the dry band close to the HV electrode. Figure 12c shows the arcs reignite at different locations due to the presence of multiple dry bands on the insulator surface. Figure 12d shows that the separated arcs were connected and led to flashover. The time frames and phenomena are consistent with the simulation results in the model.

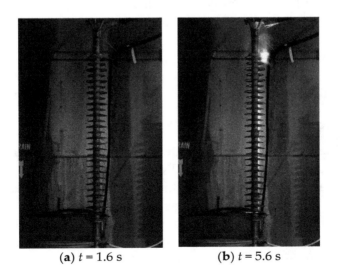

(a) $t = 1.6$ s   (b) $t = 5.6$ s

Figure 12. *Cont.*

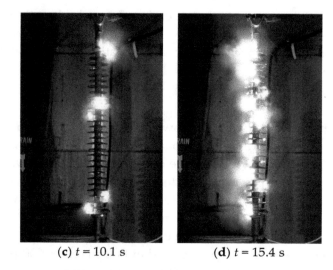

(c) $t = 10.1$ s   (d) $t = 15.4$ s

Figure 12. The experiment results at different time nodes of dry bands formation and arc propagation processes.

The time nodes during arc propagation of the experimental and simulation results are compared in Table 1 to validate the simulation model.

Table 1. Comparison of time nodes between experimental and simulation results.

Time Nodes	Simulation (s)	Experiment (s)	Error (%)
Dry band formation	1.6	1.6	0
Arc igniting	5.42	5.6	3.2
Arc extinction	5.51	5.8	5
Arc reigniting	8.57	9.7	11.6
Multiple arc occurrences	9.89	10.1	2.1
Flashover	14.64	15.4	4.9

The time node errors between simulation and experimental results were caused by the stochastic characteristics of the arc propagation.

4. Insulator Geometry Optimization

Since the experiment results verify the dry band formation and arc propagation model, three factors (creepage factor, shed angle, and alternative shed ratio) of composite insulator geometry were optimized in this section, while the creepage distance of the composite insulator remained the same as 2416 mm and the ESDD value was 0.1 mg/cm^2. Due to the stochastic property of arc propagation, the dry band formation and arcing processes were repeated 187 times to calculate the 50% flashover voltage and average duration time from $t = 0$ s to the moment of flashover. The CF is defined as the ratio of insulator creepage distance versus the arcing distance. The shed angle is the angle of the shed surface slope. The alternative shed ratio is the ratio of the small shed radius to the large shed radius.

4.1. Creepage Factor Optimization

The CF was optimized when the shed angle was 10° and the alternative shed ratio was 0.8. The range of CF was from 2.5 to 3.5 according to IEC 60815. The insulator geometry with different CF values are shown in Figure 13. 50% flashover voltage and time duration as functions of CF value are shown in Figure 14. Figure 14 indicates that 50% flashover voltage first increased then reduced with the CF values. The optimal value of CF was 2.94 to achieve the minimum flashover voltage. The average duration time decreased with the CF values because the arc had a high probability to bridge sheds as the distance between sheds reduced with the increase of CF.

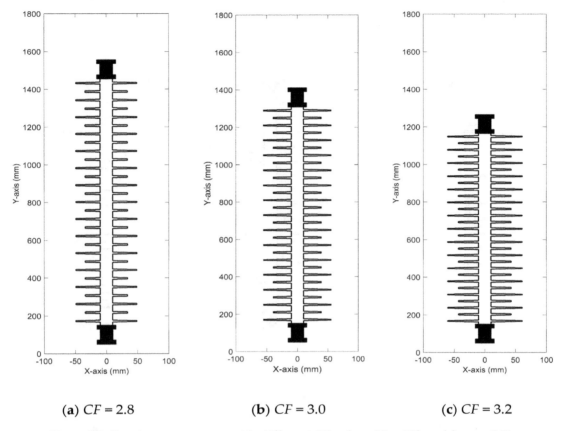

(a) $CF = 2.8$ (b) $CF = 3.0$ (c) $CF = 3.2$

Figure 13. Insulator geometry with different CF values ($\theta = 10°$ and $k_{shed} = 0.8$).

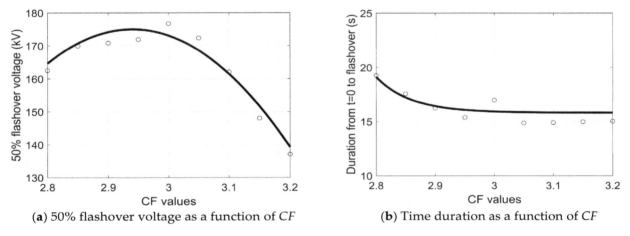

(a) 50% flashover voltage as a function of CF (b) Time duration as a function of CF

Figure 14. 50% flashover voltage and time duration as functions of the CF value.

4.2. Shed Angle Optimization

The shed angle was optimized when the CF value was 3.0 and the alternative shed ratio was 0.8. The range of shed angle was from 0° to 25° according to IEC 60815. The insulator geometry with different shed angles are shown in Figure 15. 50% flashover voltage and time duration as functions of shed angle are shown in Figure 16. Figure 16 indicates that shed angle had little impact on the 50% flashover voltage. The average duration time increased with shed angle because the average length of arc trajectories increased with shed angle.

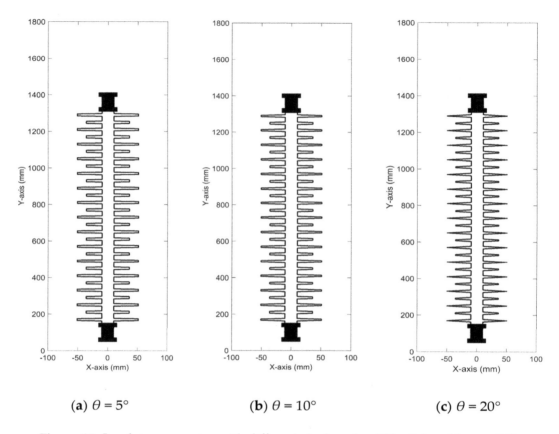

(a) $\theta = 5°$ (b) $\theta = 10°$ (c) $\theta = 20°$

Figure 15. Insulator geometry with different shed angles ($CF = 3.0$ and $k_{shed} = 0.8$).

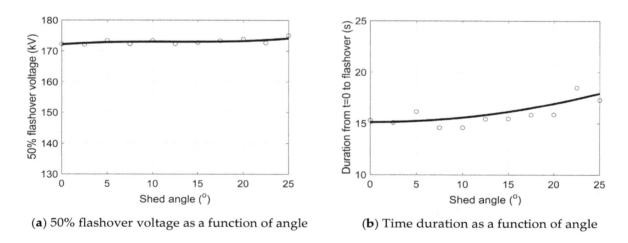

(a) 50% flashover voltage as a function of angle (b) Time duration as a function of angle

Figure 16. 50% flashover voltage and time duration as functions of the shed angle.

4.3. Alternative Shed Ratio Optimization

The alternative shed ratio was optimized when the CF value was 3.0 and the shed angle was 10°. The range of alternative shed ratio was from 0.7 to 1.0 according to IEC 60815. The insulator geometry with different alternative shed ratios are shown in Figure 17. 50% flashover voltage and time duration as functions of alternative shed ratio are shown in Figure 18. Figure 18 indicates that the 50% flashover voltage increased with alternative shed ratio because a small alternative shed ratio leads to the increasing occurrence of arc bridging between sheds. The alternative shed ratio had little impact on the average duration time.

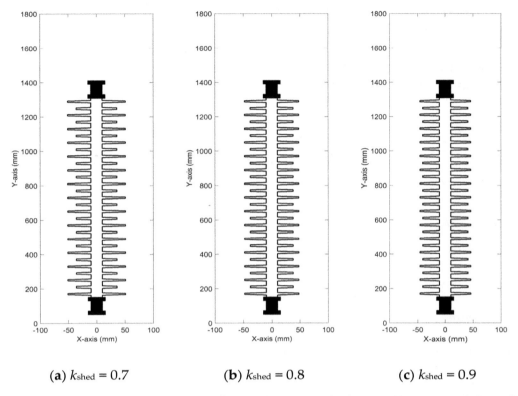

(a) $k_{shed} = 0.7$ **(b)** $k_{shed} = 0.8$ **(c)** $k_{shed} = 0.9$

Figure 17. Insulator geometry with different alternative shed ratios ($CF = 3.0$ and $\theta = 10°$).

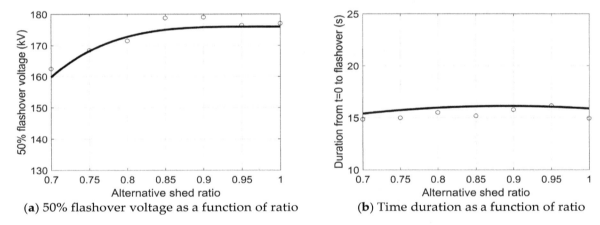

(a) 50% flashover voltage as a function of ratio   (b) Time duration as a function of ratio

Figure 18. 50% flashover voltage and time duration as functions of the alternative shed ratio.

5. Conclusions

This paper modeled the dry band formation and arcing processes of polluted composite insulators. Instantaneous electric and thermal fields were calculated by the GFDTD method to investigate the mechanism of dry band arcing and flashover. The simulation results were verified by the laboratory experiments. Insulator dimension factors were analyzed to optimize insulator geometry when the creepage distances remained the same.

1. The GFDTD method is suitable to calculate the electric and thermal fields for the insulator geometry by improving the field calculation accuracy at the high precision requirement area and reducing the computational complexity at the low precision requirement area.

2. The stochastic characteristics and the arc trajectory jumping between insulator sheds were modelled to simulate the physical phenomena of the arc.

3. The maximum electric field decreases with the expansion of the dry band. The heat transfer model demonstrates that the leakage current density is the dominant factor to affect dry band

formation before the arc initialization, while the arc radiation becomes the dominant factor to form the dry band after the arc ignition.

4. The 50% flashover voltage of composite insulators increases with the decrease of the *CF* value and the increase of the alternative shed ratio. The duration time from the pollution layer generation moment to flashover increases with the decrease of the *CF* value and the increase of the alternative shed angle.

Author Contributions: Conceptualization: J.H.; Methodology: J.H.; Software: J.H. and K.H.; Validation: K.H. and B.G.; Formal Analysis: J.H. and K.H.; Investigation: J.H.; Experiment: B.G.; Writing—Original Draft Preparation: J.H.; Writing—Review and Editing: J.H., K.H., and B.G.

Nomenclature

A	coefficient matrix multiply with $[D_\varphi]$
$(a^{-1})_{r,c}$	element at r-th row and c-th column of matrix $[A]^{-1}$
$a(E - E_c)$	step function of random walk
B	coefficient matrix multiply with $[\varphi]$
$B(u)$	residual function of two discrete points
b	constant matrix which equals to the product of $[A]$ and $[D_\varphi]$
$b_{r,c}$	element at r-th row and c-th column of matrix $[B]$
c	specific heat capacity
CF	creepage factor
D	constant matrix which equals to the product of $[A]^{-1}$ and $[B]$
D_u	partial difference column matrix
d	insulator arcing distance
$d_{1,2,3\ldots}$	distance between two discrete points
$d_{r,c}$	element at r-th row and c-th column of matrix $[D]$
E	electric field strength
E_c	RMS value of the threshold field
E_i^{tn}	electric field strength in GFDTD form
$ESDD_B$	ESDD value of bottom part of the insulator
$ESDD_T$	ESDD value of top part of the insulator
ΔH_{water}^{steam}	phase changing enthalpy of water
h	heat transfer coefficient of convection
h_{ij}	absolute value X coordinate differences between two discrete points
J	leakage current density
J_i^{tn}	leakage current density in GFDTD form
k_{ij}	absolute value Y coordinate differences between two discrete points
k_{shed}	ratio of radii of large and small sheds
l	length of insulator leakage distance
l_1, l_2	insulator leakage distance
P	probability of random walk
$P_{1,2,3\ldots}$	discrete points
p_1, p_2	saturated vapor pressure
R	universal gas constant
r_1, r_2	radius of large and small sheds
T	thermal temperature
T_0	environment temperature
T_1, T_2	thermal temperature change before and after arc initialization
t	time
t_0	time duration of insulator current leakage
u	column matrix of discrete point values
u_i	value of field at a discrete point

V	volume
$W_{arc\_radiation}$	heat radiation energy of arcs
$W_{conduction}$	energy of heat conduction
$W_{convention}$	energy of heat convention
$W_{leakage}$	energy of leakage current
W_{water_steam}	required energy for water evaporation
$w_{1,2,3\ldots}$	weight function of discrete points in residual function
Γ	field boundary
ε	permittivity
ε_i	permittivity in GFDTD form
ε_{emit}	emissivity
θ	shed angle
λ	thermal conductivity
ρ	density
ρ_c	bulk charge density
ρ_i^{tn}	bulk charge density in GFDTD form
ρ_r	resistivity
σ	Stefan-Boltzmann constant
Φ	internal heat sources
Φ_i	internal heat sources in GFDTD form
φ	electric potential
φ_i^{tn}	electric potential in GFDTD form

Appendix A

The GFDTD method is used to calculate the electric and thermal field distributions.

$$\frac{\partial u}{\partial t} = \frac{\partial^2 u}{\partial x^2} + \frac{\partial^2 u}{\partial y^2} + C. \tag{A1}$$

The advantage of GFDTD is that the density of discrete calculation points could be different in the field domain based on the precision requirement and boundary conditions.

According to the Taylor series expansion, the value of u_j at point P_j of the function at near neighborhood of P_i is expressed as follows (Figure A1) [40]:

$$u_j = u_i + h_{ij}\frac{\partial u_i}{\partial x} + k_{ij}\frac{\partial u_i}{\partial y} + \frac{1}{2}\left(h_{ij}^2\frac{\partial^2 u_i}{\partial x^2} + k_{ij}^2\frac{\partial^2 u_i}{\partial y^2}\right) + h_{ij}k_{ij}\frac{\partial^2 u_i}{\partial x \partial y} \quad i = 1,2,\ldots,m \tag{A2}$$

where h_{ij} and k_{ij} are absolute values of X and Y coordinate differences, i.e., $h_{ij} = |x_j - x_i|$, $k_{ij} = |y_j - y_i|$.

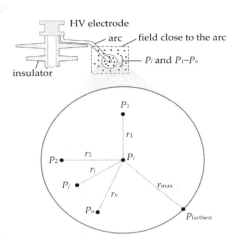

Figure A1. Generalized finite difference time domain (GFDTD) in field calculation.

P_i is the point among $P_1 \sim P_n$. The value of each point P_i and $P_1 \sim P_n$ is u_i and $u_1 \sim u_n$. The distance from each point $P_1 \sim P_n$ to P_i and is $r_1 \sim r_n$, and the farthest distance is r_{max}.

The residual function of two points $B(u)$ is defined by Equation (A3), shown as follows:

$$B(u) = \sum_{j=1}^{n} \left[\left(u_i - u_j + h_{ij}\frac{\partial u_j}{\partial x} + k_{ij}\frac{\partial u_j}{\partial y} + \frac{h_{ij}^2}{2}\frac{\partial^2 u_j}{\partial x^2} + \frac{k_{ij}^2}{2}\frac{\partial^2 u_j}{\partial y^2} \right) w_j \right]^2. \tag{A3}$$

The weight function of the j-th point w_j is calculated below:

$$w_j = 1 - 6\left(\frac{r_j}{r_{max}}\right)^2 + 8\left(\frac{r_j}{r_{max}}\right)^3 - 3\left(\frac{r_j}{r_{max}}\right)^4. \tag{A4}$$

Derive $B(u)$ for $\partial^2 u/\partial x^2$ and $\partial^2 u/\partial y^2$, then get $[A][D_u] = [b]$, where matrixes $[A]$, $[D_u]$ and $[b]$ are shown as follows,

$$[A] = \begin{bmatrix} \sum_{j=1}^{n}\frac{h_{ij}^4 w_j^2}{4} & \sum_{j=1}^{n}\frac{h_{ij}^2 k_{ij}^2 w_j^2}{4} \\ \sum_{j=1}^{n}\frac{h_{ij}^2 k_{ij}^2 w_j^2}{4} & \sum_{j=1}^{n}\frac{k_{ij}^4 w_j^2}{4} \end{bmatrix} \tag{A5}$$

$$[D_u] = \begin{bmatrix} \frac{\partial^2 u}{\partial x^2} & \frac{\partial^2 u}{\partial y^2} \end{bmatrix}^{\mathrm{T}} \tag{A6}$$

$$[b] = \begin{bmatrix} \sum_{j=1}^{n}\left(u_j - u_i\right)\frac{h_{ij}^2 w_j^2}{2} & \sum_{j=1}^{n}\left(u_j - u_i\right)\frac{k_{ij}^2 w_j^2}{2} \end{bmatrix}^{\mathrm{T}} \tag{A7}$$

Decompose the matrix $[b]$ as $[b] = [B][u]$, where

$$[B] = \begin{bmatrix} \sum_{j=1}^{n}\frac{-h_{ij}^2 w_j^2}{2} & \frac{h_{i1}^2 w_1^2}{2} & \frac{h_{i2}^2 w_2^2}{2} & \cdots & \frac{h_{ij}^2 w_j^2}{2} & \cdots & \frac{h_{in}^2 w_n^2}{2} \\ \sum_{j=1}^{n}\frac{-k_{ij}^2 w_j^2}{2} & \frac{k_{i1}^2 w_1^2}{2} & \frac{k_{i2}^2 w_2^2}{2} & \cdots & \frac{k_{ij}^2 w_j^2}{2} & \cdots & \frac{k_{in}^2 w_n^2}{2} \end{bmatrix} \tag{A8}$$

$$[u] = \begin{bmatrix} u_i & u_1 & u_2 & \cdots & u_j & \cdots & u_n \end{bmatrix}^{\mathrm{T}} \tag{A9}$$

The matrix $[D_u]$ is written in another form, i.e., $[D_u] = [A]^{-1}[b] = [A]^{-1}[B][u] = [D][u]$, where $[D] = [A]^{-1}[B]$ is a matrix with two rows and $(n + 1)$ columns shown below:

$$[D] = [A]^{-1}[B] = \begin{bmatrix} \sum_{i=1}^{2}(a^{-1})_{1,i}b_{i,1} & \sum_{i=1}^{2}(a^{-1})_{1,i}b_{i,2} & \sum_{i=1}^{2}(a^{-1})_{1,i}b_{i,3} & \cdots & \sum_{i=1}^{2}(a^{-1})_{1,i}b_{i,n+1} \\ \sum_{i=1}^{2}(a^{-1})_{2,i}b_{i,1} & \sum_{i=1}^{2}(a^{-1})_{2,i}b_{i,2} & \sum_{i=1}^{2}(a^{-1})_{2,i}b_{i,2} & \cdots & \sum_{i=1}^{2}(a^{-1})_{2,i}b_{i,n+1} \end{bmatrix} \tag{A10}$$

where $(a^{-1})_{r,c}$ and $b_{r,c}$ are the elements at r-th row and c-th column of matrix $[A]^{-1}$ and $[B]$.

Thus, $\partial^2 u/\partial x^2$ and $\partial^2 u/\partial y^2$ are written as

$$\begin{cases} \frac{\partial^2 u}{\partial x^2} = d_{1,1}u_i + \sum_{j=1}^{n} d_{1,(j+1)}u_j \\ \frac{\partial^2 u}{\partial y^2} = d_{2,1}u_i + \sum_{j=1}^{n} d_{2,(j+1)}u_j \end{cases} \tag{A11}$$

where $d_{r,c}$ is the element at r-th row and c-th column of matrix $[D]$.

After substituting Equation (A11) into Equation (A1), the PDE is written as

$$\frac{u_i^{t_{n+1}} - u_i^{t_n}}{\Delta t} = d_{1,1}u_i^{t_n} + \sum_{j=1}^{n} d_{1,(j+1)}u_j^{t_n} + d_{2,1}u_i^{t_n} + \sum_{j=1}^{n} d_{2,(j+1)}u_j^{t_n} + C_i^{t_n} \tag{A12}$$

where the superscript "t_n" and "t_{n+1}" are the present and next stages of the point u, respectively.

References

1. Gorur, R.S.; Cherney, E.A.; Burnham, J.T. *Outdoor Insulators*; Ravi, S., Ed.; Gorur Inc.: Phoenix, AZ, USA, 1999.
2. Gorur, R.S. Status Assessment of Composite Insulators for Outdoor HV Applications. In Proceedings of the 5th International Conference on Properties and Applications of Dielectric Materials, Seoul, Korea, 25–30 May 1997.
3. Headley, P. Development and Application Experience with Composite Insulators for Overhead Lines. In Proceedings of the IEE Colloquium on Non-Ceramic Insulators for Overhead Lines, London, UK, 16 October 1992.
4. Clift, S. Composite Fiber Optic Insulators and Their Application to High Voltage Sensor Systems. In Proceedings of the IEE Colloquium on Structural Use of Composites in High Voltage Switchgear/Transmission Networks, London, UK, 16 October 1992.
5. Ilomuanya, C.S.; Nekahi, A.; Farokhi, S. Acid Rain Pollution Effect on the Electric Field Distribution of a Glass Insulator. In Proceedings of the 2018 IEEE International Conference on High Voltage Engineering and Application (ICHVE), Athens, Greece, 10–13 September 2018.
6. Liu, Y.; Wang, J.; Zhou, M.; Fang, C.; Zhou, W. Research on the Silicone Rubber Sheds Performance of Composite Insulator. In Proceedings of the 2008 International Conference on High Voltage Engineering and Application, Chongqing, China, 9–12 November 2008.
7. Hussain, M.M.; Farokhi, S.; McMeekin, S.G.; Farzaneh, M. Effect of Cold Fog on Leakage Current Characteristics of Polluted Insulators. In Proceedings of the 2015 International Conference on Condition Assessment Techniques in Electrical Systems (CATCON), Bangalore, India, 10–12 December 2015.
8. Hussain, M.M.; Farokhi, S.; McMeekin, S.G.; Farzaneh, M. Mechanism of Saline Deposition and Surface Flashover on Outdoor Insulators near Coastal Areas Part II: Impact of Various Environment Stresses. *IEEE Trans. Dielectr. Electr. Insul.* **2017**, *24*, 1068–1076. [CrossRef]
9. Hussain, M.; Farokhi, S.; McMeekin, S.G.; Farzaneh, M. Effect of Uneven Wetting on E-field Distribution along Composite Insulators. In Proceedings of the 2016 IEEE Electrical Insulation Conference (EIC), Montreal, QC, USA, 19–22 June 2016.
10. Hussain, M.M.; Farokhi, S.; McMeekin, S.G.; Farzaneh, M. Dry Band Formation on HV Insulators Polluted with Different Salt Mixtures. In Proceedings of the 2015 IEEE Conference on Electrical Insulation and Dielectric Phenomena (CEIDP), Ann Arbor, MI, USA, 18–21 October 2015.
11. Nekahi, A.; McMeekin, S.G.; Farzaneh, M. Ageing and Degradation of Silicone Rubber Insulators due to Dry Band Arcing under Contaminated Conditions. In Proceedings of the 2017 52nd International Universities Power Engineering Conference (UPEC), Heraklion, Greece, 28–31 August 2017.
12. Qiao, X.; Zhang, Z.; Jiang, X.; Li, X.; He, Y. A New Evaluation Method of Aging Properties for Silicon Rubber Material Based on Microscopic Images. *IEEE Access* **2019**, *7*, 15162–15169. [CrossRef]
13. Kim, S.H.; Cherney, E.A.; Hackam, R. Effect of Dry Band Arcing on the Surface of RTV Silicone Rubber Coatings. In Proceedings of the Record of the 1992 IEEE International Symposium on Electrical Insulation, Baltimore, MD, USA, 7–10 June 1992.
14. Qiao, X.; Zhang, Z.; Jiang, X.; Liang, T. Influence of DC Electric Fields on Pollution of HVDC Composite Insulator Short Samples with Different Environmental Parameters. *Energies* **2019**, *12*, 2304. [CrossRef]
15. Hussain, M.M.; Chaudhary, M.A.; Razaq, A. Mechanism of Saline Deposition and Surface Flashover on High-Voltage Insulators near Shoreline: Mathematical Models and Experimental Validations. *Energies* **2019**, *12*, 3685. [CrossRef]
16. Hussain, M.M.; Farokhi, S.; McMeekin, S.G.; Farzaneh, M. Observation of Surface Flashover Process on High Voltage Polluted Insulators near Shoreline. In Proceedings of the 2016 IEEE International Conference on Dielectrics (ICD), Montpellier, France, 3–7 July 2016.
17. Hampton, B.F. Flashover Mechanism of Polluted Insulation. *Proc. Inst. Electr. Eng.* **1964**, *111*, 985–990. [CrossRef]
18. Salthouse, E.C. Dry-Band Formation and Flashover in Uniform-Field Gaps. *Proc. Inst. Electr. Eng.* **1971**, *118*, 630. [CrossRef]
19. Salthouse, E.C. Initiation of dry bands on polluted insulation. *Proc. Inst. Electr. Eng.* **1968**, *115*, 1707–1712. [CrossRef]
20. Löberg, J.O.; Salthouse, E.C. Dry-Band Growth on Polluted Insulation. *IEEE Trans. Electr. Insul.* **1971**, *EI-6*, 136–141.

21. Zhou, J.; Gao, B.; Zhang, Q. Dry Band Formation and Its Influence on Electric Field Distribution along Polluted Insulator. In Proceedings of the 2010 Asia-Pacific Power and Energy Engineering Conference, Chengdu, China, 28–31 March 2010.

22. Das, A.; Ghosh, D.K.; Bose, R.; Chatterjee, S. Electric Stress Analysis of a Contaminated Polymeric Insulating Surface in Presence of Dry Bands. In Proceedings of the 2016 International Conference on Intelligent Control Power and Instrumentation (ICICPI), Kolkata, India, 21–23 October 2016.

23. Benito, J.J.; Urena, F.; Gavate, L. Influence of Several Factors in the Generalized Finite Difference Method. *Appl. Math. Model.* **2001**, *25*, 1039–1053. [CrossRef]

24. Gavete, L.; Gavate, M.L.; Benito, J.J. Improvements of Generalized Finite Difference Method and Comparison with Other Meshless Method. *Appl. Math. Model.* **2003**, *27*, 831–847. [CrossRef]

25. Chen, J.; Gu, Y.; Wang, M.; Chen, W.; Liu, L. Application of the Generalized Finite Difference Method to Three-dimensional Transient Electromagnetic Problems. *Eng. Anal. Bound. Elem.* **2018**, *92*, 257–266. [CrossRef]

26. Ureña, F.; Gavete, L.; García, A.; Benito, J.J.; Vargas, M. Solving Second Order Non-linear Parabolic PDEs Using Generalized Finite Difference Method (GFDM). *J. Comput. Appl. Math.* **2019**, *354*, 211–241. [CrossRef]

27. Suchde, P.; Kuhnert, J.; Tiwari, S. On Meshfree GFDM Solvers for the Incompressible Navier–Stokes Equations. *Comput. Fluids* **2018**, *165*, 1–12. [CrossRef]

28. Yee, K. Numerical Solution of Initial Boundary Value Problems Involving Maxwell's Equations in Isotropic Media. *IEEE Trans. Antennas Propag.* **1966**, *14*, 302–307.

29. Hou, K.; Li, W.; Ma, L.; Cheng, Y.; Jin, L. Multi-Objective Structural Optimization of UHV Composite Insulators based on Pareto Dominance. In Proceedings of the 2018 12th International Conference on the Properties and Applications of Dielectric Materials (ICPADM), Xi'an, China, 20–24 May 2018.

30. Li, L. Shed Parameters Optimization of Composite Post Insulators for UHV DC Flashover Voltages at High Altitudes. *IEEE Trans. Dielectr. Electr. Insul.* **2015**, *22*, 169–176. [CrossRef]

31. Doufene, D.; Bouazabia, S.; Ladjici, A.A. Shape Optimization of a Cap and Pin Insulator in Pollution Condition Using Particle Swarm and Neural Network. In Proceedings of the 2017 5th International Conference on Electrical Engineering-Boumerdes (ICEE-B), Boumerdes, Algeria, 29–31 October 2017.

32. Liu, L. The Influence of Electric Field Distribution on Insulator Surface Flashover. In Proceedings of the 2018 IEEE Conference on Electrical Insulation and Dielectric Phenomena (CEIDP), Cancun, Mexico, 21–24 October 2018.

33. Yu, X.; Yang, X.; Zhang, Q.; Yu, X.; Zhou, J.; Liu, B. Effect of Booster Shed on Ceramic Post Insulator Pollution Flashover Performance Improvement. In Proceedings of the 2016 IEEE International Conference on High Voltage Engineering and Application (ICHVE), Chengdu, China, 19–22 September 2016.

34. Jiang, X.; Yuan, J.; Zhang, Z.; Hu, J.; Sun, C. Study on AC Artificial-Contaminated Flashover Performance of Various Types of Insulators. *IEEE Trans. Power Deliv.* **2007**, *22*, 2567–2574. [CrossRef]

35. Jiang, X.; Zhang, Z.; Hu, J. Investigation on Flashover Voltage and Non-uniform Pollution Correction Coefficient of DC Composite Insulator. In Proceedings of the 2008 International Conference on High Voltage Engineering and Application, Chongqing, China, 9–12 November 2008.

36. He, J.; Gorur, R.S. Flashover of Insulators in a Wet Environment. *IEEE Trans. Dielectr. Electr. Insul.* **2017**, *24*, 1038–1044. [CrossRef]

37. Abd Rahman, M.S.B.; Izadi, M.; Ab Kadir, M.Z.A. Influence of Air Humidity and Contamination on Electrical Field of Polymer Insulator. In Proceedings of the 2014 IEEE International Conference on Power and Energy (PECon), Kuching, Malaysia, 1–3 December 2014.

38. He, J.; Gorur, R.S. A Probabilistic Model for Insulator Flashover under Contaminated Conditions. *IEEE Trans. Dielectr. Electr. Insul.* **2016**, *23*, 555–563. [CrossRef]

39. Pushpa, Y.G.; Vasudev, N. Artificial Pollution Testing of Polymeric Insulators by CIGRE Round Robin Method -Withstand & Flashover Characteristics. In Proceedings of the 3rd International Conference on Condition Assessment Techniques in Electrical Systems (CATCON), Chandigarh, India, 16–18 November 2017.

40. Chan, H.; Fan, C.; Kuo, C. Generalized Finite Difference Method for Solving Two-Dimensional Non-Linear Obstacle Problems. *Eng. Anal. Bound. Elem.* **2013**, *37*, 1189–1196. [CrossRef]

Permissions

All chapters in this book were first published in MDPI; hereby published with permission under the Creative Commons Attribution License or equivalent. Every chapter published in this book has been scrutinized by our experts. Their significance has been extensively debated. The topics covered herein carry significant findings which will fuel the growth of the discipline. They may even be implemented as practical applications or may be referred to as a beginning point for another development.

The contributors of this book come from diverse backgrounds, making this book a truly international effort. This book will bring forth new frontiers with its revolutionizing research information and detailed analysis of the nascent developments around the world.

We would like to thank all the contributing authors for lending their expertise to make the book truly unique. They have played a crucial role in the development of this book. Without their invaluable contributions this book wouldn't have been possible. They have made vital efforts to compile up to date information on the varied aspects of this subject to make this book a valuable addition to the collection of many professionals and students.

This book was conceptualized with the vision of imparting up-to-date information and advanced data in this field. To ensure the same, a matchless editorial board was set up. Every individual on the board went through rigorous rounds of assessment to prove their worth. After which they invested a large part of their time researching and compiling the most relevant data for our readers.

The editorial board has been involved in producing this book since its inception. They have spent rigorous hours researching and exploring the diverse topics which have resulted in the successful publishing of this book. They have passed on their knowledge of decades through this book. To expedite this challenging task, the publisher supported the team at every step. A small team of assistant editors was also appointed to further simplify the editing procedure and attain best results for the readers.

Apart from the editorial board, the designing team has also invested a significant amount of their time in understanding the subject and creating the most relevant covers. They scrutinized every image to scout for the most suitable representation of the subject and create an appropriate cover for the book.

The publishing team has been an ardent support to the editorial, designing and production team. Their endless efforts to recruit the best for this project, has resulted in the accomplishment of this book. They are a veteran in the field of academics and their pool of knowledge is as vast as their experience in printing. Their expertise and guidance has proved useful at every step. Their uncompromising quality standards have made this book an exceptional effort. Their encouragement from time to time has been an inspiration for everyone.

The publisher and the editorial board hope that this book will prove to be a valuable piece of knowledge for researchers, students, practitioners and scholars across the globe.

List of Contributors

Zhenyu Li and Xuezeng Zhao
Department of Mechatronics Control and Automation, School of Mechatronics Engineering, Harbin Institute of Technology, Harbin 150001, China

Maurizio Albano, A. Manu Haddad and Nathan Bungay
School of Engineering, Cardiff University, The Parade, Cardiff CF24 3AA, UK

Alhaytham Alqudsi
Mechanical Engineering Department, École de technologie supérieure, Montréal, QC H3C 1K3, Canada

Ayman El-Hag
Electrical and Computer Engineering, University of Waterloo, Waterloo, ON N2L 3G1, Canada

Rodrigo Nuricumbo-Guillén, Fermín P. Espino Cortés and Carlos Tejada Martínez
Departamento de Ingeniería Eléctrica SEPI-ESIME ZAC, Instituto Politécnico Nacional, Mexico City 07738, Mexico

Pablo Gómez
Electrical and Computer Engineering Department, Western Michigan University, Kalamazoo, MI 49008, USA

Mohammed El Amine Slama and Abderrahmane (Manu) Haddad
Advanced High Voltage Engineering Centre, School of Engineering Cardiff University, Queen's Buildings The Parade, Cardiff, Wales CF24 3AA, UK

Abderrahmane Beroual
Laboratoire Ampère, University of Lyon, 36 Avenue Guy de Collongues, 69130 Ecully, France
École Centrale de Lyon, University of Lyon, Ampère CNRS UMR 5005, 36 Avenue Guy Collongue, 69134 Écully, France

Jian Hao and Ruijin Liao
State Key Laboratory of Power Transmission Equipment & System Security and New Technology, Chongqing University, Chongqing 400044, China

Min Dan
State Key Laboratory of Power Transmission Equipment & System Security and New Technology, Chongqing University, Chongqing 400044, China
State Grid Chongqing Nanan Power Supply Company, Chongqing 401223, China

Lin Cheng, Jie Zhang and Fei Li
Najing NARI Group Corporation, State Grid Electric Power Research Institute, Nanjing 211000, China
Wuhan NARI Co. Ltd., State Grid Electric Power Research Institute, Wuhan 430077, China

Muhammad Majid Hussain
Faculty of Computing, Engineering and Science, University of South Wales, Treforest, Cardiff CF37 1DL, UK

Muhammad Akmal Chaudhary
Department of Electrical and Computer Engineering, Ajman University, Ajman, UAE

Abdul Razaq
School of Design and Informatics, Abertay University, Dundee DD1 1HG, UK

Kazuki Komatsu, Hao Liu and Yukio Mizuno
Department of Electrical and Mechanical Engineering, Nagoya Institute of Technology, Nagoya 466-8555, Japan

Mitsuki Shimada
Information and Analysis Technologies Division, Nagoya Institute of Technology, Nagoya 466-8555 Japan

WenWei Zhu
School of Electric Power, South China University of Technology, Guangzhou 510640, China
Grid Planning Research Center, Guangdong Power Grid Co., Ltd., Guangzhou 510000, China

YiFeng Zhao, ZhuoZhan Han, Gang Liu, Yue Xie and NingXi Zhu
School of Electric Power, South China University of Technology, Guangzhou 510640, China

Xiang Bing Wang and YanFeng Wang
Grid Planning Research Center, Guangdong Power Grid Co., Ltd., Guangzhou 510000, China

Edward Gulski and Rogier Jongen
Onsite hv solutions AG, Lucerne, Toepferstrasse 5, 6004 Lucerne, Switzerland

Aleksandra Rakowska and Krzysztof Siodla
Institute of Electric Power Engineering, Poznan University of Technology, Piotrowo 3A, 60-965 Poznan, Poland

Yiming Zang, Yong Qian, Yongpeng Xu, Gehao Sheng and Xiuchen Jiang
Department of Electrical Engineering, Shanghai Jiao Tong University, 800 Dongchuan Road, Minhang, Shanghai 200240, China

Wei Liu
Key Laboratory for Sulfur Hexafluoride Gas Analysis and Purification of SGCC, Anhui Electric Power Research Institute of SGCC, Hefei 230022, China

Michail Michelarakis and Phillip Widger
Advanced High Voltage Engineering Research Centre, School of Engineering, Cardiff University, The Parade, Cardiff CF24 3AA, UK

Muhd Shahirad Reffin, Abdul Wali Abdul Ali, Normiza Mohamad Nor, Nurul Nadia Ahmad, Syarifah Amanina Syed Abdullah and Azwan Mahmud
Faculty of Engineering, Multimedia University, 63100 Cyberjaya, Malaysia

Farhan Hanaffi
Faculty of Electrical Engineering, Universiti Teknikal Malaysia Melaka, 76100 Durian Tunggal, Malaysia

Disheng Wang, Lin Du and Chenguo Yao
State Key Laboratory of Power Transmission Equipment and System Security and New Technology, Chongqing University, Chongqing 400030, China

Jiahong He, Kang He and Bingtuan Gao
School of Electrical Engineering, Southeast University, Nanjing 210096 China

Index